DIFFERENTIAL EQUATIONS

WITH APPLICATIONS AND PROGRAMS

S Balachandra Rao
Head of Department of Mathematics
National College
Basavanagudi, Bangalore 560 004

H R Anuradha
Lecturer in Mathematics
VVS Firstgrade College for Women
Basaveswaranagar, Bangalore 560 079

Universities Press

UNIVERSITIES PRESS (INDIA) PRIVATE LIMITED

Registered Office
3-6-747/1/A & 3-6-754/1, Himayatnagar, Hyderabad 500 029 (A.P.), India
info@universitiespress.com; www.universitiespress.com

Distributed by
Orient Blackswan Private Limited

Registered Office
3-6-752 Himayatnagar, Hyderabad 500 029 (A.P.) India

Other Offices
Bangalore / Bhopal / Chennai / Ernakulam / Guwahati / Hyderabad
Jaipur / Kolkata / Lucknow / Mumbai / New Delhi / Noida / Patna

Reprinted 2013

ISBN 978 81 7371 023 0

Typeset by
Vijaya Enterprises, Madras 600 006

Printed at
Printline Printers
Hyderabad 500 029

Published by
Universities Press (India) Private Limited
3-6-747/1/A & 3-6-754/1, Himayatnagar,
Hyderabad 500 029 (A.P.), India.

Contents

Preface

This book would serve as a textbook in Differential Equations for undergraduate students of science, engineering and technology. In fact, the book conforms to the prescribed syllabi of most Indian universities. It can be used for polytechnic courses also by leaving out the rigorous theoretical portions wherever these are not relevant. Students of M.Sc., M.E. and M.Tech., as also professionals and researchers who may need solutions of differential equations will greatly benefit from the book.

This book ensures that the student gets a good grounding in the fundamentals of the subject through a rigorous treatment of the theoretical concepts. Included are a large number of solved problems as also exercises (with answers) drawn from question papers of various Indian universities. Examples and exercises from real-life situations are presented to give the student a basic understanding of the practical applications of differential equations in diverse fields such as physics, chemistry, biological sciences, engineering and economics.

Numerical methods for the solution of differential equations are also presented, with the related algorithms and computer programs in two popular languages — **FORTRAN** and **Pascal**. Illustrative examples and their solutions are also provided after the program for each numerical method.

Though the book is self-sufficient insofar as differential equations are concerned, a necessary prerequisite is a working knowledge of differential and integral calculus. Students having access to a personal computer (PC) can readily use the programs (in Chapter 16) listed in **FORTRAN** and **Pascal**. The programs may be saved on a floppy and preserved for regular use. Of course, modifications may be made in the programs to suit specific needs of the user.

Organization and content

Basic concepts like the distinction between ordinary and partial differential equations (ODE and PDE), classification of ODE, general form and solution of an ODE, linear and nonlinear ODE, and initial and boundary value problems are discussed in Chapter 1.

Chapter 2 deals with ODE of the first order and first degree. Method of solution by separation of variables, logistic law of population growth, solutions of homogeneous, linear, exact, reducible to the exact and Bernoulli's forms of ODE are discussed in detail with very interesting examples.

Differential equations of first order and of higher degree constitute Chapter 3. Equations solvable for p, x, y and Clairaut's equation are discussed in this chapter.

Orthogonal trajectories, in the cartesian as well as polar forms, are discussed in Chapter 4.

Chapter 5 deals with solutions of higher order ODE with constant coefficients. Homogeneous and non-homogeneous equations are considered in detail. Application of the methods to mechanical free oscillations and forced oscillations as also to electrical circuits is discussed at length.

Cauchy–Euler differential equations are studied in Chapter 6.

Simultaneous differential equations in two variables with constant coefficients are discussed in Chapter 7. Euler's solution of homogeneous linear systems of differential equations forms an important part of this chapter.

Chapter 8 is devoted to the various methods of solving second order ODE with variable coefficients. The methods considered are (i) when a factor of the complementary function is known, (ii) variation of parameters, (iii) changing the independent variable, (iv) reduction to the normal form, and (v) exact differential equations.

Chapter 9 concentrates on power series solutions of ODE.

Legendre's differential equation and polynomials constitute Chapter 10. Important recurrence relations for the Legendre polynomials $P_n(x)$ and the orthogonality of these polynomials are discussed.

Chapter 11 deals with Bessel's differential equation. Series solution of Bessel's equation, identifier for the Bessel's functions $J_n(x)$, continued fraction expansion for $\tan x$, orthogonality of Bessel's functions and Bessel's integral formula are studied in detail.

Total differential equations of the form $Pdx + Qdy + Rdz = 0$ are discussed in Chapter 12.

Solution of simultaneous total differential equations is discussed in Chapter 13.

Chapter 14 is devoted to a detailed study of Laplace transforms. Convolution theorem and application of the Laplace transforms method to differential equations as also to problems in mechanics form an important part of this chapter.

Fourier series expansions and their applications, and finite Fourier transforms are studied in Chapter 15.

Chapter 16 is devoted to an elaborate study of numercial methods for the solution of ODE. The methods discussed in this chapter include Taylor's series method, Picard's successive approximation method, Euler's method, improved and modified Euler's methods, the second and fourth order Runge–Kutta methods, the error estimates for these methods and extensions of the Runge–Kutta methods due to Gill, Merson, Butcher and Nyström.

Computer programs in FORTRAN and Pascal are provided for all these methods along with outputs for the solution of illustrative examples.

Acknowledgements

We express our profuse thanks to Mr. S. Kedar Shankar for ably and sincerely assisting us in preparing the computer programs. We acknowledge our indebtedness to all the authors and publishers of the titles listed in the bibliography.

S. Balachandra Rao
H.R. Anuradha

CHAPTER ONE

Basic Concepts

1.1 Introduction

Most of the problems arising in various disciplines of science can be expressed in the form of differential equations. The solution of such problems, therefore, lies in solving the differential equations which govern those problems.

Various analytical methods have been developed to obtain the so-called "closed form" (i.e. precise) solutions for different types of differential equations. In the case of problems where closed form solutions cannot be obtained, *numerical methods* have been developed to provide approximate solutions.*

In this chapter we shall acquaint ourselves with the basic concepts concerning differential equations, their classification and solutions.

1.2 Ordinary and partial differential equations

An equation involving an independent variable x, a dependent variable y, which is a function of x, and the derivatives (of any order) of y with respect to x is called an *ordinary differential equation* (ODE). For example,

$$\frac{d^2y}{dx^2} - x\frac{dy}{dx} = e^x$$

is an ordinary differential equation.

Similarly, an equation involving two or more independent variables, a dependent variable and its partial derivatives is called a *partial differential equation* (PDE). For example, if z is a function of x and y, then

$$\frac{\partial^2 z}{\partial x^2} = a^2\frac{\partial^2 z}{\partial y^2}$$

is a partial differential equation.

*See *Numerical Methods* (with programs in BASIC, FORTRAN and Pascal) by S. Balachandra Rao and C.K. Shantha.

1.3 Classification of ordinary differential equations

Definition 1.1: The *order* of an ordinary differential equation is the same as that of the highest order derivative in the equation.

Definition 1.2: The *degree* of an ordinary differential equation is the positive integral power to which the highest order derivative is raised.

Consider the following example:

$$\left(\frac{d^3y}{dx^3}\right)^2 + 5\frac{d^2y}{dx^2} - x\left(\frac{dy}{dx}\right)^3 = x^2.$$

Here, the highest order derivative, d^3y/dx^3 is raised to the power 2. Therefore, the differential equation is of *order* 3 and *degree* 2.

1.4 General form and solution of a differential equation

An ordinary differential equation has the general form

$$F[x, y, y', y'', \ldots, y^{(n)}] = 0 \ldots \qquad [1.1]$$

where $y', y'', \ldots, y^{(n)}$ stand for the derivatives $\dfrac{dy}{dx}, \dfrac{d^2y}{dx^2}, \ldots, \dfrac{d^ny}{dx^n}$.

In [1.1], F is a specified function of its arguments and there exists atleast one of the derivatives of y among the arguments of F.

Definition 1.3: A *solution* of a differential equation is any function which satisfies the given equation. A function $\phi(x)$ is said to be a solution of differential equation [1.1] if

$$F[x, \phi(x), \phi'(x), \ldots, \phi^n(x)] = 0.$$

In other words, $y = \phi(x)$ satisfies differential equation [1.1].

The solution of an *n*th order differential equation containing n arbitrary constants is called the *general solution* or the *complete solution.*

In the general solution of a differential equation when particular values are specified for the arbitrary constants, the resulting solution is called a *particular solution.*

1.5 Formation of a differential equation

Differential equations arise from practical situations, following natural laws and geometrical constraints, in such diverse disciplines as physical sciences, biological sciences and even economics and statistics.

We shall illustrate the formation of differential equations from the given conditions in the following examples.

Example 1.1 The slope of a curve at any point (x, y) is proportional to x^2. Form the differential equation.

Solution 1.1 The slope of the curve at a point (x, y) is dy/dx. It is given that the slope is proportional to x^2 where x is the abscissa of the point (x, y) on the curve. This yields

$$\frac{dy}{dx} = kx^2 \qquad [1]$$

where k is a constant of proportionality. Thus the given condition is translated into differential equation [1]. The general solution of [1] can be easily obtained by directly integrating [1]. We get

$$y = \int kx^2 dx$$

or $$y = \frac{kx^3}{3} + C \qquad [2]$$

where C is an arbitrary constant. The general solution [2] of the differential equation [1] is the cartesian equation representing a cubic curve geometrically. In fact, if the coordinates of a specific point through which the curve [2] passes are given, then C is determined.

Note: From a given algebraic or transcendental equation a governing differential equation can be formed by eliminating the arbitrary constants from the given equation in the process of repeated differentiation. This method is demonstrated in Example 1.2.

Example 1.2 The equation

$$(x - h)^2 + y^2 = r^2 \qquad [1]$$

where h and r are arbitrary constants represents a family of circles of different radii and centres lying on the x-axis. Form a differential equation by eliminating h and r from [1].

Solution 1.2 Differentiating [1] with respect to x, we get

$$2(x - h) + \frac{2ydy}{dx} = 0 \quad \text{or} \quad x - h = -y\frac{dy}{dx}.$$

Differentiating the above equation with respect to x, we get

$$1 = -y\frac{d^2y}{dx^2} - \left(\frac{dy}{dx}\right)^2$$

or $yy'' + (y')^2 + 1 = 0.$ [2]

Thus, differential equation [2] is formed from algebraic equation [1] by differentiation and eliminating the arbitrary constants h and r. Note that [1] contains two arbitrary constants and correspondingly differential equation [2], got by eliminating the two arbitrary constants of [1], has the *order* 2. Hence an equation containing n arbitrary constants leads to an nth order differential equation.

Example 1.3 The motion of a particle falling freely under gravity is described by the differential equation

$$d^2y/dt^2 = -g \quad [1]$$

where y is the vertical height of the particle and g is the constant acceleration due to gravity and t is the time. Solve equation [1].

Solution 1.3 Integrating [1] with respect to t, we get

$$dy/dt = -gt + A \quad [2]$$

where A is an arbitrary constant. Now, integrating [2] with respect to t, we get

$$y = -gt^2/2 + At + B \quad [3]$$

where B is another constant of integration. Thus, [3] is a solution of the *second order* differential equation [1] and involves *two arbitrary constants* A and B.

The arbitrary constants A and B can be determined if additional conditions are imposed on the given differential equation [1]. This aspect will be discussed later in the sections on initial value and boundary value problems.

In practical problems, differential equations motivated by some law or condition require solutions which satisfy some specific conditions. These specifications yield appropriate values for the arbitrary constants in the general solutions. Such solutions, with specific values of arbitrary constants determined, are called *particular solutions*.

Example 1.4 In Example 1.3, for the problem of a particle falling freely under gravity satisfying the equation of motion

$$d^2y/dt^2 = -g \quad [1]$$

suppose we impose the following two conditions.

(1) The displacement y of the particle (i.e., the height of the particle) is 100 metres at time $t = 0$ i.e. *initially*. This means $y = 100$ when $t = 0$.

(2) The particle is at rest at $t = 0$; i.e. the velocity $dy/dt = 0$ at $t = 0$.

Solution 1.4 The general solution of differential equation [1] is

$$y = -\frac{gt^2}{2} + At + B \qquad [2]$$

Using condition (1) in [2], we get $B = 100$ so that the solution [2] takes the form

$$y = -\frac{gt^2}{2} + At + 100 \qquad [3]$$

Differentiating [3] with respect to t, we get the velocity

$$\frac{dy}{dt} = -gt + A \qquad [4]$$

Now, using condition (2) *viz.*, $dy/dt = 0$ when $t = 0$ we get $A = 0$ so that [2] becomes

$$y = -\frac{gt^2}{2} + 100 \qquad [5]$$

Thus, [5] is the *particular* solution of the differential equation [1] with the *initial conditions* (1) and (2).

1.6 Linear and nonlinear differential equations

Definition 1.4 An ordinary differential equation in which the dependent variable y and its derivatives occur as first degree terms, and no two of which are multiplied together, is called a *linear* differential equation.

A linear differential equation has the general form

$$a_0(x)\frac{d^n y}{dx^n} + a_1(x)\frac{d^{n-1}y}{dx^{n-1}} + \ldots + a_n(x)y = f(x) \qquad [1.2]$$

where $a_0(x), a_1(x), \ldots, a_n(x)y$ and $f(x)$ are, in general functions of x which are given.

A differential equation which is not linear, i.e., not of the form [1.2], is said to be *nonlinear*. There are no general methods available to solve nonlinear differential equations although some special forms of nonlinear equations can be solved.

The linear differential equation [1.2] can be written in the form

$$Ly = f(x) \tag{1.3}$$

where $$L \equiv a_0(x)\frac{d^n}{dx^n} + a_1(x)\frac{d^{n-1}}{dx^{n-1}} + \ldots + a_n(x) \tag{1.4}$$

is a *linear operator*. If $y_1(x)$ and $y_2(x)$ are any two functions that are differentiable n times and c_1 and c_2 are arbitrary constants, then we know that

$$\frac{d^m}{dx^m}[c_1y_1(x) + c_2y_2(x)] = c_1\frac{d^my_1(x)}{dx^m} + c_2\frac{d^my_2(x)}{dx^m} \quad (\text{for } 1 \leqslant m \leqslant n).$$

From the above result it follows that the linear differential operator L, defined in [1.4], has the property

$$L[c_1y_1(x) + c_2y_2(x)] = c_1Ly_1(x) + c_2Ly_2(x).$$

Definition 1.5 In the linear differential equation [1.2], if the known function $f(x) \neq 0$, then the equation is said to be *non-homogeneous*. If $f(x) = 0$, then the differential equation is called a *homogeneous* differential equation which is of the form $Ly = 0$. If $y_1, y_2, \ldots, y_n$ are solutions of [1.3], with $f(x) \equiv 0$, then $Ly_i = 0$ for $i = 1, 2, 3, \ldots, n$ or $\sum_{i=1}^{n} Lc_iy_i = 0$. Therefore $y = \sum c_iy_i$ is a solution.

1.7 Initial value problem

For a differential equation of *order* n if the values of the unknown function $y(x)$ and its first $(n-1)$ derivatives are specified at a particular point, say $x = x_0$, then the system consisting of the differential equation and the *initial conditions* (i.e. the values of y and its derivatives at $x = x_0$) is called an *initial value problem*. The initial conditions are specified in the form

$$y(x_0) = k_0, y'(x_0) = k_1, y''(x_0) = k_2, \ldots, y^{(n-1)}(x_0) = k_{n-1} \tag{1.5}$$

where $k_0, k_1, \ldots, k_{n-1}$ are the specified constants.

We have the following important theorem on the existence and uniqueness of an initial value problem.

Theorem If $a_0(x), a_1(x), \ldots, a_n(x)$ and $f(x)$ are real and continuous, with $a_0(x) \neq 0$, in an interval I, and x_0 be a point in I, then there exists one and only one solution of a linear differential equation $Ly = f(x)$ in I satisfying the initial conditions [1.5].

The proof of the theorem is omitted since it is beyond the scope of the book.

Example 1.5 Consider an electric circuit that involves a resistance R, a capacitance C and an inductance L, where R, C and L are constants, connected in series with a battery. The applied voltage is $E(t)$ which varies with time t. The charge $Q(t)$ on the capacitance and the current $I(t)$ in the loop are related by the equation

$$I = \frac{dQ}{dt}.$$

According to Kirchhoff's law, the sum of the voltage-drops around the loop is equal to the voltage applied. This results in the equation

$$L\frac{dI}{dt} + RI + \frac{1}{C}Q = E(t)$$

or $$L\frac{d^2Q}{dt^2} + R\frac{dQ}{dt} + \frac{1}{C}Q = E(t) \qquad [1]$$

which is a second-order differential equation. Now, if the charge Q and the current $I(= dQ/dt)$ are specified at time $t = t_0$ in the form

$$Q(t_0) = Q_0, Q'(t_0) = I_0 \qquad [2]$$

where Q_0 and I_0 are given constants, then the differential equation [1] along with the initial conditions is an *initial value problem.*

1.8 Boundary value problem

In problems involving second and higher order differential equations, often values of the unknown function $y(x)$ and its derivatives are specified at two or more particular points. Such specified conditions are called *boundary conditions* and the differential equation, along with these boundary conditions, is called a *boundary value problem.*

Example 1.6 Consider a second-order differential equation

$$\frac{d^2s}{dt^2} = 16 - 24t \quad \text{with} \quad s(0) = 2, s(1) = 7.$$

This is a boundary value problem in which the values of the unknown function s at two points $t = 0$ and $t = 1$ are specified.

Example 1.7 The following is a fourth-order boundary value problem specifying two conditions at each of two values of the independent variable.

The deflection y of a beam of length L, subjected to a uniform load q, is goverened by the differential equation

$$\frac{d^4y}{dx^4} + ky = q \quad \text{with} \quad y(0) = y'(0) = 0, y(L) = y''(L) = 0$$

where k is a given constant.

Exercise 1

1.1 Write down the differential equations of the curves for the following and indicate their *order* and *degree*:

(i) The slope of the curve is – 5

(ii) The length of the subtangent at any point (x, y) on the curve is equal to the sum of the coordinates of the point.

(iii) The tangent of the angle between the radius-vector and the tangent (i.e. $\tan\phi$) at any point (r, θ) of the curve is equal to the tangent of the vectorial angle.

1.2 Express the following statements in the form of differential equations:

(i) A particle moves along the x-axis such that its velocity is inversely proportional to time.

(ii) The potential difference E across an element of inductance L is equal to the product of L and the time-rate of change of current I in the inductance.

(iii) The air resistance acting on a falling body exerts a force that is proportional to the square of the velocity.

1.3 Find the general solutions of the following differential equations by *direct integration*:

(i) $\dfrac{dy}{dx} = 3x^2 - 2x$ **(ii)** $\dfrac{ds}{dt} = \cos t - \sin t$

(iii) $\dfrac{d^2y}{dx^2} = -\cos x$ **(iv)** $\dfrac{d^3y}{dx^3} = x^3 + x^2 + 1$

1.4 Find the particular solutions of the following differential equations for the specified conditions:

(i) $\dfrac{dy}{dx} = \cos x, y = 1$ when $x = \pi$ **(ii)** $\dfrac{dy}{dx} = xe^x, y(1) = 3$

(iii) $\dfrac{d^3y}{dx^3} = \cos x$ if $y = 1,\ y' = 0,\ y'' = 1$ when $x = \dfrac{\pi}{2}$.

CHAPTER TWO

Differential Equations Of The First Order And First Degree

2.1 Introduction

In this chapter we shall discuss the different techniques of solving differential equations which are of the first order and first degree. These equations contain only the first order derivative of the dependent variable with respect to the independent variable. The general solution will involve only one arbitrary constant. There are various methods of solving differential equations of this type depending on the form of the equation. We shall start with the simplest form.

2.2 Separation of variables

Consider a differential equation of the form

$$\frac{dy}{dx} = \frac{f(x)}{g(y)} \quad \text{(where } g(y) \neq 0). \tag{2.1}$$

In order to solve an equation of this type we *separate the variables* and write [2.1] in the form

$$g(y)dy = f(x)dx. \tag{2.2}$$

Integrating [2.2], we write $\int g(y)dy = \int f(x)dx$ and get the solution in the form

$$G(y) = F(x) + C \tag{2.3}$$

where $F(x)$ and $G(y)$ are functions of x and y respectively, being anti-derivatives of $f(x)$ and $g(y)$, and C is an arbitrary constant.

Example 2.1 Solve: $dy/dx = x^2y + y$.

Solution 2.1 We can rewrite the equation as $dy/dx = (x^2 + 1)y$. Separating the variables, we have $(1/y)\,dy = (x^2+1)dx$. On integration, we get $\log y = (x^3/3) + x + C$ where C is an arbitrary constant. This general solution can also be expressed in the exponential form

$$y = \exp\left(\frac{x^3}{3} + x + C\right) \quad \text{or} \quad y = \exp\left(\frac{x^3}{3} + x\right)\exp C.$$

Since C is an arbitrary constant, exp C is so and hence writing A = exp C, the general solution can be written as

$$y = A\exp\left(\frac{x^3}{3} + x\right)$$

Note: The *exponential* function e^x is also written as exp x. Sometimes it is convenient to write in the latter form, particularly when the argument involves powers and fractions.

Example 2.2 Solve: $e^y(dy/dt) - t - t^3 = 0$.

Solution 2.2 The equation can be written in the form

$$\frac{dy}{dt}(e^y) = t + t^3$$

where y is a function of t. On integration, we get

$$e^y = \frac{t^2}{2} + \frac{t^4}{4} + C \quad \text{or} \quad y = \log\left(\frac{t^2}{2} + \frac{t^4}{4} + C\right).$$

Example 2.3 Solve: $(2x - 3y + 1)dx + (6y - 4x + 3)dy = 0$.

Solution 2.3 The given equation is not in the variables separable form as such. But it can be reduced to the variables separable form and then solved as follows:

The equation can be written as

$$\frac{dy}{dx} = \frac{2x - 3y + 1}{4x - 6y - 3} = \frac{(2x - 3y) + 1}{2(2x - 3y) - 3}. \qquad [1]$$

$$z = 2x - 3y \quad \text{(say)} \qquad [2]$$

so that $$\frac{dz}{dx} = 2 - 3\frac{dy}{dx}$$

or $$\frac{dy}{dx} = \frac{1}{3}\left(2 - \frac{dz}{dx}\right). \qquad [3]$$

Using [2] and [3] in [1], the equation becomes

$$\frac{1}{3}\left(2 - \frac{dz}{dx}\right) = \frac{z+1}{2z-3}$$

or $$\frac{dz}{dx} = \frac{z-9}{2z-3} \qquad [4]$$

which is in the variables separable form. By separating the variables in [4] and integrating, we get

$$\int \frac{2z-3}{z-9} dz = \int dx \quad \text{or} \quad \int 2\frac{(z-9)+15}{z-9}\, dz = x + A$$

or $2z + 15\log(z-9) = x + A.$

Substituting for z from [2], the general solution is

$$(4x - 6y) + 15\log(2x - 3y - 9) = x + A$$

or $3x - 6y + 15\log(2x - 3y - 9) = A$ or $x - 2y + 5\log(2x - 3y - 9) = C$

where $C = (1/3)A.$

Note: Example 2.3 can also be solved by reducing it to the *homogeneous* form – see Example 2.18 .

Example 2.4 Solve: $(x+y)^2\dfrac{dy}{dx} = a^2.$

Solution 2.4 The given equation can be reduced to the variables separable form by substituting .

$$z = x + y \qquad [1]$$

so that we have $\dfrac{dz}{dx} = 1 + \dfrac{dy}{dx}$ and hence

$$\frac{dy}{dx} = \frac{dz}{dx} - 1. \qquad [2]$$

Substituting [1] and [2] in the given equation, we get

$$z^2\left[\frac{dz}{dx}-1\right]=a^2 \qquad [3]$$

or $\dfrac{z^2}{a^2+z^2}\,dz = dx$ or $\dfrac{(a^2+z^2)-a^2}{a^2+z^2}dz = dx.$

On integration we get

$$\int 1dz - a^2\int\frac{1}{a^2+z^2}dz = \int dx \quad \text{or} \quad z - a^2\left[\frac{1}{a}\tan^{-1}\left(\frac{z}{a}\right)\right] = x + C.$$

Substituting for z , from [1], we get the general solution of the given equation as

$$x+y-a\tan^{-1}\left(\frac{x+y}{a}\right) = x + C \quad \text{or} \quad y - a\tan^{-1}\left(\frac{x+y}{a}\right) = C.$$

Example 2.5 Solve: $\dfrac{dy}{dx} = \sin(x+y)+\cos(x+y).$

Solution 2.5 Let us substitute

$$z = x + y \qquad [1]$$

so that we have $\dfrac{dz}{dx} = 1 + \dfrac{dy}{dx}$ and hence

$$\frac{dy}{dx} = \frac{dz}{dx} - 1. \qquad [2]$$

Substituting [1] and [2] in the given equation, we get

$$\frac{dz}{dx} - 1 = \sin z + \cos z \quad \text{or} \quad \frac{dz}{dx} = \sin z + \cos z + 1.$$

Separating the variables, we have

$$\frac{dz}{\sin z + (1+\cos z)} = dx \quad \text{or} \quad \frac{dz}{2\sin\frac{z}{2}\cos\frac{z}{2} + 2\cos^2\frac{z}{2}} = dx.$$

Dividing the numerator and the denominator of the left-hand side by $\cos^2(\frac{z}{2})$, we get

$$\frac{\sec^2\left(\frac{z}{2}\right)dz}{2\left[\tan\left(\frac{z}{2}\right)+1\right]} = dx.$$

On integration, this yields

$$\log\tan\left[\frac{z}{2}+1\right] = x + C \quad \left[\text{by the substitution,} \quad t = \tan\left(\frac{z}{2}\right)\right].$$

Now, substituting for z from [1], the general solution is

$$\log\tan\left[\frac{x+y}{2}+1\right] = x + C \quad \text{or} \quad \tan\left[\frac{x+y+2}{2}\right] = Ae^{x}$$

where $A = e^{C}$.

Example 2.6 The rate at which a radioactive substance decomposes is proportional to the number of atoms present at any instant. If, initially, there are N_0 atoms at time t_0, find the number of atoms N at any instant t.

Solution 2.6 The rate of decomposition of the radioactive material is $[-dN/dt]$.

The negative sign indicates *decomposition* or *decay* of the material. Since this rate of decomposition is proportional to the number of atoms N present at any instant t, we have

$$-\frac{dN}{dt} = \lambda N \text{ or } \frac{dN}{dt} = -\lambda N \qquad [1]$$

where λ is the constant of proportionality.

Separating the variables in [1] and integrating, we have

$$\int \frac{dN}{N} = \int -\lambda dt$$

or $\log N = -\lambda t + C.$ [2]

Initially, at $t = t_0$ and $N = N_0$, we have

$$\log N_0 = -\lambda t_0 + C. \qquad [3]$$

From [2] and [3], we get $\log N - \log N_0 = -\lambda(t - t_0)$

or $\log\left(\frac{N}{N_0}\right) = -\lambda\left(t - t_0\right)$ [4]

or $\frac{N}{N_0} = e^{-\lambda(t-t_0)}$

so that the general solution is $N = N_0 e^{-\lambda(t-t_0)}$.

Note: Half-life and radioactive dating: The constant of proportionality λ is called *radioactive rate constant* since its value is a measure of the rate at which the reaction (or radioactive decay) takes place.

An important measure of the radioactive disintegration of a substance is its *half-life* which is defined as the time required for half of the radioactive substance to disintegrate.

If we substitute $N/N_0 = 1/2$ in [4], we get $\log 1/2 = -\lambda(t - t_0)$ so that

$$t - t_0 = \frac{\log 2}{\lambda} \simeq \frac{0.6931}{\lambda}. \tag{5}$$

The *half-lives* of many substances have been determined. For example, the half-life of Carbon-14 is 5568 years and the half-life of Uranium-238 is 4.5 billion years.

This concept of half-life of a radioactive substance can be used in *radioactive dating*. The process of radioactive dating enables us to determine how old a substance is.

The following are some well known and interesting incidents of historical and investigative importance which are determined using the technique of radioactive dating:

(1) The very famous and beautiful painting, *Disciples at Emmaus* was forged by a third-rate Dutch painter H.A. Van Meegeren. This forgery of the famous painting was conclusively established by the technique of radioactive dating from the radioactive elements, lead-210 (^{210}Pb) and radium-226 (^{226}Ra) . [Reference: *Van Meegeren's Faked Vermeers and De Hooghs* by P. Coremans, 1949].

(2) Linen wrappings from the Dead Sea scrolls of the Book of Isaiah, found in a cave in Palestine, were earlier believed to belong to the first or second century B.C. By radioactive dating it was found to be of the 18th or 19th century A.D.

(3) A site of remarkable paintings, the Lascaux cave in southern France, is found to be $15,516 \pm 900$ years old from the radioactive dating of charcoal from the site.

(4) The prehistoric monuments at Stonehenge in southern England are determined to be 3798 ± 275 years old from charcoal.

(5) Charcoal from a tree burned at the time of the volcanic explosion that formed the Crater Lake in Oregon is found to be 6453 ± 250 years old.

By using pieces of charcoal, fibre sandals, fragments of burned bison bone etc., campsites of ancient man throughout the Western world have been dated by the process of radioactive dating. The results point out that man did not arrive in that part of the world until about the period of the last Ice Age, some 11,500 years ago. During that period the level of the water in the oceans was greatly lower than it now is and man could have easily walked across the Bering Straits from Siberia to Alaska.

It is remarkable to note that great findings of historical and investigative importance, such as those mentioned above, rest on the mathematical foundation provided by [1] and its solution given by [5]. [References: (1) *Radiocarbon Dating* by Libby, 2nd Ed., University of Chicago Press, 1955. (2) *America's Buried Past* by G.C. Baldwin, Putnam, New York, 1962].

2.3 Logistic law of population growth

The growth of population of a species, including human, is studied by using the *population growth model*. It is found from experiments on bacterial growth as well as of other species that the rate of growth of a population is proportional to the current population. Thus, if P is the population of a species at time t, then

$$\frac{dP}{dt} = aP, P(t_0) = P_0 \tag{2.4}$$

where a is a constant which is found to be positive assuming that the birth rate is higher than the death rate. At time $t = t_0, P = P_0$.

Now, separating the variables in [2.4], we have $(dP/P) = adt$. On integration

$$\log P = at + \log C \quad \text{or} \quad \log(P/C) = at$$

or $P = Ce^{at}$. [2.5]

Since initially, $P = P_0$ at $t = t_0$, from [2.5] we get $P_0 = Ce^{at_0}$ so that [2.5] takes the form (see Fig.2.1)

$$P = P_0 e^{a(t-t_0)}. \tag{2.6}$$

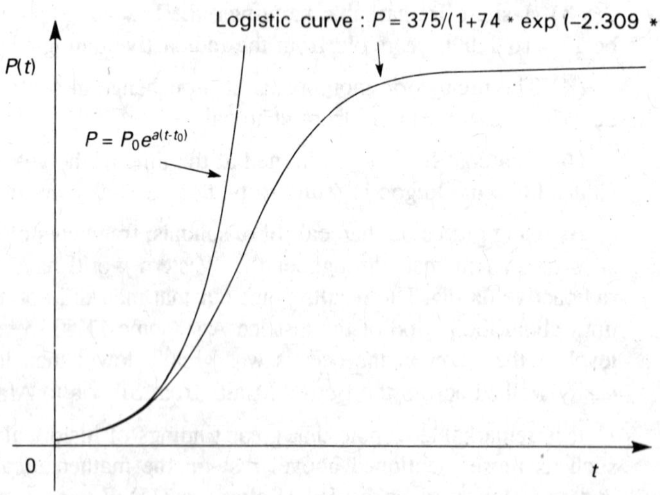

Fig.2.1 *Logistic curve and exponential growth curve*

Based on this model of *exponential growth* of population, we have Example 2.8 on bacterial population. While this model works in simplistic cases, it is found that for fairly large values of t, the figures we get for P are enormously large. For example, the world population in 2510 would be as large as 200,000 billion and in 2670 it would be 3,600,000 billion. These are indeed astronomical figures which are very unrealistic. In fact, this would mean that in 2635 each person will have only one square foot to stay on and by 2670 we will have to stand two deep on each other's shoulders.

Although the exponential growth model [2.6] is unrealistic for human population over large intervals of time, it is found that the model fits in well with the population growth of several species especially over reasonably small intervals of time.

In fact, linear model [2.4] works out satisfactorily as long as the population is not too large. When population is very large, as in the case of humans, there is need to reflect the fact that individual members compete with each other for the limited living space, natural resources and available food. Thus, there is need to add a *competition term* to the linear differential equation [2.4]. Since the statistical average of the number of encounters of two members per unit time is proportional to P^2, and the effect is to diminish the population, the term added to [2.4] is $-bP^2$ and we have

$$\frac{dP}{dt} = aP - bP^2, P(t_0) = P_0. \qquad [2.7]$$

Equation [2.7] is called the *logistic law* of population growth. This model was first introduced by the Dutch mathematical-biologist Verhulst in 1837. In general, the coefficient b is very small compared to a in [2.7]. Therefore, when P is not large, $-bP^2$ is negligible compared to aP in which case model [2.7] reduces to the exponential growth model [2.4].

The logistic equation [2.7] can be successfully used to predict the future growth of an isolated population. This initial value problem can be solved by the method of separation of variables. We have

$$\frac{dP}{P(a - bP)} = dt. \qquad [2.8]$$

By resolving into partial fractions, we can write

$$\frac{1}{P(a - bP^2)} = \frac{1}{a}\left(\frac{1}{P} + \frac{b}{a - bP}\right).$$

Hence on integration, from [2.8], we get

$$\frac{1}{a}\left(\log\frac{P}{P_0} + \log\left|\frac{a - bP_0}{a - bP}\right|\right) = t - t_0$$

or $$\frac{1}{a}\log\left(\frac{P}{P_0}\left|\frac{a - bP_0}{a - bP}\right|\right) = t - t_0.$$

Since $\dfrac{a - bP_0}{a - bP} > 0$, we have

$$a(t-t_0) = \log\left(\frac{P(a-bP_0)}{P_0(a-bP)}\right) \quad \text{or} \quad e^{a(t-t_0)} = \frac{P(a-bP_0)}{P_0(a-bP)}$$

or $P_0(a-bP)e^{a(t-t_0)} = P(a-bP_0)$. Solving for $P \equiv P(t)$, we get

$$P(t) = \frac{aP_0e^{a(t-t_0)}}{a-bP_0+bP_0e^{a(t-t_0)}}$$

$$\text{or} \quad P(t) = \frac{aP_0}{bP_0+(a-bP_0)e^{-a(t-t_0)}}. \qquad [2.9]$$

As $t \to \infty$ in [2.9], we see that $P(t) \to a/b$. This limit is independent of the initial value of P.

Further, by differentiating the differential equation in [2.7], we have

$$\frac{d^2P}{dt^2} = a\frac{dP}{dt} - 2bP\frac{dP}{dt} = (a-2bP)\left(\frac{dP}{dt}\right) = (a-2bP)P(a-bP).$$

Now, $d^2P/dt^2 > 0$ for $P < a/2b$ in which case dP/dt is *increasing*. Similarly, dP/dt is *decreasing* for $P > a/2b$. Therefore, if $P_0 < a/2b$, the graph of $P(t)$ will have the form shown in Fig. 2.1. This curve is called *logistic curve* or "S-shaped" curve.

From the shape of the curve it is clear that the time period before the population reaches half its limiting value is a period of accelerated growth. After that the rate of population growth decreases until it tends to zero eventually.

Example 2.7 The protozoa of *Paramecium Caudatum*, starting with 5, were found to increase at a rate of 230.9% per day when their numbers were low. The number increased rapidly at first and then more slowly, until towards the fourth day it attained a maximum level of 375, saturating the container with Paramecium. Obtain an expression for the Paramecium population using the logistic law.

Solution 2.7 We use the logistic law of population growth, given by

$$\frac{dP}{dt} = aP - bP^2$$

Since for small values of P, the rate of growth is 230.9%, we have $a = 2.309$. The maximum population, reached according to the logistic law, is given by

$$\frac{a}{b} = 375; \quad \text{so that} \quad b = \frac{a}{375} = \frac{2.309}{375}.$$

The solution of the logistic law of population growth is given by

$$P(t) = \frac{aP_0}{bP_0 + (a - b\ P_0)e^{-a(t-t_0)}}.$$

Now, substituting $\quad a = 2.309, b = \dfrac{2.309}{375}, P_0 = 5 \quad \text{and} \quad t_0 = 0,$

the above equation takes the form

$$P(t) = \frac{(2.309)5}{\dfrac{(2.309)5}{375} + \left[2.309 - \dfrac{(2.309)5}{375}\right]e^{-2.309t}}$$

or $\quad P(t) = \dfrac{375}{1 + 74e^{-2.309t}}.$

Note: The logistic law can be successfully applied to predict the future human population of any country or of the earth. For this it is important to estimate the important controlling coefficients a and b in the logistic equation [2.9]. Usually a and b are estimated from sets of known population over a period.

For predicting our global population some ecologists have estimated that $a = 0.029$. The earth's human population was increasing at the rate of about 2% per year when the population was 3.06×10^9. Since from [2.7],

$$\frac{1}{P}\frac{dP}{dt} = a - bP$$

we have $0.02 = a - b(3.06 \times 10^9)$ so that with 0.029, we get $b = 2.941 \times 10^{-12}$. Thus, according to the logistic law, the human population of the earth reaches its limiting value

$$\frac{a}{b} = \frac{0.029}{2.941 \times 10^{-12}} \simeq 9.86 \text{ billion.}$$

Example 2.8 In a biological problem, a bacterial population of size P is known to have a rate of growth proportional to P itself. If between 9 a.m. and 10 a.m. the population doubles, at what time will the population become 100 times what it was at 9 a.m.?

Solution 2.8 If P is the bacterial population at time t, then it is given that

$$\frac{dP}{dt} = kP \qquad [1]$$

where k is the constant of proportionality. Equation [1] is the differential equation which governs the *population growth*. Separating the variables in [1] and integrating we get

$$\int \frac{dP}{P} = \int kt \quad \text{or} \quad \log P = kt + \log A (\log A = \text{ arbitrary constant})$$

or $\log(P/A) = kt$

or $P = Ae^{kt}$ [2]

This is the general solution of the given problem. Suppose $P = P_0$ at 9 a.m. Now, taking time $t = 0$ at 9 a.m., we have $t = 60$ minutes at 10 a.m. By data, we have $P = 2P_0$ at $t = 60$ minutes. Substituting $P = P_0, t = 0$ in [2], we have

$$P_0 = Ae^0 = A \qquad [3]$$

Setting $P = 2P_0$ and $t = 60$ in [2], we have

$$2P_0 = Ae^{60k} \qquad [4]$$

From [3] and [4], we get $2 = e^{60k}$ or $\log 2 = 60k$ so that

$$k = \frac{\log 2}{60}. \qquad [5]$$

Using [5] in [2], we get

$$P = P_0\ e^{(\log 2)t/60} \qquad [6]$$

From [6] we have to find t for which $P = 100P_0$. Putting $P = 100P_0$ in [6], we have

$$100P_0 = P_0e^{(\log 2)t/60} \quad \text{or} \quad \log 100 = (\log 2)t/60$$

so that $t = 60 \log 100/ \log 2$

or $t \approx 398.63$ minutes.

Therefore, at time $t = 398.63$ minutes after 9 a.m., i.e. at about $3^{\text{h}}38^{\text{m}}38^{\text{s}}$ p.m., the bacterial population becomes 100 times what it was at 9 a.m.

Example 2.9 A piece of metal at 180°C is placed in a room at 18°C. After 10 minutes it has cooled to 126°C. After another 10 minutes the temperature is 90°C. What will the temperature of the piece of metal be at the end of one hour?

Solution 2.9 The excess temperatures over the surroundings at intervals of 0,10 and 20 minutes are

$$\theta_0 = 180° - 18° = 162°; \theta_1 = 126° - 18° = 108°; \theta_2 = 90° - 18° = 72°$$

where the temperatures are in degrees celsius. Now, by *Newton's law of cooling*, we have

$$\frac{dT}{dt} = -k(T - T_s) \qquad [1]$$

where T is the temperature of the hot metallic piece at time t, T_s is the temperature of the surroundings and k is the constant of proportionality, called *coefficient of thermal conductivity* of the material. Let T_0 be the initial temperature of the metal piece when $t = 0$. By separating the variables in [1] and integrating, we get

$$\int \frac{dT}{T - T_s} = -\int k dt$$

or $\log(T - T_s) = -kt + C$ [2]

Since $T = T_0$ when $t = 0$, from [2], we get $C = \log\,(T_0 - T_s)$. Substituting this in [2], we have

$$\log\,(T - T_s) = -kt + \log(T_0 - T_s) \quad \text{or} \quad \log\left[\frac{T - T_s}{T_0 - T_s}\right] = -kt$$

or $(T - T_s) = (T_0 - T_s)e^{-kt}$ [3]

where $(T_0 - T_s)$ is a constant being the excess temperature of the body over the surroundings initially. In the given problem, we have $T_s = 18°\text{C}$, $T_0 = 180°\text{C}$, $T_1 = 126°\text{C}$, $T_2 = 90°\text{C}$ where the temperatures T_0, T_1 and T_2 are taken respectively at timings $t_0 = 0$, $t_1 = 10$ minutes and $t_2 = 20$ minutes. Therefore, $(T_0 - T_s) = 180° - 18° = 162°$ and [3] becomes

$(T - 18) = 162e^{-kt}$ [4]

In [4], substituting $T = T_1 = 126$ at $t = t_1 = 10$, we have $126 - 18 = 162e^{-10k}$

$$\text{or} \quad e^{-10k} = \frac{108}{162} = \frac{2}{3} \quad \text{or} \quad -10k = \log\frac{2}{3}$$

$$\text{or} \quad k = \frac{1}{10}\log\frac{3}{2}.$$

Therefore, the required temperature at $t = 60$ minutes is given by using [4]:

$$T - 18 = 162e^{(-\frac{1}{10}\log\frac{3}{2})60} = 162e^{6\log(\frac{2}{3})}$$

$$= 162 \exp\log\left[\left(\frac{2}{3}\right)^6\right] = 162 \times \frac{64}{729} = \frac{2}{9} \times 64$$

$$\text{so that} \quad T = 18° + \frac{128°}{9} = 32\frac{2}{9}°\text{C}.$$

Note: Substituting T_1 and T_0 for T in [4] and dividing, we get

$$\frac{T_1 - 18}{T_0 - 18} = \frac{162e^{-kt_1}}{162e^{-kt_0}} = e^{-k(t_1 - t_0)}$$

and similarly, by putting T_2 and T_1 in [4], we have

$$\frac{T_2 - 18}{T_1 - 18} = e^{-k(t_2 - t_1)}.$$

Since $t_2 - t_1 = t_1 - t_0 = 10$ minutes, we get

$$\frac{T_1 - 18}{T_0 - 18} = \frac{T_2 - 18}{T_1 - 18}$$

which means the excess temperatures over the surroundings, $T_0 - 18, T_1 - 18$ and $T_2 - 18$ taken at equal intervals of time, are in *geometric progression.*

Therefore, problems of this type based on Newton's law of cooling can be solved using the property of geometric progression. In the given problem, let

$$\theta_0 = T_1 - 18 = 162°; \theta_1 = T_2 - 18 = 108°; \theta_2 = T_3 - 18 = 72°.$$

So $$\frac{\theta_1}{\theta_0} = \frac{108°}{162°} = \frac{2}{3} \text{ and } \frac{\theta_2}{\theta_1} = \frac{72°}{108°} = \frac{2}{3}.$$

This means the excess of temperatures over the surroundings, $\theta_0, \theta_1, \theta_2, \ldots,$ form a geometric progression with first term $a = \theta_0 = 162°$ and common ratio $r = 2/3$. Since $\theta_0, \theta_1, \theta_2, \ldots$ are at time intervals $t = 0, 10, 20, \ldots$ (in minutes), the general term of the geometric progression is given by $\theta_n = ar^n$. Therefore, at $t = 60$ minutes, the excess temperature is given by

$$\theta_6 = 162° \times \left(\frac{2}{3}\right)^6 = \frac{162 \times 64}{729} = \frac{128°}{9}.$$

The actual temperature is given by

$$T_6 = \theta_6 + 18° = 18° + \frac{128°}{9} = 32\frac{2}{9}° \text{ C}.$$

Example 2.10 Let F be the constant force generated by the motor of an automobile of mass M, and its velocity V be given by

$$M\frac{dV}{dt} = F - kV$$

where k is a constant. Express V in terms of t given that $V = 0$ when $t = 0$.

Solution 2.10 The given equation can be written as

$$M\,dV = (F - kV)dt \quad \text{or} \quad \frac{M}{F - kV}dV = dt.$$

On integration we get

$$\frac{M}{k}\log(F - kV) = C - t. \qquad [1]$$

By substituting $t = 0$ and $V = 0$ in [1], we get

$$\frac{M}{k}\log F = C. \qquad [2]$$

Substituting [2] in [1], we get the solution of the initial value problem as

$$\frac{M}{k}\log(F - kV) = \frac{M}{k}\log F - t \quad \text{or} \quad \frac{M}{k}\log\left[1 - \frac{kV}{F}\right] = -t$$

or $\log\left[1 - \dfrac{kV}{F}\right] = -\dfrac{kt}{M}$ or $1 - \dfrac{kV}{F} = e^{-kt/M}$

so that the velocity at any instant of time t is given by

$$V = \frac{F}{k}\left[1 - e^{-kt/M}\right].$$

Example 2.11 The distance x descended by a person falling by means of a parachute satisfies the equation

$$\left[\frac{dx}{dt}\right]^2 = k^2\left[1 - e^{-2gx/k^2}\right]$$

where k is a constant and g is the constant acceleration due to gravity. Obtain an expression for x in terms of time t given that $x = 0$ when $t = 0$.

Solution 2.11 The given equation is

$$\left[\frac{dx}{dt}\right]^2 = k^2\left[1 - e^{-2gx/k^2}\right]. \qquad [1]$$

Taking square root, we have

$$\frac{dx}{dt} = k\left[1 - e^{-2gx/k^2}\right]^{1/2}$$

$$= k\left[1 - \frac{1}{e^{2gx/k^2}}\right]^{1/2} = k\left[\frac{e^{2gx/k^2} - 1}{e^{2gx/k^2}}\right]^{1/2}$$

$$= k\frac{\left[e^{2gx/k^2} - 1\right]^{1/2}}{e^{gx/k^2}}$$

or $$\frac{e^{gx/k^2}\,dx}{\sqrt{e^{2gx/k^2} - 1}} = kdt. \qquad [2]$$

Substitute $e^{gx/k^2} = y$ so that $e^{gx/k^2} \cdot \frac{g}{k^2}dx = dy$

or $$e^{gx/k^2}\,dx = \frac{k^2}{g}dy. \qquad [3]$$

Using [3] in [2], we have $\frac{k^2}{g}\frac{dy}{\sqrt{y^2 - 1}} = kdt$. On integration, we get

$$\frac{k}{g}\cosh^{-1}y = t + c_1$$

or $\cosh^{-1}y = \frac{g}{k}t + c,\quad [c = gc_1/k]$ or $y = \cosh(\frac{gt}{k} + c)$

i.e. $e^{gx/k^2} = \cosh\left(\frac{gt}{k} + c\right)$ or $x = \frac{k^2}{g}\log\cosh\left(\frac{gt}{k} + c\right)$.

Example 2.12 [*Brachistochrone Problem*] Determine the shape of a wire, down which a bead slides from a point A to another point B, not directly below it, in the *shortest possible time*.

Solution 2.12 We choose A as the origin, the vertically downward direction AY (along the plumb-line) as the y-axis and the horizontal direction AX as the x-axis (Fig. 2.2).

Suppose the bead placed at A slides along a curve AB, under the influence of gravity to reach B in the *shortest possible time*. It is important to note that in this case, due to gravity, the shortest *time* does not correspond to the shortest *distance*. In fact, the shortest distance between A and B is along the line-segment AB. But the bead does not slide along this line-segment.

Let $P(x, y)$ be any point on the curve along which the bead slides and PT be the tangent thereat. Let the inclinations of PT with the x-axis and the y-axis be ψ and α

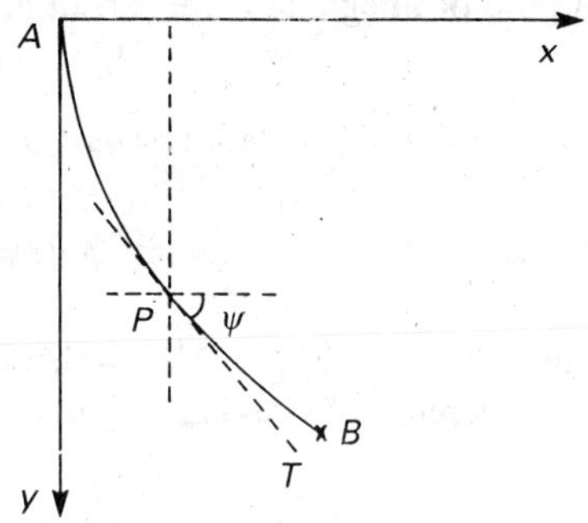

Fig. 2.2 *Brachistochrone problem*

respectively. If v is the velocity of the bead, then by *minimizing* the time taken by the bead to move from A to B, we have

$$\frac{\sin\alpha}{v} = c_0, \text{ (c_0 is a constant).} \qquad [1]$$

By the principle of conservation of energy, the velocity attained by the bead at a given height is determined by its loss of potential energy in reaching that height below the starting point and not by the path along which the bead moves. Now since the gain in kinetic energy is equal to the loss in potential energy, we have

$$\frac{1}{2}mv^2 = mgy$$

or $\quad v = \sqrt{2gy}.$ $\qquad$ [2]

From Fig. 2.2, we have $\quad \sin\alpha = \cos\psi = \dfrac{1}{\sec\psi} = \dfrac{1}{\sqrt{1+\tan^2\psi}}$

or $\quad \sin\alpha = \dfrac{1}{\sqrt{1+(y')^2}}.$ $\qquad$ [3]

From [1], [2] and [3], we get

$$y\left[1+(y')^2\right] = c \qquad [4]$$

where c is a constant. Thus, [4] is the first order differential equation governing the Brachistochrone problem. Now, [4] can be written as

$$\frac{dy}{dx} = \left[\frac{c}{y} - 1\right]^{1/2} \quad \text{or} \quad dx = \left[\frac{y}{c-y}\right]^{1/2} dy \text{ (by separation of variables).}$$

For the purpose of integration, we substitute $\left[\frac{y}{c-y}\right]^{1/2} = \tan\phi$ so that

$$y = c\sin^2\phi, dy = 2c\sin\phi\cos\phi\, d\phi$$

and $$dx = \tan\phi\; dy = 2c\sin^2\phi\; d\phi = c(1-\cos 2\phi)d\phi.$$

On integration we get $x = c/2(2\phi - \sin 2\phi) + c_1$ since the curve has to pass through the origin A. Therefore, by taking $x = 0, y = 0$, we get $c_1 = 0$ so that

$$x = \frac{c}{2}(2\phi - \sin\, 2\phi); \qquad [5]$$

$$y = c\sin^2\phi = \frac{c}{2}(1-\cos 2\phi). \qquad [6]$$

Now, by substituting $a = c/2$ and $\theta = 2\phi$, [5] and [6] take the form

$$x = a(\theta - \sin\theta), y = a(1-\cos\theta). \qquad [7]$$

Equations [7] are the standard parametric equations of a *cycloid* shown in Fig. 2.3a. For a suitable value of a, the first arch of this cycloid passes through A and B of Fig. 2.2. The cycloid [7] is generated by a fixed point on a circle of radius a rolling along the x-axis [Fig. 2.3b].

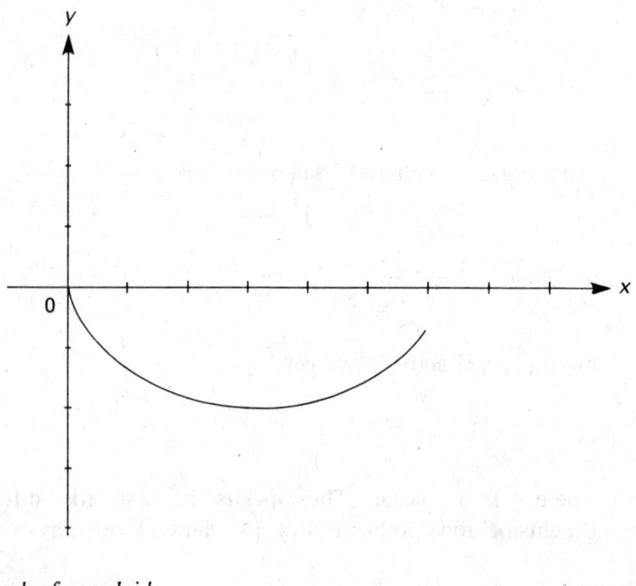

Fig. 2.3a *Arch of a cycloid*

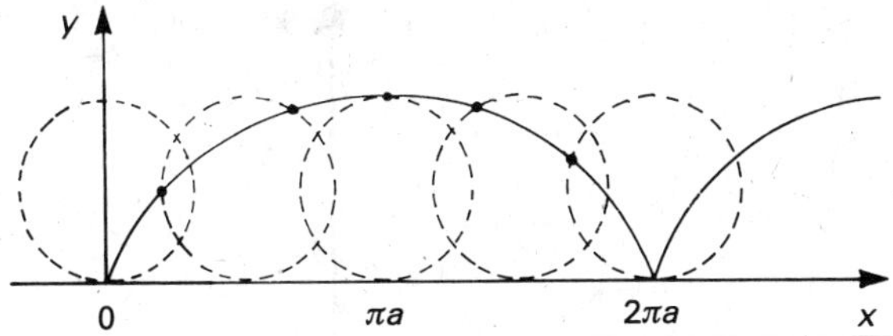

Fig. 2.3b *Cycloid generated by a fixed point on a rolling circle*

Note: In June 1696, Johann Bernoulli posed the *brachistochrone* problem as a challange to mathematicians. *Brachistochrone* means "shortest time" (from the Greek *brachistos* — shortest — and *chronos* — time). This problem was solved by Newton, Leibniz, L'Hospital, Johann Bernoulli himself and his elder brother Jakob Bernoulli. This problem was formulated and incorrectly solved by Galileo much earlier who gave the arc of a circle as the answer.

The brachistochrone problem, apart from its importance as a problem in mechanics, has a larger significance — historically, it was the source of a very important and useful branch of mathematical analysis called *calculus of variations*.

Example 2.13 [*Hanging chain problem*] Find the shape of an inextensible and thin chain suspended from two fixed points, not in a vertical line, and hanging under its own weight.

Solution 2.13 We choose the vertical line through the lowest point A of the chain as the y-axis (Fig. 2.4) and the direction perpendicular to the y-axis at a fixed distance below A, to be determined later, as the x-axis.

Let $\omega(s)$ be the linear density of the chain, i.e. the weight per unit length of the chain and s be the length of the arc AP along the curve. Now, the portion AP of the chain is in *equilibrium* under the action of three forces namely, (i) the horizontal tension T_0 at the lowest point A, (ii) the variable tension T at $P(x, y)$ which acts along the tangent to the curve at P and (iii) the downward weight of the chain between A and P.

The horizontal and vertical components of T are respectively $T\cos\psi$ and $T\sin\psi$. Due to the equilibrium of the chain AP, the horizontal component $T\cos\psi$ is equal and opposite to the horizontal tension T_0. Similarly, the vertical component $T\sin\psi$ is equal and opposite to the downward weight. In otherwords,

$$T\cos\psi = T_0 \quad \text{and} \quad T\sin\psi = \int_0^s \omega(s)ds. \qquad [1]$$

From the first of equations [1], we have $T\sin\psi = T\cos\psi\cdot\tan\psi = T_0\tan\psi = T_0 y'$ so that from the second of equations [1],

$$T_0 y' = \int_0^s \omega(s)ds. \qquad [2]$$

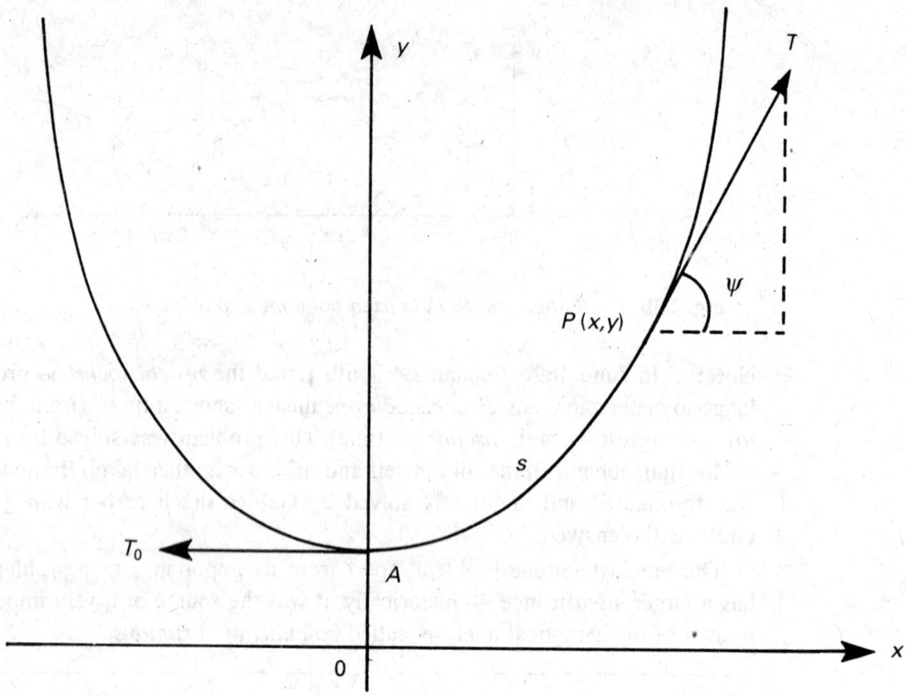

Fig. 2.4 *Hanging chain problem*

Differentiating [2] with respect to x, we get

$$T_0 y'' = \frac{d}{dx}\int_0^s \omega(s)ds = \frac{d}{ds}\int_0^s \omega(s)ds \cdot \frac{ds}{dx}$$

or $\quad T_0 y'' = \omega(s) \cdot \sqrt{1+(y')^2} \qquad [3]$

since $\quad \dfrac{ds}{dx} = \sqrt{1+(y')^2}.$

Now, [3] is the differential equation of the curve whose shape the suspended chain takes. Although [3] is a differential equation of the second order, we reduce it to the first order by substituting $y' = p$ and $y'' = dp/dx$ so that [3] reduces to

$$\frac{dp}{\sqrt{1+p^2}} = a\,dx \qquad [4]$$

where $\quad a = \dfrac{\omega(s)}{T_0} = \dfrac{\omega}{T_0}$ assuming that the linear density of the chain $\omega(s)$ is a constant, ω. By integrating [4], we get

$$\sinh^{-1} p = ax + c \quad \text{or} \quad p = \sinh\,(ax + c)$$

or $$\frac{dy}{dx} = \sinh\,(ax + c). \qquad [5]$$

At $x = 0$, the tangent is parallel to the x-axis and hence $dy/dx = 0$; putting these in [5], we get $c = 0$ so that

$$dy/dx = \sinh\,(ax) \quad \text{or} \quad y = (1/a)\cosh\,(ax) + c_1 \text{ (on integration)}$$

where c_1 is an arbitrary constant. Now, taking $y = 1/a$ when $x = 0$, we get $c_1 = 0$. Thus, the solution of the differential equation [3] is

$$y = \frac{1}{a}\cosh\,(ax). \qquad [6]$$

Equation [6] represents a curve called *catenary.*

Note: The lowest point A of the chain has the coordinates $(0, 1/a)$. By taking $c = 1/a$, [6] takes the form: $y = c\cosh\,(x/c)$ with $A \equiv (0, c)$.

Example 2.14 Find the curve for which the length of the normal at any point is proportional to the square of the ordinate of the point.

Solution 2.14 Let $P(x, y)$ be any point on the required curve. We know, from calculus, that the length of the normal to the curve at $P(x, y)$ is given by

$$PN = y\sqrt{1 + \left(\frac{dy}{dx}\right)^2}.$$

It is given that $PN \propto y^2$ or

$$y\sqrt{1 + \left(\frac{dy}{dx}\right)^2} = ay^2 \qquad [1]$$

where a is the constant of proportionality. Cancelling y in [1] and squaring, we get

$$1 + \left(\frac{dy}{dx}\right)^2 = a^2y^2 \quad \text{or} \quad \frac{dy}{dx} = \sqrt{a^2y^2 - 1}$$

or $$\frac{dy}{\sqrt{a^2y^2 - 1}} = dx \quad \text{or} \quad \frac{dy}{a\sqrt{y^2 - (1/a)^2}} = dx.$$

Also $$\int \frac{dy}{\sqrt{y^2 - (1/a)^2}} = \int a\,dx \text{ (on integration)}$$

$$\text{or}\quad \cosh^{-1}\left(\frac{y}{1/a}\right) = ax + c \quad \text{or} \quad y = \frac{1}{a}\cosh(ax + c).$$

This equation represents a *catenary* for any specified constant c.

Example 2.15 Find the equation to the curve whose polar subnormal is proportional to $\sin\theta$ where θ is the polar angle at a point (r, θ).

Solution 2.15 Let (r, θ) be the polar coordinates of any point on the required curve. We know, from calculus, that the polar subnormal (PSN) at (r, θ) is given by

$$PSN = \frac{dr}{d\theta}.$$

Since the polar subnormal is proportional to $\sin\theta$ at a point (r, θ) of the curve,

$$\frac{dr}{d\theta} = a\sin\theta; (a = \text{ constant of proportionality}) \qquad [1]$$

or $dr = a\sin\theta\, d\theta$. Integrating, we get $r = -a\cos\theta + c$ or $r = c - a\cos\theta$. In particular, if the curve is taken to pass through the point (0, 0) i.e. the pole, then $c = a$ so that the equation to the curve becomes

$$r = a(1 - \cos\theta).$$

This equation represents the famous curve, *cardioid* (Fig. 2.5).

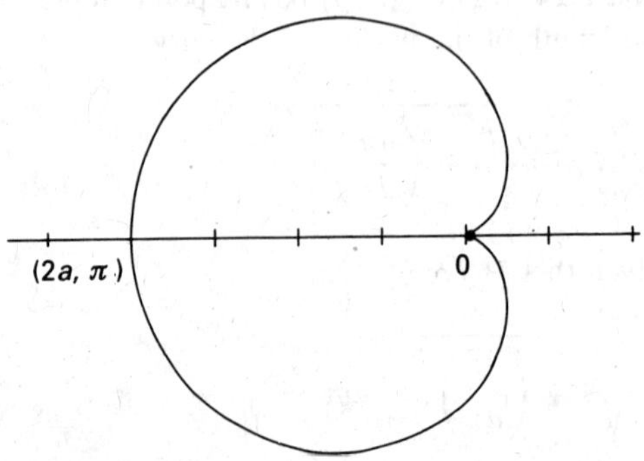

Fig. 2.5 *Cardioid* $r = a(1 - \cos\theta)$

Exercise 2A

I Find the general solutions of the following equations:

2.1 $(1+y^2)x\ dx = (1+x^2)ydy$

2.2 $y\dfrac{dy}{dx} + x^2 = 0$

2.3 $xy^2(1+x^2)\dfrac{dy}{dx} = y^3 + 1$

2.4 $(\cos x)\dfrac{dy}{dx} = (\sin x + x\sec x)\cot y$

2.5 $x\dfrac{dy}{dx} = y\log x$

2.6 $\sec y dx + \operatorname{cosec} x\ dy = 0$

2.7 $\cos^2 x\dfrac{dy}{dx} + y = 2$

2.8 $\dfrac{dy}{dx} + \sqrt{\dfrac{1-y^2}{1-x^2}} = 0$

2.9 $y\sqrt{1+x^2}dx + x\sqrt{1+y^2}\ dy = 0$

2.10 $\dfrac{ds}{dt} + t^2 = t^2e^{3t}$

II Reduce the following to the variables separable form and solve:

2.11 $(x-y)^2\dfrac{dy}{dx} = a^2$

2.12 $\dfrac{dy}{dx} = (x+y)^2$

2.13 $\dfrac{dy}{dx} = \cos(x+y)$

2.14 $(x+2y)(dx-dy) = dx+dy$

2.15 $\dfrac{dy}{dx} = \tan^2(x+y)$

2.16 $\dfrac{dy}{dx} = \dfrac{x-2y+3}{x-2y+5}$

2.17 $dy - (x+4y)^2\ dx = 0$ if $y = \dfrac{1}{8}$ when $x = 0$.

2.18 $(2x-2y+3)dx - (x-y+1)\ dy = 0$; given that $y = 1$ when $x = 0$.

2.19 $\dfrac{dy}{dx} = (3x+2y+4)^2$

2.20 $\dfrac{dy}{dx} = \dfrac{6x-2y-7}{3x-y+4}$

2.21 $\dfrac{dy}{dx} = e^{(x+y)} - 1$

2.22 $x^4\dfrac{dy}{dx} + x^3y + \operatorname{cosec}(xy) = 0$

[**Hint:** Substitute $v = xy$]

III Find the particular solutions of the following:

2.23 $x\dfrac{dy}{dx} = y(1-y)$given $y = 2$ when $x = -4$

2.24 $\dfrac{dy}{dx} = e^y\sin x$ given $y = 0$ when $x = 0$

2.25 $\dfrac{dy}{dx} = \dfrac{xe^y}{\sqrt{1+x^2}}, y\left(\dfrac{3}{4}\right) = 0$

2.26 $e^x \dfrac{dy}{dx} = xe^y, y(0) = 0$

2.27 $\dfrac{dy}{dx} = \dfrac{4xy}{x^2+1}, \; y(0) = 1$

IV Solve the following application-oriented problems:

2.28 A particle moves along a straight line so that at time t seconds its distance from a fixed point O is x cm and its velocity v cm/sec. If the acceleration of the particle is $(v^3 - v)\text{cm/sec}^2$ directed away from O and $v = 3$ at $x = 0$, show that

$$x = \frac{1}{2}\log 2 - \coth^{-1} v.$$

2.29 A particle moves along the x-axis so that its velocity is proportional to the product of its instantaneous position x measured from $x = 0$, and the time t measured from $t = 0$. If the particle is located at $x = 54$ when $t = 0$ and at $x = 36$ when $t = 1$, where will it be when $t = 2$?

2.30 The current i in an electric circuit comprising an inductance L and a resistance R, where L and R are constants, satisfies the equation

$$L\frac{di}{dt} + Ri = E$$

in the presence of a constant electromotive force (emf) E, t being calculated from the moment E is applied. Show that $i \to E/R$ as $t \to \infty$.

2.31 A copper ball is heated to a temperature of 100° C. At time $t = 0$, it is placed in water which is maintained at a temperature of 30° C. At the end of 3 minutes, the temperature of the ball is reduced to 70° C. Find the time at which the temperature of the ball is reduced to 31° C.

2.32 A naphthalene ball loses its volume by evaporation at a rate proportional to its instantaneous surface area. If the diameter of the ball decreases from 2 cm to 1 cm in 3 months, how long will it take for the ball to evaporate until its diameter is reduced 1 mm?

2.33 The population of the world has been estimated at 700 million in 1700 and as 1800 million in 1900. Assuming that the rate of population growth is proportional to the current population at the moment, estimate the population of the world in 1960 (using a *billion* people as the population unit and a *century* as the time unit). What will be the world population in the year 2000 according to this model? [**Note:** In USA, 1 billion = 1 thousand million = 10^9].

2.34 A family of salmon fish living off the Alaskan Coast obeys the *Malthusian law* of population growth, $dp/dt = 0.003p$ where t is in minutes. At time $t = 0$, a group of sharks establishes residences in these waters and begins attacking the salmon fish. The rate at which the fish are killed by the sharks is $0.001\,p^2$ where $p = p(t)$ is the population of the fish at time t. Further, since an enemy has entered into their jurisdiction, 0.002 salmon fish per minute leave the Alaskan waters. (i) Modify the Malthusian law to take into account these two factors; (ii) Assuming that at $t = 0$ there are one million fish, find the population $p(t)$. What happens as $t \to \infty$?

2.4 Homogeneous equations

A differential equation of the type

$$\frac{dy}{dx} = f\left(\frac{y}{x}\right) \qquad [2.10]$$

is called a *homogeneous* differential equation of the first degree. In [2.10] it is understood that $f(y/x)$ is a function depending only on the quotient y/x of the variables. For example,

$$\frac{dy}{dx} = \frac{y}{x} + \sqrt{1 + \left(\frac{y}{x}\right)^2}$$

is a homogeneous equation.

A function $\phi(x, y)$ is called a *homogeneous function* of degree n if

$$\phi(tx, ty) = t^n \phi(x, y) \qquad [2.11]$$

for all suitably restricted values of x, y and t. For example, $x^3 - 3xy^2$, $\sqrt{2x^2 - 3y^2}$ and $\cos(x/y)$ are homogeneous functions of degrees 3, 1 and 0 respectively.

If a function $\phi(x, y)$ is a homogeneous function of degree 0, it can always be expressed as another function of the quotient y/x. Now consider a differential equation of the form as follows

$$M(x, y)dx + N(x, y)\, dy = 0. \qquad [2.12]$$

If $M(x, y)$ and $N(x, y)$ are homogeneous functions of the *same order*, then the function $M(x, y)/N(x, y)$ is a homogeneous function of degree 0 and hence [2.12] takes the form

$$\frac{dy}{dx} = -\frac{M(x, y)}{N(x, y)} = f\left(\frac{y}{x}\right) \qquad [2.13]$$

where $f(y/x)$ is some function of y/x. For example, consider the differential equation

$$(x^2 + 3xy + 2y^2)dx + (2x^2 - xy + 3y^2)dy = 0.$$

Here, the functions $M(x,y) = x^2 + 3xy + 2y^2$ and $N(x,y) = 2x^2 - xy + 3y^2$ are homogeneous functions of the *same degree* 2.

The given equation can be written in the form

$$\frac{dy}{dx} = -\frac{x^2 + 3xy + 2y^2}{2x^2 - xy + 3y^2} \equiv \phi(x,y), \text{ say.}$$

Now, on dividing the numerator and the denominator of the right-hand side by x^2,

$$\phi(x,y) = -\frac{1 + 3(y/x) + 2(y/x)^2}{2 - (y/x) + 3(y/x)^2} \equiv f(y/x), \text{ say.}$$

Thus, the given differential equation has taken the form

$$\frac{dy}{dx} = -\frac{M(x,y)}{N(x,y)} = f\left(\frac{y}{x}\right)$$

where $M(x,y)$ and $N(x,y)$ are homogeneous functions of the *same degree* in x and y.

2.5 Method of solving homogeneous equations

A homogeneous differential equation [2.10], say

$$\frac{dy}{dx} = f\left(\frac{y}{x}\right)$$

can be solved by changing it to the *variables separable* form on substituting

$$\frac{y}{x} = v \text{ or } y = vx. \qquad [2.14]$$

that is, by changing the dependent variable from y to v, keeping the same independent variable x. Differentiating [2.14] with respect to x, we get

$$\frac{dy}{dx} = v + x\frac{dv}{dx}. \qquad [2.15]$$

Using [2.9a] in [2.4a], we get $v + x\dfrac{dv}{dx} = f(v)$ so that, on separation of variables,

$$\frac{dx}{x} = \frac{dv}{f(v) - v}, \qquad [2.16]$$

By integrating [2.16], we get the general solution of [2.10]. However, it should be noted that this method breaks down if $f(v) = v$ i.e. $f(y/x) = y/x$.

Example 2.16 Solve $(x - y)dy - (x + y)dx = 0.$

Solution 2.16 The given equation can be written as $\dfrac{dy}{dx} = \dfrac{x+y}{x-y}$.

Since the function in the right-hand side is a homogeneous function, we substitute $y = vx$ so that

$$\frac{dy}{dx} = v + x\frac{dv}{dx}.$$

The equation becomes

$$v + x\frac{dv}{dx} = \frac{1+v}{1-v} \quad \text{or} \quad x\frac{dv}{dx} = \frac{1+v}{1-v} - v = \frac{(1+v) - v(1-v)}{1-v}$$

or $$\frac{1-v}{1+v^2}dv = \frac{dx}{x}.$$

On integration, we have $$\int \frac{1}{1+v^2}\,dv - \frac{1}{2}\int \frac{2v}{1+v^2}\,dv = \int \frac{dx}{x}$$

or $$\tan^{-1} v - \frac{1}{2}\log(1+v^2) = \log x + c$$

or $$\tan^{-1}\frac{y}{x} - \frac{1}{2}\log\left(1 + \frac{y^2}{x^2}\right) = \log x + c$$

or $$\tan^{-1}\frac{y}{x} - \log\sqrt{x^2+y^2} = c$$

or $$\tan^{-1}\frac{y}{x} = \log\sqrt{x^2+y^2} + c.$$

Example 2.17 Solve $\dfrac{dy}{dx} = \dfrac{x^2 - 2xy + 5y^2}{x^2 + 2xy + y^2}$ given that $y = -1$ when $x = 0$.

Solution 2.17 The equation can be written as $\dfrac{dy}{dx} = \dfrac{1 - 2(y/x) + 5(y/x)^2}{1 + 2(y/x) + (y/x)^2}$.

Substituting $\dfrac{y}{x} = v$ and $\dfrac{dy}{dx} = v + x\dfrac{dv}{dx}$ the equation becomes

$$x\frac{dv}{dx} = \frac{1-2v+5v^2}{1+2v+v^2} - v$$

or $$x\frac{dv}{dx} = \frac{(1-2v+5v^2) - v(1+2v+v^2)}{(1+v)^2} = \frac{(1-v)^3}{(1+v)^2}$$

so that $$\int \frac{dx}{x} = \int \frac{(1+v)^2}{(1-v)^3}dv.$$

In order to perform integration, we substitute $1 - v = t, dv = -dt$, and the integral takes the form

$$\log x = \frac{2}{t^2} - \frac{4}{t} - \log t + c.$$

Substituting back $t = 1 - v = 1 - y/x$ the general solution of the equation is

$$\log(x-y) = \frac{2x^2}{(y-x)^2} - \frac{4x}{(y-x)} + c.$$

The given initial condition is $y = -1$ when $x = 0$ so that we get $c = 0$. Therefore the solution becomes

$$\log(x-y) = \frac{2x}{y-x}\left[\frac{x}{y-x} - 2\right].$$

Example 2.18 Solve: $\dfrac{dy}{dx} = \dfrac{x+2y-3}{2x+y-3}$.

Solution 2.18 The equation is *not* homogeneous. But it can be reduced to the homogeneous type by substituting

$$x = X + h \text{ and } y = Y + k$$

where h and k are constants to be determined later. From the substitution, we have

$$\frac{dy}{dx} = \frac{dY}{dX}$$

and the equation takes the form

$$\frac{dY}{dX} = \frac{(X+2Y)+(h+2k-3)}{(2X+Y)+(2h+k-3)} \qquad [1]$$

Now, h and k are chosen such that $h+2k-3=0$ and $2h+k-3=0$, solving which we get $h = 1$ and $k = 1$. Therefore [1] takes the form

$$\frac{dY}{dX} = \frac{X + 2Y}{2X + Y} \qquad [2]$$

which is a homogeneous equation. By substituting

$$Y = VX \text{ and } \frac{dY}{dX} = V + X\frac{dV}{dX},$$

equation [2] takes the form $X\dfrac{dV}{dX} = \dfrac{1+2V}{2+V} - V$ or $\dfrac{dX}{X} = \dfrac{2+V}{1-V^2}\, dV$

so that $\displaystyle\int \frac{dX}{X} = \int \left[\frac{1/2}{1+V} + \frac{3/2}{1-V}\right] dV$

or $\log X = \dfrac{1}{2}\log(1+V) - \dfrac{3}{2}\log(1-V) + \log C$

where C is an arbitrary constant. The general solution simplifies to

$$X^2(1-V)^3 = C^2(1+V). \qquad [3]$$

Now, substituting $V = \dfrac{Y}{X} = \dfrac{y-k}{x-h} = \dfrac{y-1}{x-1}$ in [3], we get the general solution

$$(x-y)^3 = C^2(x+y-2).$$

Note: A differential equation of the form

$$\frac{dy}{dx} = \frac{ax + by + c}{lx + my + n}$$

can be reduced to the *homogeneous* form if and only if the determinant

$$\begin{vmatrix} a & b \\ l & m \end{vmatrix} \neq 0, \text{i.e. } \frac{a}{l} \neq \frac{b}{m}.$$

On the other hand, if

$$\frac{a}{l} = \frac{b}{m} = k,$$

then the equation takes the form $\dfrac{dy}{dx} = \dfrac{k(lx+my)+c}{(lx+my)+n}$ which can be solved by the separation of variables by substituting

$$z = lx + my.$$

Example 2.19 Solve: $\dfrac{dy}{dx} = \dfrac{x + 2y + 1}{2x + 4y + 3}$.

Solution 2.19 This equation is *not* homogeneous and belongs to the type mentioned in the above note. The equation can be written as

$$\frac{dy}{dx} = \frac{(x + 2y) + 1}{2(x + 2y) + 3}.$$

We substituting $z = x + 2y$ so that on differentiation with respect to x,

$$\frac{dy}{dx} = \frac{1}{2}\left(\frac{dz}{dx} - 1\right).$$

Then the differential equation becomes

$$\frac{1}{2}\left(\frac{dz}{dx} - 1\right) = \frac{z + 1}{2z + 3} \quad \text{or} \quad \frac{dz}{dx} = \frac{2z + 2}{2z + 3} + 1 = \frac{4z + 5}{2z + 3}.$$

Separating the variables, we have $\dfrac{2z + 3}{4z + 5}\, dz = dx$. Integrating, we get

$$\int \frac{1/2(4z + 5) + 1/2}{4z + 5}\, dz = \int dx \quad \text{or} \quad \frac{1}{2}\int \left[1 + \frac{1}{4z + 5}\right] dz = x + c_1$$

$$\text{or} \quad \frac{1}{2}\left[z + \frac{1}{4}\log(4z + 5)\right] = x + c_1 \quad \text{or} \quad 4z + \log(4z + 5) = 8x + c$$

where $c = 8c_1$, an arbitrary constant. Substituting back for z, the solution is

$$4(x + 2y) + \log(4x + 8y + 5) = 8x + c$$

$$\text{or} \quad 4(2y - x) + \log(4x + 8y + 5) = c.$$

This is the general solution of the given differential equation.

Example 2.20 Find the shape of a reflector which is such that the light rays emitted by a point source are reflected parallel to a fixed line.

Solution 2.20 Suppose the origin O of a rectangular coordinate system represents a point source of light and CD is the cross-section of the reflector (Fig. 2.6). Let OP be

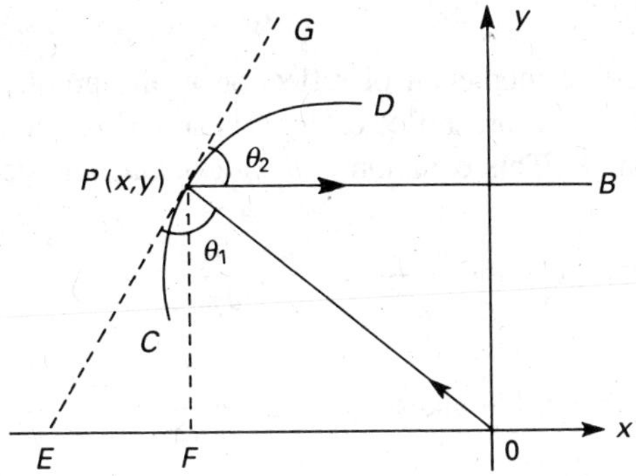

Fig. 2.6 *Parabolic reflector*

an incident ray and PB be the ray reflected by the reflector CD such that PB is parallel to the x-axis.

We need to find the shape of the reflector such that all the rays emanating from O and hitting the reflector are reflected parallel to a fixed line, say, the x-axis.

Let $P(x, y)$ be any point on the reflector and let

$$O\hat{P}E = \theta_1 \quad \text{and} \quad B\hat{P}G = \theta_2$$

where EG is the tangent to the curve CD at P and cutting the x-axis at E. Actually, the angles of reflection and incidence are respectively $90° - \theta_2$ and $90° - \theta_1$. Now, by the law of reflection

$$90° - \theta_1 = 90° - \theta_2 \quad \text{or} \quad \theta_1 = \theta_2.$$

Since PB is parallel to EX, we have

$$O\hat{E}P = \theta_2, X\hat{O}P = O\hat{E}P + O\hat{P}E = \theta_2 + \theta_1 = 2\theta_1$$

since $\theta_2 = \theta_1$. The slope of OP is $\tan(2\theta_1) = (y/x)$.

Therefore, $$\frac{2\tan\theta_1}{1 - \tan^2\theta_1} = \frac{y}{x}.$$

But $\tan\theta_1 = \tan\theta_2 = (dy/dx) \equiv y'$ so that we have

$$\frac{2y'}{1 - (y')^2} = \frac{y}{x}.$$

Simplifying and solving the resulting quadratic equation in y', we get

$$\frac{dy}{dx} = \frac{-x \pm \sqrt{x^2+y^2}}{y} \tag{1}$$

Thus, the phenomenon of reflection in the problem, in which the rays emanating from a point source are reflected as a beam of parallel rays is governed by the differential equation [1]. This equation is *homogeneous* and hence solved by the substitution

$$y = vx \quad \text{and} \quad \frac{dy}{dx} = v + x\frac{dv}{dx}.$$

Then equation [1] takes the form $x\dfrac{dv}{dx} = \dfrac{-1 - v^2 \pm \sqrt{v^2+1}}{v}$.

On integration, after separating the variables,

$$\int \frac{dx}{x} = -\int \frac{v\,dv}{(v^2+1) \pm \sqrt{v^2+1}} \tag{2}$$

Now, substituting $v^2 + 1 = u^2$ and $vdv = udu$ in [2], we get

$$\log x + c_1 = -\int \frac{du}{u \pm 1} \quad \text{or} \quad \log x + c_1 = -\log(u \pm 1)$$

or $\log x + c_1 = -\log[\sqrt{v^2+1} \pm 1]$ from which it follows that

$$x\left[\sqrt{v^2+1} \pm 1\right] = c \quad \text{or} \quad \sqrt{x^2+y^2} = c \pm x$$

since $v = y/x$. Squaring and simplifying this equation, we get

$$y^2 = c^2 \pm 2cx. \tag{3}$$

Now, for a given value of $c(\neq 0)$, [3] represents two *parabolas* which are symmetrical with respect to the x-axis as shown in Fig. 2.7. The two parabolas have the separate equations

$$y^2 = 2cx + c^2, c > 0 \tag{4}$$

$$y^2 = -2cx + c^2, c > 0 \tag{5}$$

which are respectively "open to the right" and "open to the left" (see Fig. 2.7). Both the parabolas have the same focus viz., the origin O. By considering different rays emanating from the point source at O, we get a family of parabolas.

If, for example, the parabola [4] is revolved about the x-axis, we obtain (a *paraboloid of revolution*). An electric bulb placed at the origin, i.e. the focus, sends rays of light "bouncing off" the reflector to produce a direct beam of light rays which is the most efficient way of lighting. This important property accounts for the paraboloid shape of automobile headlights.

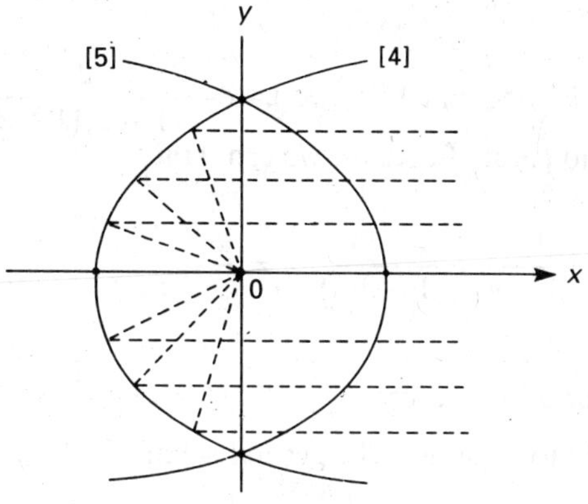

Fig. 2.7 *Parabolic reflector*

Example 2.21 Find the curve for which the subnormal at any point on it is equal to half the sum of the coordinates of the point in a rectangular cartesian system.

Solution 2.21 Let $P(x, y)$ be any point on the required curve and SN be the length of the subnormal at P.

We know that the length of the subnormal at a point (x, y) is given by

$$SN = y\frac{dy}{dx}.$$

It is given that $SN = \frac{1}{2}(x + y)$ so that we have

$$y\frac{dy}{dx} = \frac{1}{2}(x + y). \quad [1]$$

Equation [1] is the differential equation of the required curve. It can be written as

$$\frac{dy}{dx} = \frac{1}{2}\,\frac{x + y}{y}$$

which is a homogeneous equation (of degree 0). We substitute

$$y = vx \quad \text{and} \quad \frac{dy}{dx} = v + x\frac{dv}{dx}$$

so that $v + x\dfrac{dv}{dx} = \dfrac{1}{2}\,\dfrac{1+v}{v}$ or $x\dfrac{dv}{dx} = \dfrac{1+v}{2v} - v = \dfrac{1+v-2v^2}{2v}$

or $x\dfrac{dv}{dx}=\dfrac{(1-v)(1+2v)}{2v}$.

Separating the variables, we have $\dfrac{2v}{(1-v)(1+2v)}\,dv=\dfrac{dx}{x}$. Resolving the left-hand side into partial fractions, we can write

$$2\left[\frac{(1/3)}{1-v}+\frac{(-1/3)}{1+2v}\right]dv=\frac{dx}{x}.$$

On integration, we get $\dfrac{2}{3}\left[-\log(1-v)-\dfrac{1}{2}\log(1+2v)\right]=\log x+\log c_1$, where c_1 is an arbitrary constant. The general solution gets simplified as

$$-\frac{1}{3}\left[\log(1-v)^2+\log(1+2v)\right]=\log(c_1x)$$

or $\log\left[(1-v)^2(1+2v)\right]=\log\,(c_1x)^{-3}$ or $(1-v)^2(1+2v)=\dfrac{1}{c_1^3x^3}$.

Substituting back for v as $v=y/x$, the general solution takes the form

$$\left(1-\frac{y}{x}\right)^2\left(1+\frac{2y}{x}\right)=\frac{1}{c_1^3x^3}\quad\text{or}\quad\frac{(x-y)^2}{x^2}\frac{(x+2y)}{x}=\frac{1}{c_1^3x^3}$$

or $(x-y)^2(x+2y)=c$

where $c=c_1^3$, an arbitrary constant.

Exercise 2B

Solve the following equations:

2.35 $xy\dfrac{dy}{dx}=x^2+y^2$

2.36 $(x^2+y^2)\dfrac{dy}{dx}=xy$

2.37 $(x^2-2y^2)dx+xydy=0$

2.38 $x^2\dfrac{dy}{dx}-3xy-2y^2=0$

2.39 $x^2\dfrac{dy}{dx}=3(x^2+y^2)\tan^{-1}\dfrac{y}{x}+xy$

2.40 $x\sin\left(\dfrac{y}{x}\right)\dfrac{dy}{dx}=y\sin\left(\dfrac{y}{x}\right)+x$

2.41 $(x^2+y^2)dy = 2(x^2+xy-y^2)dx, y(0)=2$

2.42 $t\dfrac{dx}{dt} = 2x+\sqrt{x^2+t^2}, x(1)=0$

2.43 $(y^2-3x^2)ydy = (x^2-3y^2)xdx, y(0)=2$

2.44 $\dfrac{dy}{dx} = \dfrac{y(x+2y)}{x(2x+y)}, y(0)=1$

2.45 $xy\dfrac{dy}{dx} = x^2+2y^2, y(1)=0$

2.46 $\dfrac{dy}{dx} = \dfrac{x+y+4}{x-y-6}$

2.47 $(x+y+4)dx = (x+y-6)dy$

2.48 $(x-y+1)\dfrac{dy}{dx} = (x+y+1)$

2.49 $\dfrac{dy}{dx} = \dfrac{2(x+y-1)}{3x+y-5}$

2.50 $(t-8x+7)dx = (t-x)dt$

2.51 $(4x+y-3)dy = (2x+3y-4)dx, y(1)=1$

2.52 $\dfrac{dx}{dt} = \dfrac{2t-5x+3}{2t+4x-6}, x(0)=-1$

2.53 Find a curve for which the subnormal at any point is equal to half the sum of the cartesian coordinates of the point.

2.54 Find the equation to the curve for which the tangent at any point on it cuts the x-axis at a point which is equidistant from the origin and the point on the curve.

2.6 Linear differential equations

A first order *linear* differential equation is one which is linear with respect to an unknown variable and its derivative. If y is the unknown variable and x is the known (or independent) variable, then a first order linear differential equation is of the form

$$\frac{dy}{dx} + P(x)y = Q(x) \qquad [2.17]$$

where $P(x)$ and $Q(x)$ are independent of y. By multiplying [2.17] throughout by $e^{\int P dx}$ we get

$$e^{\int P dx}\frac{dy}{dx} + e^{\int P dx}Py = Qe^{\int P dx} \qquad [2.18]$$

which can be written as $\frac{d}{dx}\left[ye^{\int P dx}\right] = Qe^{\int P dx}$. On integration, we get

$$ye^{\int P dx} = \int Qe^{\int P dx} dx \qquad [2.19]$$

The integral in the right-hand side of [2.19], being an indefinite integral, contains an arbitrary constant C. The factor [2.18] with which the differential equation [2.17] is multiplied to render it integrable is called the *integrating factor* of [2.17].

Note: (i) From the preceding discussion one might get an impression that for a linear differential equation of the form [2.17] an integrating factor can be easily determined and hence the equation can be solved. But, in many cases, it may not be possible to find an integrating factor easily even if one exists. For example, consider the equation

$$\frac{dy}{dx} + (\sqrt{1+x^3})y = 1$$

for which an integrating factor is $e^{\int \sqrt{1+x^3} dx}$. But this is a difficult integral which would make a solution by this method unattainable.
(ii) Even if an integrating factor can be easily determined, there are many instances where the integration in the right-hand side of [2.19] cannot be carried out. Consider, for example,

$$\frac{dy}{dx} + 2xy = 1.$$

The integrating factor of this equation is $e^{\int 2x dx} = e^{x^2}$. Multiplying the equation by the integrating factor we get $d[ye^{x^2}]/dx = e^{x^2}$. On integration, we have $ye^{x^2} = \int e^{x^2} dx$.

The integral in the right-hand side cannot be expressed in a "closed form" in terms of the familiar functions. These circumstances necessitate the use of numerical methods to solve such differential equations over a given finite interval (see Chapter 16).

Example 2.22 Solve: $\frac{dy}{dx} + \frac{y}{x} = 1$.

Solution 2.22 This is a linear differential equation of the type $dy/dx + Py = Q$. Here $P = 1/x$ and $Q = 1$. The integrating factor is

$$e^{\int P dx} = e^{\int (1/x) dx} = e^{\log x} = x$$

so that the general solution is $ye^{\int P dx} = \int Qe^{\int P dx} dx$, i.e.,

$$yx = \int (1)x dx \quad \text{or} \quad xy = \frac{x^2}{2} + c_1 \quad \text{or} \quad 2xy = x^2 + 2c_1$$

or $x^2 - 2xy = c$ (where $c = -2c_1$ is an arbitrary constant).

Example 2.23 Solve: $y^2\dfrac{dx}{dy} + xy = 2y^2 + 1.$

Solution 2.23 The above equation, similar to

$$\frac{dx}{dy} + \frac{x}{y} = 2 + \frac{1}{y^2} \qquad [1]$$

is not linear with respect to y as the unknown function. However, [1] can be considered as linear with respect to x being treated as the unknown function and y the independent variable. In that case [1] is a *linear* equation of the form

$$\frac{dx}{dy} + P(y)x = Q(y). \qquad [2]$$

The integrating factor of [2] is $e^{\int P(y)dy} = e^{\int (1/y)dy} = e^{\log y} = y$, so that the general solution is given by

$$xe^{\int P(y)dy} = \int Q(y)e^{\int P(y)dy}dy$$

that is $xy = \displaystyle\int (2 + \frac{1}{y^2})ydy$ or $xy = y^2 + \log y + c.$

Example 2.24 Solve: $\dfrac{dI}{dt} + \dfrac{10I}{2t+5} = 10$ given that $I = 0$ when $t = 0.$

Solution 2.24 The equation is in the form

$$\frac{dI}{dt} + P(t)I = Q(t)$$

where $P(t) = 10/(2t+5)$ and $Q(t) = 10$ and I is the unknown function, t being the independent variable. The integrating factor is

$$e^{\int P(t)dt} = e^{\int 10/(2t+5)dt} = e^{5\log(2t+5)} = (2t+5)^5.$$

Therefore, the general solution is

$$I(2t+5)^5 = \int 10(2t+5)^5 dt \quad \text{or} \quad (2t+5)^5 I = \frac{5}{6}(2t+5)^6 + c.$$

Now, substituting the initial values, $I = 0$ and $t = 0$ in the general solution, we get

$$c = -\frac{5}{6}(5)^6 = -\frac{78,125}{6}.$$

Therefore, the solution of the given initial value problem is

$$I = \frac{5}{6}(2t+5) - \frac{78,125}{6(2t+5)^5}.$$

Example 2.25 A tank contains S_0 kilograms (mass) of salt dissolved in 1000 litres of water. Starting from time $t = 0$, water containing 100 gm (mass) of salt per litre enters the tank at the rate of 20 litres/minute and the well-stirred solution leaves the tank at the same rate. Find the concentration of salt in the tank at any time, t (minutes).

Solution 2.25 Let $S(t)$ denote the amount of salt in the tank at time t. Then the rate of change of salt content $S'(t)$ in the tank at time t must be equal to the rate at which salt enters the tank *minus* the rate at which it leaves the tank.

The rate at which salt enters the tank :

(100 gm/litre) × (20 litres/min) = 2 kg/min.

The rate at which salt leaves the tank :

(20 litres/min) $(S(t)/1000)$kg/litre = $S(t)/50$ kg/min.

This results in the differential equation

$$\frac{dS}{dt} = 2 - \frac{S}{50}, \; S(0) = S_0. \qquad [1]$$

The integrating factor of [1] is $e^{\int(1/50)dt} = e^{t/50}$ so that the general solution is

$$Se^{t/50} = \int 2e^{t/50}dt \quad \text{or} \quad Se^{t/50} = 2\frac{e^{t/50}}{(1/50)} + c$$

or $Se^{t/50} = 100e^{t/50} + c.$ [2]

Initially at $t = 0$, $S = S_0$ so that we get $c = S_0 - 100$. Substituting for c in [2], the solution is

$$Se^{t/50} = 100e^{t/50} + S_0 - 100 \quad \text{or} \quad S = S_0e^{-t/50} + 100(1 - e^{-t/50}).$$

Therefore, the concentration $C(t)$ of salt in the tank at any time $t > 0$ is

$$C(t) = \frac{S(t)}{200} = \frac{S_0}{200}e^{-t/50} + \frac{1}{2}(1 - e^{-t/50}).$$

As $t \to \infty$, we see that $S \to 100$ kg and $C(t) \to (1/2)$kg/litre as the limiting values of the amount of salt in the tank and its concentration.

Example 2.26 The current i flowing in an electrical circuit is governed by the equation

$$L\frac{di}{dt} + Ri = E_0 \sin(\omega t)$$

where L is the constant inductance, R is the constant resistance and $E_0 \sin(\omega t)$ is the voltage at time t, E_0 and ω being constants. Find the current at any time t assuming that initially it is zero.

Solution 2.26 The given equation can be written as

$$\frac{di}{dt} + \frac{R}{L}i = \frac{E_0}{L}\sin(\omega t)$$

which is *linear* in i. The integrating factor is $e^{\int (R/L)dt} = e^{Rt/L}$.

The general solution, therefore, is $ie^{Rt/L} = (E_0/L)\int \sin(\omega t)\cdot e^{Rt/L}dt$

$$\text{or} \quad ie^{Rt/L} = \left(\frac{E_0}{L}\right)\left(\frac{e^{(Rt/L)}[(R/L)\sin(\omega t) - \omega \cos \omega t)]}{(R^2/L^2) + \omega^2}\right) + c$$

$$\text{or} \quad i = \frac{E_0}{R^2 + \omega^2 L^2}\left[R\sin(\omega t) - \omega L\cos(\omega t)\right] + ce^{-Rt/L}.$$

From the given initial condition, we have $i = 0$ when $t = 0$. This gives

$$c = \frac{\omega E_0 L}{R^2 + \omega^2 L^2}.$$

Therefore, the current i at any time t is given by

$$i = \frac{E_0}{R^2 + \omega^2 L^2}\left[R\sin(\omega t) - \omega L\cos(\omega t) + \omega L e^{-Rt/L}\right].$$

Example 2.27 A body of mass m is projected with speed V on a rough table whose coefficient of friction is μ. If the air resistance is proportional to the square of the speed, find the distance travelled by the body before it comes to rest.

Solution 2.27 Let m be the mass of the body and R_1 be the air resistance on the body when it is in motion (Fig. 2.8). Since R_1 is proportional to v^2, we have

$$R_1 = mkv^2 \qquad [1]$$

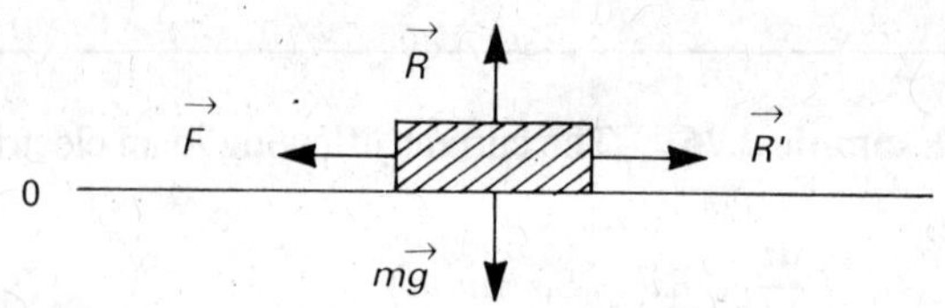

Fig. 2.8 *Motion on a rough surface*

where k is a constant and v is the velocity.

From the laws of mechanics, we know that the normal reaction $\vec{R}$ on the body of mass m, offered by the table, is equal and opposite to the weight $m\vec{g}$ so that the body does not have vertical motion, where $\vec{g}$ is the constant acceleration due to gravity. That is

$$R = mg. \tag{2}$$

The frictional force $\vec{F}$ is proportional to the normal reaction $\vec{R}$, in magnitude, so that we have

$$F = \mu R \tag{3}$$

where μ is the coefficient of friction. From [2] and [3], we have

$$F = \mu mg. \tag{4}$$

Therefore, the total force resisting the motion of the body is given by

$$F + R_1 = \mu mg + mkv^2 \quad (\text{using}[1]\text{and}[4]).$$

Since, by Newton's second law of motion, the force of motion is $m(dv/dt)$, the equation of motion is

$$m\frac{dv}{dt} = -m(\mu g + kv^2). \tag{5}$$

The acceleration term dv/dt can be written as

$$\frac{dv}{dt} = \frac{dv}{dx}\cdot\frac{dx}{dt} = \frac{dv}{dx}\cdot v = \frac{1}{2}\frac{d}{dx}(v^2).$$

Hence [5] becomes $\frac{1}{2}m\frac{d}{dx}(v^2) = -m(\mu g + kv^2)$, or

$$\frac{d}{dx}(v^2) + 2kv^2 = -2\mu g. \tag{6}$$

Comparing [6] with the standard form of the linear differential equation

$$\frac{dy}{dx} + P(x)y = Q(x)$$

we have $y = v^2$, $P = 2k$ and $Q = -2\mu g$ where P and Q are constants.

The integrating factor is $e^{\int P dx} = e^{\int 2k dx} = e^{2kx}$ so that the general solution is

$$v^2 e^{2kx} = -2\mu g \int e^{2kx} dx = -\frac{\mu g}{k} e^{2kx} + c. \qquad [7]$$

Initially, when $x = 0$, it is given that the velocity $v = V$ so that on substitution in the general solution, we get

$$c = V^2 + \frac{\mu g}{k}. \qquad [8]$$

Using [8] in [7], the solution becomes $v^2 e^{2kx} = [-\mu g/k e^{2kx}] + [V^2 + \mu g/k]$

$$\text{or} \quad v^2 = -\frac{\mu g}{k} + \left[V^2 + \frac{\mu g}{k}\right] e^{-2kx} \qquad [9]$$

The second term in [9] signifies an exponential decay of the velocity-square with distance.

Now, the body comes to rest when the velocity $v = 0$. Then, by substituting $v = 0$ in [9], we get

$$\frac{-\mu g}{k} = \left[V^2 + \frac{\mu g}{k}\right] e^{-2kx} \quad \text{or} \quad e^{2kx} = \frac{[V^2 + \mu g/k]}{(\mu g/k)} = \frac{k}{\mu g} V^2 + 1.$$

Therefore, the distance x travelled by the body before it comes to rest is

$$x = \frac{1}{2k} \log\left[1 + \frac{k}{\mu g} V^2\right].$$

Example 2.28 The slope of the tangent at any point of a curve is twice the abscissa together with thrice the ordinate of the point. Obtain the equation to the curve if it passes through the origin.

Solution 2.28 Let (x, y) be any point on the required curve. It is given that

$$\frac{dy}{dx} = 2x + 3y \quad \text{or} \quad \frac{dy}{dx} - 3y = 2x$$

which is a *linear* equation in y. The integrating factor is $e^{\int -3dx}$, i.e. e^{-3x}. Therefore, the general solution is

$$ye^{-3x} = \int 2x \cdot e^{-3x} dx \quad \text{or} \quad ye^{-3x} = 2\left[x\frac{e^{-3x}}{-3} - \int \frac{e^{-3x}}{-3} dx\right]$$

$$\text{or} \quad ye^{-3x} = \frac{-2e^{-3x}}{9}\left[3x+1\right] + c.$$

Since the curve passes through (0,0), $c = 2/9$ so that the equation to the curve is

$$y = \frac{2}{9}\left[e^{3x} - 3x - 1\right].$$

Exercise 2C

I Find the general solutions of the following equations:

2.55 $\dfrac{dy}{dx} + \dfrac{y}{x} = 1$ **2.56** $xy' + 3y = x^2$ **2.57** $y^2\dfrac{dy}{dx} + xy = 2y^2 + 1$

[**Hint** for **2.57**: Rewrite the equation to contain (dx/dy), treat y as the independent variable; the integrating factor is $e^{\int Pdy}$ where P is the coefficient of x].

2.58 $\dfrac{dy}{dx} - \dfrac{2y}{x} = x^2 \sin 3x$ **2.59** $y' + y\cot x = \cos x$

II Find the particular solutions of the following:

2.60 $\dfrac{dI}{dt} + 3I = e^{-2t}, I(0) = 5$ **2.61** $x\dfrac{dy}{dx} + 7y = 2x + 9, y(1) = 1$

2.62 $x\dfrac{dy}{dx} - y = x^3 \cos x, y(\pi) = 0$

2.63 $(x+1)\dfrac{dy}{dx} + (1+x)y = 2\sin x, y(0) = 2$

2.64 $x\cos x\dfrac{dy}{dx} + (x\sin x - \cos x)y - x^2 = 0, y\left(\dfrac{\pi}{4}\right) = 0$

2.65 $\dfrac{dy}{dx} + \dfrac{3x+2}{x^2+3x-4}\,y = \dfrac{1}{x+4},\ y(0) = \dfrac{1}{4}$

2.66 The velocity V of a rocket in motion obeys the equation of motion

$$(M - mt)\frac{dv}{dt} - mu = -(M - mt)g - kmV$$

where M, m, u, g and k are all constants. Show that the velocity of the rocket at any instant t is given by

$$V = \frac{u}{k} + \frac{g(M - mt)}{m(1-k)} - \left[\frac{Mg}{m(1-k)} + \frac{u}{k}\right]\left[1 - \frac{mt}{M}\right]^k$$

2.67 The velocity v of a solid body falling under gravity in a liquid satisfies the equation

$$\frac{dv}{dt} + kv = g\left[1 - \frac{\rho_2}{\rho_1}\right]$$

where k is a constant, g is the acceleration due to gravity and ρ_1 and ρ_2 are the densities respectively of the solid and the liquid. Find the velocity at any time t if $v \to v_1$ as $t \to \infty$.

2.68 The slope at any point (x, y) of a curve is $3x - 2y$. If the curve passes through the origin, determine the equation to the curve.

2.7 Bernoulli's equation

Certain equations of the first order which are not linear can be reduced to the linear form and hence solved by change of variables with suitable substitutions. One such form is called *Bernoulli's equation*, named after the Swiss mathematician, Jakob Bernoulli (1654–1705). A Bernoulli's equation is of the form

$$\frac{dy}{dx} + P(x)y = Q(x)y^n \qquad [2.20]$$

where $n(\neq 1)$ is a constant.

Equation [2.20] is clearly linear for $n = 0$ or $n = 1$. Otherwise, the equation can be reduced to a linear form by the change of variable with the substitution

$$u = y^{1-n} \qquad [2.21]$$

so that $\dfrac{du}{dx} = (1-n)y^{-n}\dfrac{dy}{dx}$. Then [2.21] becomes

$$\frac{du}{dx} + (1-n)P(x)u = (1-n)Q(x) \qquad [2.22]$$

which is a *linear* equation in u. The integrating factor of [2.22] is $e^{\int (1-n)P(x)dx}$ and hence the general solution of [2.22] is

$$ue^{(1-n)\int P(x)dx} = (1-n)\int \left[Q(x)e^{\int (1-n)P(x)dx}\right]dx.$$

Substituting $u = y^{(1-n)}$ from [2.21], we get the general solution of the Bernoulli equation given by [2.20].

Example 2.29 Solve: $dx - (xy + x^2y^3)dy = 0.$

Solution 2.29 The equation can be written in the form

$$\frac{dx}{dy} - yx = y^3x^2 \qquad [1]$$

which is a Bernoulli equation of the form (with $n = 2$): $\dfrac{dx}{dy} + P(y)x = Q(y)x^n$ in the unknown function x, treating y as the independent variable. We substitute

$$u = x^{1-n} = x^{1-2} = 1/x$$

so that $\dfrac{du}{dy} = -\dfrac{1}{x^2}\dfrac{dx}{dy}$ or $\dfrac{dx}{dy} = -x^2\dfrac{du}{dy} = -\dfrac{1}{u^2}\dfrac{du}{dy}.$

Therefore, [1] takes the form $-\dfrac{1}{u^2}\dfrac{du}{dy} - \dfrac{y}{u} = y^3 \cdot \dfrac{1}{u^2}$ or

$$\frac{du}{dy} + uy = -y^3. \qquad [2]$$

Equation [2] is *linear* in u. The integrating factor is

$$e^{\int y dy} = e^{(y^2/2)}$$

so that the general solution of [2] is $ue^{(y^2/2)} = \int (-y^3)e^{(y^2/2)}dy$, which yields

$$u = 2 - y^2 + ce^{-y^2/2} \qquad [3]$$

(after integrating with the substitution $t = y^2/2$). Substituting back for u in terms of x viz., $u = 1/x$, the general solution of the given equation is

$$x = \frac{1}{2 - y^2 + ce^{-y^2/2}}.$$

Example 2.30 Solve: $x\dfrac{dy}{dx} + (1 - x)y = x^2y^2.$

Solution 2.30 The equation can be written as

$$\frac{dy}{dx} + \left[\frac{1}{x} - 1\right]y = xy^2 \qquad [1]$$

which is a Bernoulli equation of the form [2.20] with $n = 2$. We substitute

$$u = y^{1-2} = \frac{1}{y} \quad \text{so that} \quad \frac{du}{dx} = -\frac{1}{y^2}\frac{dy}{dx}.$$

Equation [1] then becomes

$$\frac{du}{dx} - \left(\frac{1}{x} - 1\right)u = -x \qquad [2]$$

which is a *linear* differential equation. The integrating factor of [2] is

$$e^{\int (1-1/x)dx} = e^{x - \log x} = e^x \cdot e^{-\log x} = e^x/x.$$

Therefore, the general solution of [2] is

$$u\left(\frac{e^x}{x}\right) = \int (-x)\frac{e^x}{x}\, dx \quad \text{or} \quad \frac{ue^x}{x} = -e^x + c.$$

Substituting $u = 1/x$, the general solution of [1] is

$$xy(ce^{-x} - 1) = 1.$$

Example 2.31 Solve: $\dfrac{dy}{dx} - \dfrac{y}{x} = \dfrac{1}{2y}$ with $y = 0$ when $x = 1$.

Solution 2.31 The given equation is

$$\frac{dy}{dx} + \left(-\frac{1}{x}\right)y = \frac{1}{2}y^{-1} \qquad [1]$$

which is a Bernoulli equation with $n = -1$. Substituting $u = y^{1-(-1)} = y^2$ we get $du/dx = 2ydy/dx$ so that [1] becomes

$$\frac{du}{dx} + \left(\frac{-2}{x}\right)u = 1 \qquad [2]$$

which is a linear equation. The integrating factor of [2] is

$$e^{\int(-2/x)dx} = e^{-2\log x} = \frac{1}{x^2}.$$

Therefore, the general solution of [2] is

$$u\left(\frac{1}{x^2}\right) = \int (1)\frac{1}{x^2}\,dx \quad \text{or} \quad \frac{u}{x^2} = -\frac{1}{x} + c.$$

Substituting $u = y^2$, the general solution of [1] is

$$y^2 = -x + cx^2.$$

From the initial condition, putting $x = 1$ and $y = 0$, we get $c = 1$ so that the solution becomes

$$y^2 = -x + x^2 \quad \text{or} \quad x^2 - y^2 - x = 0.$$

Example 2.32 A chain coiled up near the edge of a smooth table begins to fall over the edge. When a length x of the chain has fallen, the equation of motion is given by

$$\frac{d}{dt}(mxv) = mxg$$

where m is the mass of the chain per unit length, v is the speed, g is the acceleration due to gravity and t is the time. Show that the speed is given by

$$v = \sqrt{(2/3)gx}.$$

Solution 2.32 The given equation of motion can be rewritten as

$$x\frac{dv}{dt} + v\frac{dx}{dt} = xg. \qquad [1]$$

Since $\dfrac{dx}{dt} = v$ and $\dfrac{dv}{dt} = \dfrac{dv}{dx}\cdot\dfrac{dx}{dt} = \dfrac{dv}{dx}\,v$, [1] becomes

$$\frac{dv}{dx} + \left(\frac{1}{x}\right)v = gv^{-1} \qquad [2]$$

which is a Bernoulli equation with $n = -1$. Substituting $u = v^{1-(-1)} = v^2$, we get the relation $du/dx = 2v dv/dx$ so that equation [2] becomes

$$\frac{du}{dx} + \left(\frac{2}{x}\right)u = 2g \qquad [3]$$

which is a linear differential equation in u. The integrating factor of [3] is

$$e^{\int (2/x)dx} = e^{2\log x} = e^{\log x^2} = x^2.$$

Therefore, the general solution of [3] is $ux^2 = \int (2g)x^2 dx$ or

$$ux^2 = \frac{2}{3}gx^3 + c.$$

Substituting $u = v^2$, we get $v^2x^2 = \frac{2}{3}gx^3 + c.$

Since the chain *begins* to fall initially, we have $v = 0$ when $x = 0$ so that $c = 0$. Therefore, the solution of [1] is

$$v^2x^2 = \frac{2}{3}gx^3 \quad \text{or} \quad v = \sqrt{(2/3)gx}.$$

From this solution we note that the velocity profile with respect to distance is a *parabola*.

Note: Another type of equation which can be reduced to the linear form is

$$f'(y)\frac{dy}{dx} + P(x)f(y) = Q(x). \qquad [2.23]$$

An equation of this type can be reduced to the linear form by taking $u = f(y)$.

Example 2.33 Solve: $\dfrac{dy}{dx} - \dfrac{\tan y}{1+x} = (1+x)e^x \sec y.$

Solution 2.33 Multiplying the equation by $\cos y$, we get

$$\cos y\frac{dy}{dx} - \frac{\sin y}{1+x} = (1+x)e^x. \qquad [1]$$

Equation [1] is of the form [2.23] where

$$f(y) = \sin y, P(x) = -\frac{1}{1+x}, \quad \text{and} \quad Q(x) = (1+x)e^x.$$

Substituting $u = \sin y$ so that $\cos y(dy/dx) = (du/dx)$, equation [1] takes the form

$$\frac{du}{dx} - \frac{1}{1+x}u = (1+x)e^x \qquad [2]$$

which is linear. The integrating factor is $e^{\int -(1/1+x)dx} = \dfrac{1}{1+x}$.

Therefore, the general solution of [2] is

$$u\left(\frac{1}{1+x}\right) = e^x + c$$

and hence the general solution of [1] is $\sin y = (1+x)(e^x + c)$.

Exercise 2D

Solve the following equations:

2.69 $xy' = y - \dfrac{1}{y}$

2.70 $3y' + xy = \dfrac{x}{y^2}$

2.71 $(e^y - 2xy)y' = y^2$

2.72 $y - xy' = y'y^2e^y$

2.73 $x\dfrac{dy}{dx} + y = y^2 \log x$

2.74 $\dfrac{dy}{dx} + y \tan x = y^2 \sec^3 x$

2.75 $xy^2\dfrac{dy}{dx} + y^3 = x \cos x$

2.76 $(x + \sin y) \sin y dx + 2x^2 \cos y dy = 0$

2.77 $6y^2dx - x(2x^3 + y)dy = 0, y(1) = 1$

2.78 $y(1 + x^4y^3)dx + xdy = 0, y(1) = -1$

2.79 $\dfrac{dy}{dx} - \dfrac{y}{x} = \dfrac{1}{2y}, y(1) = 0$

2.80 The motion of a particle along the x-axis is governed by the equation

$$v\frac{dv}{dx} + v^2 + 2x = 0$$

where v is the speed at a point x. If the particle starts from rest at a unit distance from the origin, show that it reaches the origin with a speed equal to $\sqrt{1-e^2}$.

2.8 Exact differential equations

A differential equation of the type

$$M(x,y)dx + N(x,y)dy = 0 \tag{2.24}$$

is said to be an *exact* (or *total*) differential equation if the left-hand side of [2.24] is an exact (or total) differential of some function, say $u(x,y)$. That is

$$M(x,y)dx + N(x,y)dy = du.$$

In that case, on integration, $u(x,y) = c$ is the general solution of [2.24] (where c is an arbitrary constant). But then, how do we confirm if [2.24] is an *exact* (or *total*) differential? This question is answered in the following theorem.

Theorem 2.1 A necessary and sufficient condition for the differential equation

$$M(x,y)dx + N(x,y)dy = 0$$

to be *exact* is $\dfrac{\partial M}{\partial y} = \dfrac{\partial N}{\partial x}$.

Proof: *(i) Condition is necessary* Suppose the left-hand side of the equation

$$M\,dx + N\,dy = 0$$

is exact (or total) i.e., it is the differential of some function $u(x,y)$. Therefore,

$$du = M\,dx + N\,dy. \tag{2.25}$$

But $$du = \frac{\partial u}{\partial x}dx + \frac{\partial u}{\partial y}dy. \tag{2.26}$$

From [2.25] and [2.26], we see that [2.24] is *exact* if there exists a function $u(x,y)$ such that

$$\frac{\partial u}{\partial x} = M;\ \frac{\partial u}{\partial y} = N. \tag{2.27}$$

Assuming that M and N are defined and have continuous first-order partial derivatives in a certain region in the xy-plane whose boundary is a closed curve, we have from [2.27]

$$\frac{\partial M}{\partial y} = \frac{\partial^2 u}{\partial y \partial x} \quad \text{and} \quad \frac{\partial N}{\partial x} = \frac{\partial^2 u}{\partial x \partial y}$$

so that $$\frac{\partial M}{\partial y} = \frac{\partial N}{\partial x} \qquad [2.28]$$

thus proving that the condition is *necessary*.

(ii) Condition is sufficient Suppose the condition [2.28] is satisfied. That is

$$\frac{\partial M}{\partial y} = \frac{\partial N}{\partial x}$$

We shall find a function $u(x, y)$ such that $du = M(x, y)dx + N(x, y)dy$ or [2.27] holds. That is

$$\frac{\partial u}{\partial x} = M(x, y) \quad \text{and} \quad \frac{\partial u}{\partial y} = N(x, y).$$

We shall obtain $u(x, y)$ which satisfies the first condition in [2.27]. Integrating this equation with respect to x, keeping y constant, we obtain

$$u = \int M(x, y)dx + \phi(y) \qquad [2.29]$$

where $\phi(y)$ is an arbitrary function of y. Now, we choose $\phi(y)$ such that the second condition in [2.27] is also satisfied, namely

$$\frac{\partial u}{\partial y} = N(x, y).$$

Now, under the given condition $\partial M/\partial y = \partial N/\partial x$, such a choice of the function $\phi(y)$ is always possible. Indeed, from [2.29] we have

$$\frac{\partial u}{\partial y} = \frac{\partial}{\partial y}\left[\int M(x, y)dx\right] + \phi'(y).$$

Equating the right-hand side of the equation thus obtained to $N(x, y)$, we find that

$$\phi'(y) = N(x, y) - \frac{\partial}{\partial y}\int M(x, y)dx. \qquad [2.30]$$

The left-hand side of [2.30] is independent of x. Let us verify that, under the condition [2.27], x does not enter into the right hand side of [2.30] either. For that purpose, we shall show that the partial derivative of the right-hand side of [2.30] is identically zero.

$$\text{Now}\quad \frac{\partial}{\partial x}\left[N - \frac{\partial}{\partial y}\int M(x,y)dx\right] = \frac{\partial N}{\partial x} - \frac{\partial}{\partial x}\left[\frac{\partial}{\partial y}\int M dx\right]$$

$$= \frac{\partial N}{\partial x} - \frac{\partial}{\partial y}\left[\frac{\partial}{\partial x}\int M dx\right].$$

But $\dfrac{\partial}{\partial x}\displaystyle\int M dx = M(x,y)$, and hence

$$\frac{\partial}{\partial x}\left[N - \frac{\partial}{\partial y}\int M(x,y)dx\right] = \frac{\partial N}{\partial x} - \frac{\partial M}{\partial y} = 0$$

since it is given that $\partial N/\partial x = \partial M/\partial y$. Now, integrating [2.30] with respect to y we get

$$\phi(y) = \int\left[N - \frac{\partial}{\partial y}\int M dx\right]dy + c \qquad [2.31]$$

where c is an arbitrary constant. Substituting [2.31] in [2.29], we get the required function

$$u(x,y) = \int M dx + \int\left[N - \frac{\partial}{\partial y}\int M dx\right]dy + c$$

whose exact differential is $M dx + N dy$ as can be verified easily.

Note: The technique of constructing $u(x,y)$, indicated in the above proof, in fact, constitutes the method of finding the general solution of an exact differential equation.

A "*working rule*" convenient to apply for solving an exact differential equation [2.24] is as follows:

$$\underset{(y\text{const.})}{\int M dx} + \int[\text{Terms of } N \textit{ not} \text{ containing } x]dy = c. \qquad [2.32]$$

Example 2.34 Solve: $(2xy + 3y)dx + (x^2 + 3x)dy = 0$.

Solution 2.34 Here, $M = 2xy + 3y$ and $N = x^2 + 3x$. We have

$$\frac{\partial M}{\partial y} = 2x + 3 \quad\text{and}\quad \frac{\partial N}{\partial x} = 2x + 3 \quad\text{so that}\quad \frac{\partial M}{\partial y} = \frac{\partial N}{\partial x}$$

and hence the equation is *exact*. The general solution is given by

$$\int_{(y\,\text{const.})} M\,dx + \int [\text{Terms of } N \text{ not containing } x]\,dy = c,$$

that is $\displaystyle\int_{(y\,\text{const.})} (2xy + 3y)dx + \int (0)dy = c$ or $x^2y + 3xy = c.$

Note: Sometimes it is possible, by inspection, to rewrite the left-hand side of the differential equation as an exact differential or as a sum or difference of exact differentials of simple functions. Then, direct integration gives the general solution. In Example [2.34], the equation can be rewritten as

$$(2xydx + x^2dy) + 3(xdy + ydx) = 0$$

which is equivalent to $d(x^2y) + 3d(xy) = 0$. Therefore, on integration, the general solution is

$$x^2y + 3xy = c.$$

Example 2.35 Verify that the equation $(2x + 3x^2y)dx + x^3dy = 0$ is exact and solve it.

Solution 2.35 Here, we have $M = 2x + 3x^2y$ and $N = x^3$. Therefore

$$\frac{\partial M}{\partial y} = 3x^2 \quad \text{and} \quad \frac{\partial N}{\partial x} = 3x^2 \quad \text{so that} \quad \frac{\partial M}{\partial y} = \frac{\partial N}{\partial x}$$

and hence the equation is exact. The general solution of the equation is

$$\int_{(y\,\text{const.})} (2x + 3x^2y)dx + \int (0)dy = c$$

or $x^2 + x^3y = c.$

Example 2.36 Solve: $(ax + hy + g)dx + (hx + by + f)dy = 0.$

Solution 2.36 Here, we have $M = ax + hy + g$ and $N = hx + by + f$. Therefore

$$\frac{\partial M}{\partial y} = h \quad \text{and} \quad \frac{\partial N}{\partial x} = h \quad \text{so that} \quad \frac{\partial M}{\partial y} = \frac{\partial N}{\partial x}$$

and hence the equation is *exact*. The general solution of the equation is

$$\underset{(y\text{const.})}{\int} M\,dx + \int (\text{Terms of } N \textit{ not } \text{containing } x)dy = c_1,$$

that is $$\underset{(y\text{const.})}{\int} (ax + hy + g)dx + \int (by + f)dy = c_1$$

or $$\left[\frac{ax^2}{2} + hxy + gx\right] + \left[\frac{by^2}{2} + fy\right] = c_1.$$

Multiplying by 2, we get

$$ax^2 + 2hxy + by^2 + 2gx + 2fy + c = 0 \quad (\text{where } c = -2c_1).$$

Example 2.37 Verify for exactness and solve the equation

$$(2x \log y)dx + \left[\frac{x^2}{y} + 3y^2\right] dy = 0.$$

Solution 2.37 Here, we have $M = 2x \log y$ and $N = (x^2/y) + 3y^2$. Therefore,

$$\frac{\partial M}{\partial y} = \frac{2x}{y} \quad \text{and} \quad \frac{\partial N}{\partial x} = \frac{2x}{y} \quad \text{so that} \quad \frac{\partial M}{\partial y} = \frac{\partial N}{\partial x}$$

and hence the equation is exact. The general solution is

$$\underset{(y\text{const.})}{\int} (2x \log y)dx + \int (3y^2)dy = c \quad \text{or} \quad x^2 \log y + y^3 = c.$$

Example 2.38 Solve: $(xy - y^2)dx - x^2dy = 0$.

Solution 2.38 With $M = xy - y^2$ and $N = -x^2$ we have

$$\frac{\partial M}{\partial y} = x - 2y \quad \text{and} \quad \frac{\partial N}{\partial x} = -2x$$

and hence the equation is *not* exact.

But, if we divide the equation by xy^2, it becomes

$$\left(\frac{1}{y} - \frac{1}{x}\right)dx - \frac{x}{y^2}dy = 0.$$

Taking $M = \frac{1}{y} - \frac{1}{x}$ and $N = -\frac{x}{y^2}$, we have

$$\frac{\partial M}{\partial y} = -\frac{1}{y^2} \quad \text{and} \quad \frac{\partial N}{\partial x} = -\frac{1}{y^2}$$

so that the two are equal and hence the equation is rendered exact. The general solution is given by

$$\underset{(y\text{const.})}{\int}\left(\frac{1}{y} - \frac{1}{x}\right)dx + \int (0)dx = c \quad \text{or} \quad \frac{x}{y} - \log x = c$$

or $\frac{x}{y} = \log x + c$ so that

$$y = \frac{x}{\log x + c}.$$

Note: While the equation in Example [2.38] is *not* exact, we converted it into another equation which is exact by multiplying with the factor $1/xy^2$. This factor is called an *integrating factor* of the given equation. However, in the process, we have excluded $x = 0$ and $y = 0$ in the subsequent steps, as a solution of the given equation. In fact, both the y-axis and the x-axis are particular integral curves of the equation but neither is included in the final general solution.

Although [2.32] can always be applied to an equation [2.24] which is verified to be exact by the criterion [2.28], it is often easier to use the fact that if

$$du = M\,dx + N\,dy \quad \text{then} \quad (\partial u/\partial x) = M \quad \text{and} \quad (\partial u/\partial y) = N.$$

This method is demonstrated in the following example.

Example 2.39 Solve: $(x^2 - 2xy)dx + (\sin y - x^2)dy = 0$.

Solution 2.39 Here, we have $M = x^2 - 2xy$ and $N = \sin y - x^2$. Therefore

$$\frac{\partial M}{\partial y} = -2x \quad \text{and} \quad \frac{\partial N}{\partial x} = -2x \quad \text{so that} \quad \frac{\partial M}{\partial y} = \frac{\partial N}{\partial x}$$

and hence the equation is exact. We have

$$\frac{\partial u}{\partial x} = M = x^2 - 2xy.$$

On integration, with y regarded as constant, we get

$$u = \frac{x^3}{3} - x^2 y + \phi(y) \qquad [1]$$

where $\phi(y)$, a function of y, has to be determined. We also have, by differentiating [1]

$$\frac{\partial u}{\partial y} = -x^2 + \phi'(y). \qquad [2]$$

$$\frac{\partial u}{\partial y} = N = \sin y - x^2 \qquad [3]$$

so that from [2] and [3], we get $\dfrac{d\phi}{dy} = \sin y$ which on integration yields

$$\phi(y) = -\cos y + c_1. \qquad [4]$$

Substituting [4] in [1], the general solution is

$$u(x, y) \equiv \frac{x^3}{3} - x^2 y - \cos y + c_1 = c_2$$

or $\dfrac{x^3}{3} - x^2 y - \cos y = c$ (where $c = c_2 - c_1$ is an arbitrary constant).

Example 2.40 Solve: $e^{-y}dx - (2y + xe^{-y})dy = 0.$

Solution 2.40 It can be verified that the equation is exact. We have here

$$M = e^{-y} \quad \text{and} \quad N = -(2y + xe^{-y}).$$

We shall find $u(x, y)$ such that the left-hand side of the given equation is du. We have

$$\frac{\partial u}{\partial x} = M = e^{-y} \quad \text{and} \quad \frac{\partial u}{\partial y} = -(2y + xe^{-y}) \equiv N. \qquad [1]$$

Integrating the first equation in [1], treating y as constant, we get

$$u(x, y) = \int e^{-y} dx + \phi(y)$$

or $u(x, y) = xe^{-y} + \phi(y).$ [2]

Differentiating [2] with respect to y, we get

$$\frac{\partial u}{\partial y} = -xe^{-y} + \frac{d\phi}{dy}. \qquad [3]$$

From [3] and the second equation in [1], we have $\dfrac{d\phi}{dy} = -2y$ which on integration yields

$$\phi(y) = -y^2 + c_1. \qquad [4]$$

Substituting [4] in [2], we have $u(x, y) = xe^{-y} - y^2 + c_1$, so that the general solution is given by

$$u(x, y) = c_2 \quad \text{or} \quad xe^{-y} - y^2 = c$$

where $c = c_2 - c_1$ is an arbitrary constant.

Example 2.41 Solve: $2x \sin 3y dx + 3x^2 \cos 3y dy = 0$.

Solution 2.41 It can be verified that the equation is exact. We have

$$M = 2x \sin 3y \quad \text{and} \quad N = 3x^2 \cos 3y.$$

The general solution is given by

$$\underset{(y\text{const.})}{\int M\,dx} + \int [\text{Terms of } N \textit{ not } \text{containing } x]dy = c$$

that is $$\underset{(y\text{const.})}{\int 2x \sin 3y dx} + \int (0)dy = c \quad \text{or} \quad x^2 \sin 3y = c.$$

Example 2.42 Solve: $y' = (xy^2 - 1)/(1 - x^2y)$.

Solution 2.42 The equation can be rewritten in the form

$$(1 - x^2y)dy - (xy^2 - 1)dx = 0 \quad \text{or} \quad (xy^2 - 1)dx + (x^2y - 1)dy = 0.$$

By regrouping the terms, we have

$$(xy^2 dx + x^2y dy) - dx - dy = 0$$

or $d\left(\dfrac{x^2y^2}{2}\right) - d(x+y) = 0$

so that on integration we get the general solution

$$\frac{x^2y^2}{2} - (x+y) = c.$$

Exercise 2E

Verify the following equations for *exactness* and solve:

2.81 $xdy + 2y^2dx = 0$

2.82 $4\sin y dx + x\cos y dy = 0$

2.83 $\left[3x^2y + \dfrac{y}{x}\right]dx + \left[x^3 + \log x\right]dy = 0$

2.84 $(xe^{xy} + 2y)y' + ye^{xy} = 0$

2.85 $\left[x + \dfrac{2}{y}\right]dy + ydx = 0$

2.86 $(\sin x \tan y + 1)dx + \cos x \sec^2 y\ dy = 0$

2.87 $(y - x^3)dx + (x + y^3)dy = 0$

2.88 $(2y^2 - 4x + 5)dx = (4 - 2y + 4xy)dy$

2.89 $(y + y\cos xy)dx + (x + x\cos xy)dy = 0$

2.90 $9(y-1)dy + (x-3)dx = 0, y(3) = 0$

2.91 $(\cos x + y\sin x)dx = \cos x dy, y(\pi) = 0$

2.92 $y' = \dfrac{y - 2x}{2y - x}, y(1) = 2$

2.93 $2xydx + (x^2+1)dy = 0, y(1) = -3$

2.94 $y' = \dfrac{2x - \sin y}{x\cos y}, y(2) = 0$

2.95 $\dfrac{dx}{dy} = \dfrac{x \sec^2 y}{\sin 2x - \tan y}, \quad y(\pi) = \dfrac{\pi}{4}$

2.96 $(x^2 + 2ye^{2x})y' + 2xy + 2y^2e^{2x} = 0$ given that $y = 1$ when $x = 0$.

2.9 Equations reducible to the exact form

As mentioned in section 2.8, sometimes a given differential equation of the form [2.18], that is

$$M(x, y)dx + N(x, y)dy = 0$$

may *not* be *exact*, but can be reduced to an exact form by multiplying by a suitable function $f(x, y)$ which is not identically zero.

This function $f(x, y)$ is called an *integrating factor* of [2.24]. With some experience and practice integrating factors can be found by "inspection". However, in some special cases integrating factors can be found systematically. For finding integrating factors (I.F.) by "inspection", it is worth remembering the following frequently occuring *differentials*:

(i) $d\left(\dfrac{x}{y}\right) = \dfrac{ydx - xdy}{y^2}$; I.F. $= \dfrac{1}{y^2}$

(ii) $d\left(\log \dfrac{y}{x}\right) = \dfrac{xdy - ydx}{xy}$; I.F. $= \dfrac{1}{xy}$

(iii) $d\left(\tan^{-1} \dfrac{y}{x}\right) = \dfrac{xdy - ydx}{x^2 + y^2}$; I.F. $= \dfrac{1}{x^2 + y^2}$

The following theorem ensures the existence of an integrating factor for an equation which has a general solution.

Theorem 2.2 If the differential equation [2.24] viz.,

$$M(x, y)dx + N(x, y)dy = 0$$

is *not exact* and has a general solution

$$u(x, y) = c \qquad [2.33]$$

then there exists *an* integrating factor.

Proof: By differentiating the general solution [2.33], we have [2.26], namely

$$du = \frac{\partial u}{\partial x}dx + \frac{\partial u}{\partial y}dy = 0.$$

From [2.24] and [2.26], it follows that

$$\frac{\partial u/\partial x}{\partial u/\partial y} = \frac{M(x,y)}{N(x,y)}$$

holds identically. Hence a function $f \equiv f(x,y)$ exists such that

$$\frac{\partial u}{\partial x} = fM \quad \text{and} \quad \frac{\partial u}{\partial y} = fN$$

so that $f(x,y)\left[M\,dx + N\,dy\right] = du.$ [2.34]

This proves that $f(x,y)$ is ***an integrating factor.***

Note: Further, multiplication of [2.34] by a function $h(u)$ yields

$$h(u)f(x,y)\left[M\,dx + N\,dy\right] = h(u)du$$

which is also *exact*. Hence $h(u)f(x,y)$ is another integrating factor. Since $h(u)$ is any arbitrary function of u, there are infinitely many such integrating factors.

In order to find an integrating factor, the following important results are very useful:

(i) If $\frac{1}{N}\left[\frac{\partial M}{\partial y} - \frac{\partial N}{\partial x}\right] \equiv \phi(x)$ is a function of x only, it has an integrating factor

$$e^{\int \phi(x)dx}$$

(ii) If $\frac{1}{M}\left[\frac{\partial N}{\partial x} - \frac{\partial M}{\partial y}\right] \equiv \phi(y)$ is a function of y only, it has an integrating factor

$$e^{\int \phi(y)dy}$$

(iii) If $[(\partial M/\partial y) - (\partial N/\partial x)]/Ny - Mx \equiv \phi(z)$ is a function of z where

$$z = xy$$

then it has an integrating factor

$$e^{\int \phi(z)dz}.$$

Example 2.43 Find an integrating factor of the equation

$$ydx + (x^2y - x)dy = 0.$$

Solution 2.43 Here, we have $M = y$ and $N = x^2y - x$. It can be verified that the equation is not exact. Consider

$$\frac{1}{N}\left[\frac{\partial M}{\partial y} - \frac{\partial N}{\partial x}\right] = \frac{1 - (2xy - 1)}{x^2y - x} = \frac{-2(xy - 1)}{x(xy - 1)} = \frac{-2}{x}$$

which is a function of x only. Let

$$\phi(x) = -\frac{2}{x}.$$

Therefore an integrating factor is

$$e^{\int \phi(x)dx} = e^{\int(-2/x)dx} = e^{\log(1/x^2)} = \frac{1}{x^2}.$$

Example 2.44 Solve: $ydx + (3 + 3x - y)dy = 0.$

Solution 2.44 Here, $M = y$ and $N = 3 + 3x - y$ so that

$$\frac{\partial M}{\partial y} = 1 \quad \text{and} \quad \frac{\partial N}{\partial x} = 3$$

and hence the equation is not exact. Consider

$$\frac{1}{M}\left[\frac{\partial N}{\partial x} - \frac{\partial M}{\partial y}\right] = \frac{1}{y}(3 - 1) = \frac{2}{y} \equiv \phi(y)$$

a function of y only. Therefore an integrating factor is

$$e^{\int \phi(y)dy} = e^{\int (2/y)}dy = e^{2\log y} = y^2.$$

Multiplying the given equation by the integrating factor, i.e. by y^2, we have

$$y^3dx + y^2(3 + 3x - y)dy = 0.$$

Now, the modified M and N are $M = y^3$ and $N = 3y^2 + 3xy^2 - y^3$ so that

$$\frac{\partial M}{\partial y} = 3y^2 \quad \text{and} \quad \frac{\partial N}{\partial x} = 3y^2$$

so that the modified equation is exact. Therefore, the general solution is

$$\underset{(y\text{const.})}{\int y^3 dx} + \int (3y^2 - y^3) dx = c_1 \quad \text{or} \quad xy^3 + y^3 - \frac{y^4}{4} = c_1$$

or $y^3(y - 4x - 4) = c$

where $c = -4c_1$ is an arbitrary constant.

Example 2.45 Solve: $y' = x - y$ given that $y = 2$ when $x = 0$.

Solution 2.45 The equation can be rewritten as $(x - y)dx - dy = 0$. Here, we have $M = x - y$ and $N = -1$ so that

$$\frac{\partial M}{\partial y} = -1 \quad \text{and} \quad \frac{\partial N}{\partial x} = 0$$

and hence the equation is not exact. Consider

$$\frac{1}{N}\left[\frac{\partial M}{\partial y} - \frac{\partial N}{\partial x}\right] = -1[-1 - 0] = 1 \equiv f(x)$$

which is a function of x only with $f(x) = x^0 = 1$. Therefore, an integrating factor is

$$e^{\int f(x)dx} = e^{\int (1)dx} = e^x.$$

Multiplying the given equation by e^x, the equation becomes

$$e^x(x - y) - e^x dy = 0$$

with modified expressions $M = e^x(x - y)$ and $N = -e^x$ so that

$$\frac{\partial M}{\partial y} = -e^x \quad \text{and} \quad \frac{\partial N}{\partial x} = -e^x$$

and hence the modified equation is exact. The general solution is

$$\underset{(y\text{const.})}{\int e^x(x - y)dx} + \int (0)dy = c \quad \text{or} \quad (x - y)e^x - \int e^x dx = c$$

or $e^x(x-y-1)=c.$

Note: The above example can be solved as a linear differential equation also.

Example 2.46 A curve having a slope given by

$$\frac{dy}{dx}=\frac{2xy}{x^2-y^2}$$

passes through the point (2,1). Find the equation to the curve.

Solution 2.46 The differential equation can be rewritten as

$$2xydx+(y^2-x^2)dy=0$$

with $M=2xy$ and $N=y^2-x^2$. We have

$$\frac{\partial M}{\partial y}=2x \quad \text{and} \quad \frac{\partial N}{\partial x}=-2x$$

and hence the equation is not exact. Consider

$$\frac{1}{M}\left[\frac{\partial N}{\partial x}-\frac{\partial M}{\partial y}\right]=\frac{1}{2xy}\left[-2x-2x\right]=\frac{-2}{y}$$

which is a function of y only. Therefore, an integrating factor is

$$e^{\int(-2/y)dx}=e^{-2\log y}=\frac{1}{y^2}.$$

Multiplying the given equation by the integrating factor, we have the modified form of the equation

$$\frac{2x}{y}dx+\left(1-\frac{x^2}{y^2}\right)dy=0$$

with modified expressions for M and N given by

$$M=(2x/y) \quad \text{and} \quad N=1-(x^2/y^2)$$

so that $\dfrac{\partial M}{\partial y}=-\dfrac{2x}{y^2}$ and $\dfrac{\partial N}{\partial x}=\dfrac{-2x}{y^2}.$

Since $\partial M/\partial y = \partial N/\partial x$, the equation, in the modified form, is exact. The general solution, therefore, is

$$\underset{(y\text{const.})}{\int \frac{2x}{y}dx} + \int (1)dy = c \quad \text{or} \quad \frac{x^2}{y} + y = c$$

or $x^2 + y^2 - cy = 0.$

Since the curve passes through the point (2,1), by substituting $x = 2$ and $y = 1$ in the general solution we get $c = 5$. Therefore the solution is

$$x^2 + y^2 - 5y = 0.$$

The equation represents a *circle*, whose centre is (0,5/2) and radius is 5/2 passing through the origin.

Note: The equation can also be solved as a homogeneous equation.

Example 2.47 If the z-axis of a cartesian coordinate system in space carries electric current I, the work done by the magnetic field in moving a hypothetical unit pole in the xy-plane is given by

$$\int \frac{xdy - ydx}{x^2 + y^2}.$$

Find the family of *equipotential* curves on which the integrand vanishes.

Solution 2.47 The equipotential curves are governed by the equation

$$\frac{xdy - ydx}{x^2 + y^2} = 0. \qquad [1]$$

The equation [1] can be written as

$$d\left[\tan^{-1}\left(\frac{y}{x}\right)\right] = 0.$$

On integration, we get $\tan^{-1}\left(\frac{y}{x}\right) = c_1$ or $\frac{y}{x} = \tan c_1 \equiv c$

or $y = cx$ [2]

where $c = \tan c_1$ is an arbitrary constant. Along with the family of lines [2] passing through the origin, $x = 0$ is also a solution.

Exercise 2F

Solve the following equations by reducing them to the *exact* form, if necessary:

2.97 $(3x+2y^2)dx+2xydy=0$

2.98 $(2x^3-y)dx+xdy=0, y(1)=1$

2.99 $(y^2\cos x-y)dx+(x+y^2)dy=0$

2.100 $(x+x^3\sin 2y)dy-2ydx=0$

2.101 $\dfrac{dy}{dx}=\dfrac{\sin y}{x\cos y-\sin^2 y}, y(0)=\dfrac{\pi}{2}$

2.102 $(2y\sin x-\cos^3 x)dx+\cos x\,dy=0$

2.103 $\dfrac{dy}{dx}+\dfrac{4y}{x}=x$

2.104 $\dfrac{dx}{dy}=\dfrac{y^3-3x}{y}$

2.105 $2y^2dx+3xydy=0$

2.106 $ydx+x\log x\,dy=0$

2.107 $ydx-2xdy=0$

2.108 $x\cos ydy-\sin ydx=0$

2.109 $2ydx+xdy=0, y(2)=-1$

2.110 $\cos ydx=\sin ydy, y(0)=\pi$

2.111 $(3x^2-y^2)dy-2xydx=0$

2.112 $(xy-1)dx+(x^2-xy)dy=0$

2.113 $xdy+ydx+3x^3y^4dy=0$

2.114 $xdy-ydx=(1+y^2)dy$

2.115 $ydx-xdy=xy^3dy$

2.116 $xdy=(x^5+x^3y^2+y)dx$

CHAPTER THREE

Differential Equations Of The First Order And Higher Degree

Differential equations of the *first order* but of a *higher degree* can be solved by one or more of the following four methods:

Type I: Equation solvable for p where $p = dy/dx$
Type II: Equation solvable for x
Type III: Equation solvable for y
Type IV: Clairaut's equation

We shall discuss each of these four types in the following sections.

3.1 Type I: Equation solvable for *p*

Suppose a first order differential equation

$$f(x, y, p) = 0 \tag{3.1}$$

where $p \equiv dy/dx$ is solvable algebraically for p. Then equation [3.1] can be put in the form

$$[p - f_1(x, y)][p - f_2(x, y)] \cdots [p - f_n(x, y)] = 0 \tag{3.2}$$

if the equation is of *degree n,* a positive integer greater than 1.

Equating each of the factors in [3.2] to zero, we get n differential equations of the first order and first degree, namely

$$\frac{dy}{dx} = f_1(x, y), \frac{dy}{dx} = f_2(x, y), \ldots, \frac{dy}{dx} = f_n(x, y).$$

Suppose the general solutions of these n component equations are

$$\phi_1(x, y, c_1) = 0, \quad \phi_2(x, y, c_2) = 0, \ldots, \phi_n(x, y, c_n) = 0$$

where there are apparently n arbitrary constants, $c_1, c_2, \ldots, c_n$.

But, the given equation [3.1] is of the *first order* and hence its general solution has only one arbitrary constant, say c. Therefore, taking

$$c_1 = c_2 = \ldots = c_n = c$$

the general solution of equation [3.1] is

$$\phi_1(x, y, c).\phi_2(x, y, c)\ldots\phi_n(x, y, c) = 0.$$

Example 3.1 Solve: $p^2 + 2py\cot x = y^2$. If an integral curve passes through the point $(\pi/2, 1)$, find the equation to the curve.

Solution 3.1 The given equation is

$$p^2 + 2py\cot x - y^2 = 0. \qquad [1]$$

Solving [1] as a quadratic equation in p, we have

$$p = \left[-2y\cot x \pm \sqrt{4y^2\cot^2 x + 4y^2}\right]/2 \quad \text{or} \quad p = -y(\cot x \pm \operatorname{cosec} x).$$

Considering the positive and negative signs separately, we have the two corresponding first order and first degree equations respectively as

$$\frac{dy}{dx} = -y(\cot x + \operatorname{cosec} x) \qquad [2]$$

$$\frac{dy}{dx} = -y(\cot x - \operatorname{cosec} x). \qquad [3]$$

Equation [2] can be written as

$$\frac{dy}{dx} = -\frac{y(\cos x + 1)}{\sin x} = -\frac{y[2\cos^2(x/2)]}{2\sin(x/2)\cos(x/2)} \quad \text{or} \quad \frac{dy}{dx} = -y\cot\frac{x}{2}.$$

Separating the variables, we can write

$$\frac{dy}{y} = -\cot\frac{x}{2}\,dx.$$

On integration, the general solution of [2] is $\log y = -2\log\sin\left(\frac{x}{2}\right) + \log c$

or $y = c\,\text{cosec}^2\left(\frac{x}{2}\right)$. [4]

Equation [3] can be written as

$$\frac{dy}{dx} = \frac{y(1-\cos x)}{\sin x} = \frac{y(2\sin^2\frac{x}{2})}{2\sin\frac{x}{2}\cos\frac{x}{2}} \quad \text{or} \quad \frac{dy}{dx} = y\tan\left(\frac{x}{2}\right).$$

Separating the variables, we can write

$$\frac{dy}{y} = \tan\left(\frac{x}{2}\right)dx.$$

On integration, the general solution of [3] is $\log y = 2\log\sec\left(\frac{x}{2}\right) + \log c$

or $y = c\sec^2\left(\frac{x}{2}\right)$. [5]

Hence, from [4] and [5], the combined general solution of [1] is

$$\left[y - c\,\text{cosec}^2\left(\frac{x}{2}\right)\right]\left[y - c\sec^2\left(\frac{x}{2}\right)\right] = 0. \qquad [6]$$

It is given that a curve of the family [6] passes through the point $(\pi/2, 1)$ for some value of c. Therefore, on putting $x = \pi/2$ and $y = 1$ in [6], we get

$$(1-2c)(1-2c) = 0$$

which gives $c = 1/2$. Now, substituting $c = 1/2$ in [6], we get the equation for the required integral curve as

$$\left[2y - \text{cosec}^2\left(\frac{x}{2}\right)\right]\left[2y - \sec^2\left(\frac{x}{2}\right)\right] = 0.$$

Example 3.2 Solve: $p^2 + p(x+y) + xy = 0$.

Solution 3.2 The equation can be factorized as

$$(p+x)(p+y) = 0 \quad \text{or} \quad p+x=0 \quad \text{and} \quad p+y=0.$$

(i) Consider $p + x = 0$, that is

$$\frac{dy}{dx} = -x \quad \text{or} \quad dy = -x\,dx$$

which on integration gives $y + x^2/2 = c_1$ or

$$x^2 + 2y - c = 0 \quad (\text{where} \quad c = 2c_1, \text{ an arbitrary constant}).$$

(ii) Consider $p + y = 0$, that is

$$\frac{dy}{dx} = -y \quad \text{or} \quad \frac{dy}{y} = -dx$$

which on integration gives

$$\log y + x - c = 0. \qquad [2]$$

Therefore, from [1] and [2], the complete general solution of the given equation is

$$(x^2 + 2y - c)\,(\log y + x - c) = 0.$$

Note: Since the equation is of the *first order* and has only one arbitrary constant, the same c must be used in [1] and [2].

Example 3.3 Solve: $p^2y^2 \cos^2 \alpha - 2pxy\ \sin^2 \alpha + y^2 - x^2 \sin^2 \alpha = 0$.

Solution 3.3 The equation can be written as

$$(py)^2 - 2(py)x \tan^2 \rho + (y^2 \sec^2 \alpha - x^2 \tan^2 \alpha) = 0 \qquad [1]$$

after dividing throughout by $\cos^2 \alpha (\neq 0)$.
Equation [1] being a quadratic equation in py gives

$$py = \left[2x \tan^2 \alpha \pm \sqrt{4x^2 \tan^2 \alpha - 4(y^2 \sec^2 \alpha - x^2 \tan^2 \alpha)}\right]/2$$

$$\text{or} \quad y\frac{dy}{dx} = x \tan^2 \alpha \pm \sqrt{x^2 \tan^2 \alpha (\tan^2 \alpha + 1) - y^2 \sec^2 \alpha}$$

$$\text{or} \quad y\frac{dy}{dx} = x \tan^2 \alpha \pm \sec \alpha \sqrt{x^2 \tan^2 \alpha - y^2} \qquad [2]$$

$$\text{or} \quad y\ dy - x \tan^2 \alpha dx = \pm\ \sec \alpha \sqrt{x^2 \tan^2 \alpha - y^2}\ dx$$

$$\text{or} \quad \pm \frac{(x \tan^2 \alpha dx - y dy)}{\sqrt{x^2 \tan^2 \alpha - y^2}} = -\sec\ \alpha\ dx.$$

Integrating, we get

$$\pm\sqrt{x^2\tan^2\alpha - y^2} = c - (\sec\alpha)x.$$

Squaring both sides and rearranging the terms, the general solution takes the form

$$x^2 + y^2 - 2cx\sec\alpha + c^2 = 0. \qquad [3]$$

The solution [3] represents a system of circles with centres $\equiv [c\sec\alpha, 0]$ all lying on the $x-$ axis and

$$\text{radii} = \sqrt{c^2\sec^2\alpha - c^2} = c\tan\alpha.$$

Exercise 3A

Find the general solutions of the following equations:

3.1 $xp^2 + (y - x)p - y = 0$

3.2 $xyp^2 - (x^2 + y^2)p + xy = 0$

3.3 $x^2p^2 + xyp - 6y^2 = 0$

3.4 $yp^2 + (x - y)p - x = 0$

3.5 $x^2p^2 + 3xyp + 2y^2 = 0$

3.6 $yp^2 - (1 + xy)p + x = 0$

3.7 $p^2 - 3p\cosh x + 1 = 0$

3.8 $x^2p^2 - 2xyp + 2y^2 - x^2 = 0$

3.9 $xyp^2 + p(3x^2 - 2y^2) - xy\ 6x\ y = 0$

3.10 $xy^2(p^2 + 2) = 2py^3 + x^3$

3.11 $p^3 - p(x^2 + xy + y^2) + xy(x + y) = 0.$

[**Hint:** $p = x$ and $p = y$ are two of the roots]

3.12 $y = x\left[p + \sqrt{1 + p^2}\right]$

3.13 $p^2 + px + py + xy = 0$

3.14 $p^3(x + 2y) + 3p^2(x + y) + p(y + 2x) = 0$

3.2 Type II: Equation solvable for x

If a first order differential equation $f(x, y, p) = 0$, where $p = dy/dx$ is solvable for the independent variable x, then it can be put in the explicit form

$$x = \phi(y, p). \tag{3.3}$$

Differentiating [3.3] with respect to y, we get

$$\frac{dx}{dy} \equiv \frac{1}{p} = F\left(y, p, \frac{dp}{dy}\right) \tag{3.4}$$

which is a differential equation of the first order involving two variables y and p. Suppose the solution of [3.4] is

$$\phi(y, p, c) = 0 \tag{3.5}$$

where c is an arbitrary constant. Now, eliminating p between [3.3] and [3.5], we obtain the general solution of the given equation. If p cannot be easily eliminated, then the solutions of [3.3] and [3.5] for x and y, separately, in terms of p provide *parametric equations* to the integral curves in the form

$$x = f_1(p, c) \quad \text{and} \quad y = f_2(p, c)$$

where p is the parameter. These two equations together constitute the general solution of the given differential equations.

Note: If [3.4] viz., $\frac{1}{p} = F\left(y, p, \frac{dp}{dy}\right) = 0$ can be factorized to the form

$$F_1(y, p).F_2\left(y, p, \frac{dp}{dy}\right) = 0$$

then the factor $F_1(y, p)$ can be cancelled and the equation

$$F_2\left(y, p, \frac{dp}{dy}\right) = 0 \tag{3.6}$$

can be solved. Let the solution of the equation [3.6] be

$$\psi_1(y, p, c) = 0. \tag{3.7}$$

Then, eliminating p between [3.3] and [3.7], we get the general solution.

Example 3.4 Solve: $y = 2px + y^2p^3$

Solution 3.4 Solving the equation for x, we have

$$x = \frac{y}{2p} - \frac{y^2}{2}p^2 \tag{1}$$

Differentiating [1] with respect to y, we get

$$\frac{dx}{dy} \equiv \frac{1}{p} = \frac{1}{2p} - \frac{y}{2p^2}\frac{dp}{dy} - yp^2 - py^2\frac{dp}{dy}$$

$$\text{or}\quad \frac{1}{2p} + p^2y + \left(\frac{y}{2p^2} + py^2\right)\frac{dp}{dy} = 0$$

$$\text{or}\quad \frac{(1+2p^3y)}{2p} + \frac{y(1+2p^3y)}{2p^2}\frac{dp}{dy} = 0$$

$$\text{or}\quad \left[\frac{1+2p^3y}{2p^2}\right]\left[p + y\frac{dp}{dy}\right] = 0.$$

Cancelling the first factor not containing (dp/dy), we get $p+y(dp/dy)=0$. Separating the variables, we have $(dy/y)+(dp/p)=0$. On integration, we get log $(py) = \log c$ or $p = c/y$. Substituting this expression for p in the given equation we get the general solution

$$y = 2x\left(\frac{c}{y}\right) + y^2\left(\frac{c^3}{y^3}\right) \quad \text{or} \quad y^2 = 2cx + c^3$$

where c is an arbitrary constant. The solution represents a family of *parabolas* symmetrical about the x-axis.

Example 3.5 Solve: $y = 3px + 6p^2y^2$.

Solution 3.5 Solving the equation for x, we have

$$x = \frac{y}{3p} - 2py^2. \qquad [1]$$

Differentiating [1] with respect to y, we get

$$\frac{dx}{dy} \equiv \frac{1}{p} = \frac{1}{3p} - \frac{y}{3p^2}\left(\frac{dp}{dy}\right) - 4py - 2y^2\left(\frac{dp}{dy}\right)$$

or multiplying by $3p^2$, we have

$$3p = p - y\frac{dp}{dy} - 12p^3y - 6p^2y^2\frac{dp}{dy} \quad \text{or} \quad 2p + 12p^3y + y\frac{dp}{dy} + 6p^2y^2\frac{dp}{dy} = 0$$

$$\text{or}\quad 2p(1+6p^2y) + y\frac{dp}{dy}(1+6p^2y) = 0 \quad \text{or} \quad (1+6p^2y)\left(2p + y\frac{dp}{dy}\right) = 0.$$

Cancelling the first factor not involving dp/dy, we get $2p + y(dp/dy) = 0$. By separating the variables, we have $dp/p + 2dy/y = 0$.

On integration, we obtain $\log p + 2\log y = \log c$ or $\log(py^2) = \log c$

or $py^2 = c.$ [2]

Substituting for p using [2] in the given equation

$$y = 3\left(\frac{c}{y^2}\right)x + 6\left(\frac{c^2}{y^4}\right)y^2$$

or $y = \dfrac{3cx}{y^2} + \dfrac{6c^2}{y^2}$ or $y^3 = 3c(x + 2c)$.

Example 3.6 Solve: $y^2 \log y = xpy + p^2$.

Solution 3.6 Solving the equation for x, we have

$$x = \frac{y\log y}{p} - \frac{p}{y}. \qquad [1]$$

Differentiating [1] with respect to y, we get

$$\frac{dx}{dy} \equiv \frac{1}{p} = \frac{1}{p}\left(\log y + y.\frac{1}{y}\right) - y\log y.\frac{1}{p^2}\frac{dp}{dy} - \left[-p.\frac{1}{y^2} + \frac{1}{y}\frac{dp}{dy}\right]$$

or $$\frac{1}{p} = \frac{\log y}{p} + \frac{1}{p} + \frac{p}{y^2} - \left[\frac{y\log y}{p^2} + \frac{1}{y}\right]\frac{dp}{dy}$$

or $$\frac{p}{y}\left[\frac{y\log y}{p^2} + \frac{1}{y}\right] - \frac{dp}{dy}\left[\frac{y\log y}{p^2} + \frac{1}{y}\right] = 0 \quad \text{or} \quad \left[\frac{y\log y}{p^2} + \frac{1}{y}\right]\left[\frac{p}{y} - \frac{dp}{dy}\right] = 0.$$

Cancelling the first factor, we have

$$\frac{p}{y} - \frac{dp}{dy} = 0 \quad \text{or} \quad \frac{dy}{y} = \frac{dp}{p}.$$

On integrating, we get $\log p = \log y + \log c$

or $p = cy.$ [2]

Substituting [2] in the given equation, we get

$$y^2 \log y = cxy^2 + c^2y^2$$

or cancelling y^2 ($\neq 0$), we have

$$\log y = c(x + c).$$

Example 3.7 Solve: $x = y + p^2$.

Solution 3.7 Differentiating the given equation with respect to y, we get

$$\frac{dx}{dy} \equiv \frac{1}{p} = 1 + 2p\frac{dp}{dy}$$

or $$\frac{dp}{dy} = \frac{1-p}{2p^2}. \qquad [1]$$

On separating the variables in [1], we have

$$dy = -\frac{2p^2}{p-1}\,dp \quad \text{or} \quad dy = -2\left[\frac{(p^2-1)+1}{p-1}\right]dp = -2\left[p+1+\frac{1}{p-1}\right]dp.$$

On integration, we get

$$y = c - 2\left[\frac{1}{2}\,p^2 + p + \log(p-1)\right]. \qquad [2]$$

Using [2] in the given equation we have

$$x = c - 2\,[p + \log(p-1)]. \qquad [3]$$

Equations [2] and [3] together constitute the general solution of the given equation the given equation in the parametric form.

Exercise 3B

Solve the following equations

3.15 $yp^2 - 2xp + y = 0$

3.16 $ap^2 + py - x = 0$

3.17 $y = p^2y + 2px$

3.18 $4p^3 + 3xp = y$

3.19 $y^2 \log y = pxy + p^2$

3.20 $p^3y^2 - 2xp + y = 0$

3.21 $p^3 - 4xyp + 8y^2 = 0$

3.22 $x + \dfrac{p}{\sqrt{1+p^2}} = a$

3.23 $p = \tan\left[x + \dfrac{p}{1+p^2}\right]$

3.24 $x = y + p^2$

3.3 Type III: Equation solvable for *y*

In case a first order differential equation is solvable for y, we can express y explicitly in terms of x and p. Thus, the equation can be put in the form

$$y = \phi(x, p). \tag{3.8}$$

Differentiating [3.8] with respect to x, we get

$$\frac{dy}{dx} \equiv p = F\left(x, p, \frac{dp}{dx}\right) \tag{3.9}$$

which is a differential equation of the first order involving the variables x and p. Suppose the solution of [3.9] is

$$\psi(x, p, c) = 0. \tag{3.10}$$

Then, by eliminating p between [3.8] and [3.10], we obtain the general solution of the given differential equation. However, if p cannot be eliminated from [3.8] and [3.10] easily, then these two equations are solved for x and y in terms of p. The resulting parametric equations in the form

$$x = f_1(p, c) \quad \text{and} \quad y = f_2(p, c)$$

constitute the general solution where p is the parameter.

Note: If [3.9] viz., $p - F\left(x, p, \dfrac{dp}{dx}\right) = 0$ can be factorized to the form

$$F_1(x, p).F_2\left(x, p, \frac{dp}{dx}\right) = 0$$

then by cancelling the factor not containing dp/dx, we have

$$F_2\left(x, p, \frac{dp}{dx}\right) = 0. \tag{3.11}$$

Suppose the solution of this equation is

$$\psi_1(x, p, c) = 0. \tag{3.12}$$

Then, by eliminating p between [3.8] and [3.12], we get the required general solution of the given differential equation.

Example 3.8 Solve: $y = x + p^3$.

Solution 3.8 Differentiating the given equation with respect to x, we get

$$\frac{dy}{dx} \equiv p = 1 + 3p^2 \frac{dp}{dx}$$

or $p - 1 = 3p^2 \dfrac{dp}{dx}$.

Separating the variables, we have

$$dx = \frac{3p^2}{p-1}\, dp = \frac{3[(p^2-1)+1]}{(p-1)}\, dp \quad \text{or} \quad dx = 3\left[p + 1 + \frac{1}{p-1}\right] dp.$$

On integration, we get

$$x = 3\left[\frac{p^2}{2} + p + \log(p-1)\right] + c. \tag{1}$$

Substituting [1] in the given equation, we have

$$y = p^3 + 3\left[\frac{p^2}{2} + p + \log(p-1)\right] + c. \tag{2}$$

Equations [1] and [2] together constitute the general solution of the given equation in the parametric form with p as the parameter.

Example 3.9 Solve: $y = -px + x^4p^2$.

Solution 3.9 Differentiating the given equation with respect to x, we get

$$\frac{dy}{dx} \equiv p = -p - x\frac{dp}{dx} + 4x^3p^2 + 2x^4p\,\frac{dp}{dx}$$

or $x\dfrac{dp}{dx}[2px^3 - 1] + 2p[2px^3 - 1] = 0$ or $[2px^3 - 1]\left[x\dfrac{dp}{dx} + 2p\right] = 0.$

Cancelling the first factor not containing dp/dx, we get $x\dfrac{dp}{dx}+2p=0$

or $2\dfrac{dx}{x}+\dfrac{dp}{p}=0.$

On integration, we get $2\log x+\log p=\log c$ or

$$p=\frac{c}{x^2}. \qquad [1]$$

Eliminating p from [1] and the given equation, we get the general solution as

$$y=-\frac{c}{x}+x^4\left(\frac{c^2}{x^4}\right) \quad \text{or} \quad y=-\frac{c}{x}+c^2.$$

Example 3.10 Solve: $y=2px-p^2$.

Solution 3.10 Differentiating the given equation with respect to x, we get

$$\frac{dy}{dx}\equiv p=2p+2x\frac{dp}{dx}-2p\frac{dp}{dx} \quad \text{or} \quad 2(x-p)\,\frac{dp}{dx}+p=0$$

or $p\dfrac{dx}{dp}+2x-2p=0$

or $\dfrac{dx}{dp}+\left(\dfrac{2}{p}\right)x=2.$ [1]

Equation [1] is a *linear* differential equation in x, with p as the independent variable. The integrating factor of [1] is

$$e^{\int(2/p)dp}=e^{2\log p}=p^2.$$

Therefore, the solution of [1] is

$$xp^2=\int 2p^2dp+c \quad \text{or} \quad xp^2=\frac{2p^3}{3}+c$$

or $x=\dfrac{2}{3}\,p+cp^{-2}.$ [2]

Using [2] in the given equation, we have $y=2p\left[\frac{2}{3}\,p+cp^{-2}\right]-p^2$ or

$$y = \frac{1}{3}p^2 + 2cp^{-1}. \quad [3]$$

Hence, the general solution consists of the parametric equations [2] and [3] together where p is the parameter.

Exercise 3C

Find the general solutions of the following equations:

3.25 $y^2(1-p^2) = a$

3.26 $4p^3 + 3xp = y$

3.27 $y = p\sin p + \cos p$

3.28 $y = x + 2\tan^{-1} p$

3.29 $xp^2 - 2yp + ax = 0$

3.30 $16x^2 + 2p^2y - p^3x = 0$

3.31 $y = 2px + y^2p^3$

3.32 $y = xp^2 + p$

3.33 $xp^3 = a + bp$

3.34 $y = 3x + \log p$

3.4 Type IV : Clairaut's equation

A first order differential equation of the form

$$y = px + f(p) \quad [3.13]$$

where $p \equiv dy/dx$ and $f(p)$ is some function of dy/dx is called Clairaut's equation, named after the French mathematician, Alexis Claude Clairaut (1713–1765).

The degree of equation [3.13] may be higher than one. Differentiating [3.13] with respect to x, we get

$$\frac{dy}{dx} = p + x\frac{dp}{dx} + f'(p)\frac{dp}{dx} \quad \text{or} \quad p = p + [x + f'(p)]\frac{dp}{dx}$$

or $[x + f'(p)]\dfrac{dp}{dx} = 0$. This yields $\dfrac{dp}{dx} = 0$ or $x + f'(p) = 0$.

(i) For $dp/dx = 0$, we get $p = c$ where c is an arbitrary constant. Substituting $p = c$ in [3.13], we obtain one of its solutions as

$$y = cx + f(c). \qquad [3.14]$$

(ii) Consider the following equation:

$$x + f'(p) = 0. \qquad [3.15]$$

We can solve this equation for p as an explicit function of x in the form $p = g(x)$. Substituting this in [3.13], we get a solution

$$y = xg(x) + f[g(x)]. \qquad [3.16]$$

Thus, there are two solutions, given by [3.14] and [3.16], for Clairaut's equation [3.13]. The solution [3.14] involves one arbitrary constant c and that is the *general solution* of [3.13]. The other solution, [3.16], which cannot be obtained from the general solution [3.14] by assigning any value for c, is called the *singular solution* of [3.13]. The *singular solution* is defined as the c-eliminant of the general solution and its partial derivative with respect to the arbitrary constant as the parameter. The geometrical significance is discussed in section 3.5.

It will be shown in the next section that the singular solution of a Clairaut equation represents the *envelope* of the family of curves represented by the general solution. We recall that the singular solution [3.16] is the *p-eliminant* of [3.15] and the given differential equation [3.13]. Let

$$g(x, y, c) = y - cx - f(c).$$

Then, treating c as the parameter, we have $\partial g/\partial c = -x - f'(c)$. The equations $g = 0$ and $\partial g/\partial c = 0$ yield

$$y = cx + f'(c); \qquad [3.17]$$

$$x + f'(c) = 0. \qquad [3.18]$$

Now, comparing equations [3.13] and [3.15] with [3.17] and [3.18], we find that the p-eliminant of [3.13] and [3.15] is the same as the c-eliminant of [3.17] and [3.18].

Therefore, the *singular solution* may be obtained either as the p-eliminant of

$$y = px + f(p) \quad \text{and} \quad x + f'(p) = 0$$

or as the c-eliminant of $g = 0$ and $\partial g/\partial c = 0$ where $g = 0$ is the *general solution* of the given Clairaut's equation [3.13].

Example 3.11 Solve: $y = px + 2p^2$.

Solution 3.11 The equation is a Clairaut equation in its standard form. Therefore, by substituting $p = c$ in the example, its general solution is

$$y = cx + 2c^2. \qquad [1]$$

Now, writing $g(x, y, c) = y - cx - 2c^2$, we have $g = 0$ as the general solution. Then, $\partial g/\partial c = 0$ yields $-x - 4c = 0$ so that $c = -x/4$. Substituting for c in $g = 0$, we get

$$y = -\frac{x^2}{4} + 2\left(\frac{x^2}{16}\right)$$

or $x^2 = -8y.$ [2]

The singular solution [2] represents a *parabola,* with (0, 0) as the vertex and symmetrical about the negative direction of the y-axis. This parabola is the *envelope* of the family of curves represented by the general solution [2] viz., a family of straight lines having their slopes c and the y-intercepts $2c^2$ for different values of c.

Example 3.12 Solve: $y = 3px + 6p^2y^2$.

Solution 3.12 The equation is not in the standard form of Clairaut's equation. However, it can be transformed to Clairaut's form by substituting $v = y^3$ after multiplying the equation by y^2. The given equation, on multiplication by y^2 can be written as

$$y^3 = (3y^2p)x + 6p^2y^4. \qquad [1]$$

Now, by substituting $v = y^3$ and differentiating with respect to x, we have

$$\frac{dv}{dx} = 3y^2\frac{dy}{dx} \quad \text{or} \quad P = 3y^2p \quad \text{where} \quad P \equiv dv/dx \quad \text{and} \quad p \equiv dy/dx.$$

Then, [1] takes the form

$$v = Px + \frac{2}{3}P^2 \qquad [2]$$

which is a Clairaut equation. The general solution of [2] is

$$v = cx + (2/3)c^2. \qquad [3]$$

Writing $g(x, v, c) = v - cx - (2/3)c^2$ and treating c as the parameter, we get

$$\frac{\partial g}{\partial c} = -x - \frac{4}{3}c.$$

Now, eliminating c between the equations $g = 0$ and $\partial g/\partial c = 0$, we obtain

$$v = -(3/8)x^2.$$

Substituting $v = y^3$ in [3], the general solution is $3y^3 = 3cx + 2c^2$.

The singular solution is

$$8y^3 + 3x^2 = 0$$

which is a *semi-cubical parabola,* passing through the point (0, 0) and symmetrical about the y-axis.

Example 3.13 Solve: $y = xy' - (y')^2$.

Solution 3.13 The given equation is

$$y = px - p^2 \qquad [1]$$

where $p \equiv y' \equiv dy/dx$. Equation [1] is a Clairaut equation and its general solution is

$$y = cx - c^2. \qquad [2]$$

Differentiating [1] with respect to the parameter p, we get

$$x - 2p = 0 \quad (\text{or} \quad x = 2p). \qquad [3]$$

Now, eliminating p between [1] and [3], the singular solution of the given equation is

$$y = \frac{x^2}{2} - \frac{x^2}{4}$$

or $x^2 = 4y$. [4]

In fact, the singular solution [4] represents a parabola with (0, 0) as the vertex and symmetrical about the positive direction of the y-axis (Fig. 3.1). This parabola is the *envelope* of the family of straight lines [2].

Example 3.14 Solve: $x^2(y - px) = p^2y$.

Solution 3.14 The equation is not of Clairaut's form but can be reduced to that form. We substitute $x^2 = u$ and $y^2 = v$. Therefore, $2xdx = du$ and $2ydy = dv$, so that

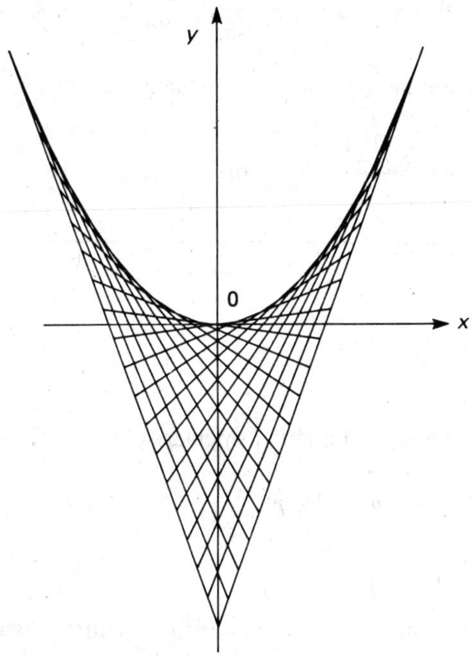

Fig. 3.1 *Parabola as the envelope of a family of straight lines*

$$\frac{y}{x}\frac{dy}{dx} = \frac{dv}{du} \quad [1]$$

or $p = \dfrac{x}{y}\dfrac{dv}{du}$ where $p \equiv dy/dx$.

Substituting for p from [1] in the given example, we have

$$x^2\left[y - \frac{x^2}{y}\frac{dv}{du}\right] = \frac{x^2}{y}\left[\frac{dv}{du}\right]^2 \quad \text{or} \quad y^2 - x^2\frac{dv}{du} = \left[\frac{dv}{du}\right]^2$$

or $v = uP + P^2$ (where $P \equiv dv/du$).

Equation [2] is of Clairaut's form and hence its general solution is

$$v = cu + c^2$$

or substituting $x^2 = u$ and $y^2 = v$, it becomes

$$y^2 = cx^2 + c^2. \quad [3]$$

Differentiating [3] with respect to c, we get

$$0 = x^2 + 2c \quad \text{or} \quad c = -\frac{x^2}{2}.$$

Substituting for c in [3], we obtain the singular solution

$$y^2 = -\frac{x^4}{2} + \frac{x^4}{4} \quad \text{or} \quad x^4 + 4y^2 = 0.$$

Exercise 3D

Find the general and the singular solutions of the following equations:

3.35 $y = px + \log p$ **3.36** $(y - px)(p - 1) = p$

3.37 $y = px + p^2$ **3.38** $y = px + e^p$

3.39 $(px - y)(py + x) = a^2 p$ [**Hint:** Substitute $u = x^2, v = y^2$]

3.40 $y + p^2 = px + 1$ **3.41** $y = 2px + y^2 p^3$ [**Hint:** $y^2 = v$]

3.42 $p^2 \cos^2 y + p \sin x \cos x \cos y = \sin y \cos^2 x$

[**Hint:** Substitute $u = \sin x, v = \sin y$]

3.5 Geometrical significance of solutions of first order differential equations

In Chapter 2 as well as in the preceding sections of the present chapter, we considered various methods of finding the solutions of the first order differential equations.

The general solution of a first order differential equation involves one arbitrary constant, say c, and the solution may be written in the form

$$g(x, y, c) = 0. \qquad [3.19]$$

Equation [3.19] represents a curve in the xy-plane for a given value of c. Now, if we treat c as a parameter and allow it to run through all real values, [3.19] yields a family of infinitely many curves. These are called *integral curves*. Thus, the general solution of a first order differential equation

$$f(x, y, dy/dx) = 0$$

geometrically represents an infinite set of curves in the xy-plane. Now, differentiating [3.19] partially with respect to c, as the parameter, we get

$$\frac{\partial g}{\partial c} = 0. \qquad [3.20]$$

By eliminating c from [3.19] and [3.20], we get an equation involving only x and y in the form $h(x, y) = 0$. This solution is the *singular solution* of the differential equation

$$f(x, y, dy/dx) = 0.$$

It is a well-known result in differential calculus * that the c-eliminant of the equation of a family of curves given by $g(x, y, c) = 0$ and its partial derivative equation

$$\frac{\partial}{\partial c}[g(x, y, c)] = 0$$

represents a curve Γ which *touches every member of the family* C of curves represented by the equation $g(x, y, c) = 0$. The curve Γ is called the *envelope* of the family of curves C. Therefore, it follows that the singular solution of a differential equation

$$f(x, y, dy/dx) = 0$$

represents the envelope of the family of curves represented by the general solution of the differential equation $g(x, y, c) = 0$.

Note: The c-eliminant of $g(x, y, c) = 0$ and $\partial g/\partial c = 0$ does *not* exist when $\partial g/\partial c$ is independent of c. In such a case, it is important to note that the *singular solution* does *not* exist. For each value of c in the general solution, we get a particular solution representing the corresponding *integral curve*. The singular solution (which cannot be obtained from the general solution for any value of c), if it exists, represents the *envelope* of the family of integral curves of the given differential equation.

Example 3.15 Show that the envelope of the integral curves of the equation

$$(x^2 - 1)p^2 - 2xyp + y^2 = 1$$

is the unit circle with its centre at the origin.

Solution 3.15 The equation can be rewritten as

$$(y - px)^2 = 1 + p^2$$

---

*See *Differential Calculus* (Wiley Eastern Ltd., Chapter 11) by S. Balachandra Rao & C. K. Shantha.

so that we have $y = px \pm \sqrt{1+p^2}$. The general solution, therefore, is given by

$$y = cx \pm \sqrt{1+c^2}$$

which represents a family of pairs of straight lines. Now, substituting

$$g(x, y, c) = (y - cx)^2 - (1 + c^2)$$

we get, on differentiation: $\partial g/\partial c = -2x(y - cx) - 2c$. Eliminating the parameter c in the following equations

$$g(x, y, c) = 0 \quad \text{and} \quad \frac{\partial g}{\partial c} = 0,$$

we get the equation

$$x^2 + y^2 = 1$$

as the *singular solution*. This proves the result.

Example 3.16 Find the integral curves and their envelope for the equation

$$p^2x^2 + py(2x + y) + y^2 = 0. \qquad [1]$$

Solution 3.16 We use the substitutions $u = y$ and $v = xy$ so that $y = u$ and $x = v/y = v/u$. Therefore

$$\frac{dy}{du} = 1 \quad \text{and} \quad \frac{dx}{du} = \frac{1}{u^2}\left[u\frac{dv}{du} - v\right]$$

$$\text{so that} \quad p \equiv \frac{dy}{dx} = \frac{dy/du}{dx/du} = \left[\frac{u^2}{u(dv/du) - v}\right].$$

Substituting these in the given equation, we obtain

$$\left[u\frac{dv}{du} - v\right]^2 + (u^2 + 2v)\left[u\frac{dv}{du} - v\right] + v^2 = 0$$

$$\text{or} \quad v = u\frac{dv}{du} + \left(\frac{dv}{du}\right)^2$$

$$\text{or} \qquad v = Pu + P^2 \qquad [2]$$

where $P \equiv dv/du$. The general solution of Clairaut's equation [2] is

$$v = cu + c^2 \tag{3}$$

or $xy = cy + c^2$. [4]

Differentiating [3] partially with respect to c, we get $0 = u + 2c$, or

$$y = -2c. \tag{5}$$

Now, eliminating c between [4] and [5], we get the singular solution

$$y(y + 4x) = 0.$$

Here, both $y = 0$ and $4x + y = 0$ represent straight lines passing through the origin. This pair of lines is the *envelope* of the integral curves [4].

Example 3.17 Find the curve for which the sum of the intercepts of the tangent on the cartesian coordinate axes is unity.

Solution 3.17 Suppose the equation of the tangent at any point (x, y) to the required curve is of the form

$$y = px + f(p) \tag{1}$$

where $p \equiv dy/dx$ is the slope of the tangent.

Equation [1] can be written in the intercepts-form as

$$\frac{x}{-f(p)/p} + \frac{y}{f(p)} = 1$$

so that the x-intercept and the y-intercepts are $-f(p)/p$ and $f(p)$ respectively. Since it is given that the sum of these intercepts is unity, we have

$$-\frac{f(p)}{p} + f(p) = 1 \quad \text{or} \quad -f(p) + pf(p) = p$$

or $f(p) = \dfrac{p}{p-1}$. [2]

Therefore, using [2] in [1], the equation to the tangent satisfying the given condition becomes

$$y = px + \frac{p}{p-1}. \tag{3}$$

Now, considering p as the parameter, [3] represents a family of tangents to the required curve. Hence, the curve is the *envelope* of the family of straight lines represented by [3]. Since the equation is in Clairaut's form, its general solution is

$$y = cx + \frac{c}{c-1}.$$

Let $g(x, y, c) = y - cx - [c/(c-1)]$, so that

$$\frac{\partial g}{\partial c} = -x + \frac{1}{(c-1)^2}.$$

Eliminating c from $g = 0$ and $\partial g/\partial c = 0$ we get the singular solution

$$(x + y - 1)^2 = 4xy$$

as the equation to the required curve.

Note: As pointed out earlier, it is possible that a differential equation may *not* have a *singular solution* at all. Geometrically, the corresponding integral curve will not have an envelope. This possibility is illustrated in the following example.

Example 3.18 Show that the integral curves of the equation

$$\frac{dy}{dx} = 2x$$

have no envelope.

Solution 3.18 The general solution of the equation, on integration, is

$$y = x^2 + c \qquad [1]$$

which represents a family of *parabolas* which are symmetrical about the positive direction of the y-axis and have their vertices $(0, c)$ all lying along the y-axis. Let

$$g(x, y, c) = y - x^2 - c. \qquad [2]$$

So $$\frac{\partial g}{\partial c} = -1. \qquad [3]$$

The envelope, if it exists, is given by eliminating c from $g(x, y, c) = 0$ and $\partial g/\partial c = 0$; but the latter contradicts [3]. Hence, the integral curves [1] do not have an envelope.

CHAPTER FOUR

Orthogonal Trajectories

4.1 Family of curves and trajectories

Consider a cartesian equation

$$F(x, y, c) = 0. \qquad [4.1]$$

For a specified value of the constant c, [4.1] represents a curve in the xy-plane assuming that the function F is continuous in a suitable interval. For each value of c in [4.1] we get a curve and hence treating c as a parameter, we get a family of curves. Since there is one parameter in [4.1], the equation represents a *one-parameter family of curves.*

A curve which cuts every member of a family of curves according to some specified property is called a *trajectory* of the given family of curves.

A trajectory which cuts every member of a family of curves at right angles (i.e. orthogonally) is called an *orthogonal trajectory* of the given family of curves.

4.2 Orthogonal trajectories in cartesian form

Suppose the differential equation governing the family of curves [4.1] is

$$F(x, y, dy/dx) = 0. \qquad [4.2]$$

If the slope of any member curve (of the family of curves [4.1]) at (x, y) on the curve is dy/dx, then the slope of a curve passing through (x, y) and cutting the member curve *orthogonally* is $(-dx/dy)$. Therefore, the slope of a member of the family of orthogonal trajectories of [4.1] is $(-dx/dy)$. Hence, the differential equation of the orthogonal trajectories may be obtained by replacing (dy/dx) by $(-dx/dy)$ in the differential equation [4.2]. We get the orthogonal trajectories of [4.1] by solving

$$F\left(x, y, -\frac{dx}{dy}\right) = 0. \qquad [4.3]$$

Example 4.1 Obtain the orthogonal trajectories of the family of circles

$$x^2 + y^2 = c^2. \quad [1]$$

Solution 4.1 Equation [1] represents a family of circles which are concentric with (0, 0) as the centre and c as their radii. Differentiating [1] with respect to x, we have $2x + 2y(dy/dx) = 0$, or

$$x + y\frac{dy}{dx} = 0. \quad [2]$$

Equation [2] is the differential equation representing the family of concentric circles [1]. Now, replacing (dy/dx) by $(-dx/dy)$ in [2], the differential equation governing the family of orthogonal trajectories of [1] is given by

$$x + y\left(-\frac{dx}{dy}\right) = 0 \quad \text{or} \quad \frac{dy}{y} = \frac{dx}{x}.$$

On direct integration, we get $\log y = \log x + \log m$, or

$$y = mx \quad [3]$$

where m is an arbitrary constant. Equation [3] represents a family of straight lines, passing through the origin and having different slopes m, and forms the orthogonal trajectories of [1] (see Fig. 4.1).

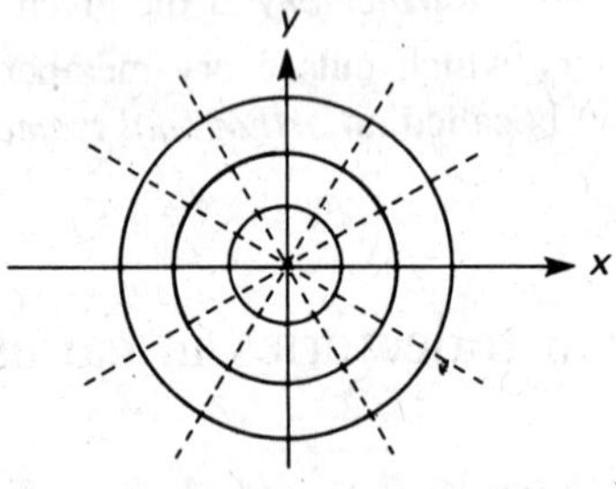

Fig. 4.1 *Orthogonal trajectories of concentric circles.*

Example 4.2 Find the orthogonal trajectories of the family of parabolas

$$y = cx^2. \quad [1]$$

Solution 4.2 Differentiating [1] with respect to x, we get

$$\frac{dy}{dx} = 2cx. \quad [2]$$

From [1] we have $c = y/x^2$. Substituting for c in [2], we get

$$\frac{dy}{dx} = \frac{2y}{x}. \qquad [3]$$

The family of orthogonal trajectories of [1] is given by the differential equation obtained by replacing (dy/dx) by $(-dx/dy)$ in [3]. We get

$$-\frac{dx}{dy} = \frac{2y}{x} \quad \text{or} \quad xdx + 2ydy = 0.$$

On integration, we have $x^2/2 + y^2 = a$, or

$$\frac{x^2}{(\sqrt{2a})^2} + \frac{y^2}{(\sqrt{a})^2} = 1 \qquad [4]$$

where $a > 0$ is an arbitrary constant. From [4], we get a family of ***ellipses*** with semi-major axes $\sqrt{2a}$ and semi-minor axes $\sqrt{a}$, as the required orthogonal trajectories of the given family of *parabolas* [1] (see Fig. 4.2).

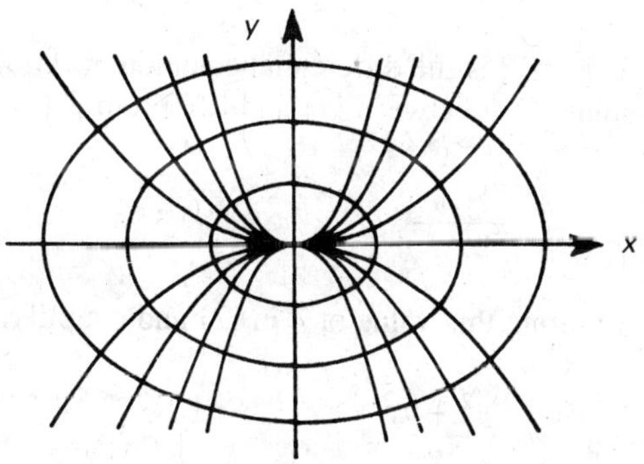

Fig. 4.2 *Ellipses as orthogonal trajectories of parabolas.*

Example 4.3 Find the orthogonal trajectories of the family of circles

$$x^2 + y^2 = 2cx. \qquad [1]$$

Solution 4.3 Equation [1] represents a family of circles with centres $(c, 0)$ lying on the x-axis and having radii c. This means all the members of the family of circles touch the y-axis at the origin. Differentiating [1] with respect to x, we get

$$2x + 2y\frac{dy}{dx} = 2c$$

$$\text{or} \quad x + y\frac{dy}{dx} = c. \tag{2}$$

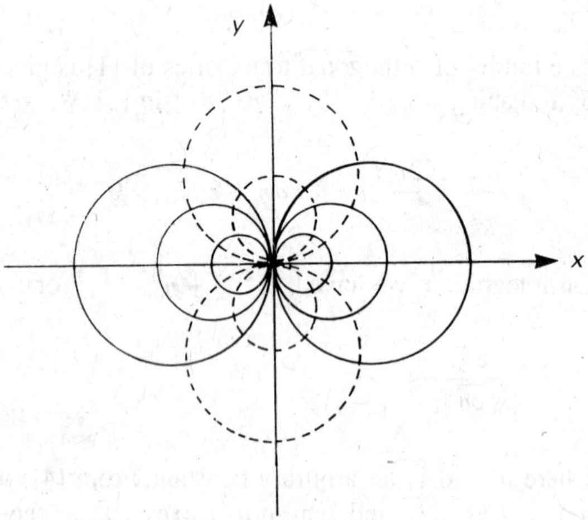

Fig. 4.3 *Circles as orthogonal trajectories of another family of circles.*

Equation [2] is the differential equation governing the family of circles [1]. We have now to eliminate c between [1] and [2]. From [1], we have

$$c = \frac{x^2 + y^2}{2x}.$$

Substituting this value of c in [2] and simplifying, we get

$$\frac{dy}{dx} = \frac{y^2 - x^2}{2xy}.$$

Replacing (dy/dx) by $(-dx/dy)$ in the above equation, we get the differential equation of the orthogonal trajectories as

$$-\frac{dx}{dy} = \frac{y^2 - x^2}{2xy}$$

$$\text{or} \quad \frac{dy}{dx} = \frac{2xy}{x^2 - y^2}. \tag{3}$$

Equation [3] is of the *homogeneous* type. On substituting $y = vx$ we obtain the general solution of [3] as

$$x^2 + y^2 = 2ay. \tag{4}$$

Equation [4], representing the family of orthogonal trajectories of [1] is also a family of circles but having their centres $(0, a)$ lying on the y- axis and with radii a, where a is the parameter. In this case, all the circles of this orthogonal family (Fig. 4.3) touch the x-axis.

Example 4.4 Show that the family of ellipses

$$\frac{x^2}{a^2+\lambda}+\frac{y^2}{b^2+\lambda}=1 \qquad [1]$$

(where a and b are constants and λ is the parameter) is *self-orthogonal.*

Solution 4.4 Differenting [1] with respect to x, we get

$$\frac{x}{a^2+\lambda}+\frac{y}{b^2+\lambda}\frac{dy}{dx}=0 \quad \text{or} \quad x(b^2+\lambda)+y(a^2+\lambda)\frac{dy}{dx}=0$$

$$\text{or} \quad \lambda\left(x+y\frac{dy}{dx}\right)=-\left(b^2x+a^2y\frac{dy}{dx}\right) \quad \text{so that} \quad \lambda=-\frac{[b^2x+a^2y(dy/dx)]}{[x+y(dy/dx)]}.$$

$$\text{Therefore} \quad a^2+\lambda=\frac{(a^2-b^2)x}{x+y(dy/dx)} \quad \text{and} \quad b^2+\lambda=\frac{-(a^2-b^2)y(dy/dx)}{x+y(dy/dx)}.$$

Substituting for $a^2+\lambda$ and $b^2+\lambda$ in [1], the equation becomes

$$\left[x+y\frac{dy}{dx}\right]\left[x-\frac{y}{(dy/dx)}\right]=a^2-b^2. \qquad [2]$$

Equation [2] is the differential equation governing the given family of ellipses [1].

Now, replacing (dy/dx) by $(-dx/dy)$ in [2], we get the differential equation of the orthogonal trajectories

$$\left[x-\frac{y}{(dy/dx)}\right]\left[x+y\frac{dy}{dx}\right]=a^2-b^2. \qquad [3]$$

But, we observe that equations [2] and [3] are identical. Therefore, the family of the orthogonal trajectories is the same as that of the given curves. In other words, the given family of ellipses is "*self-orthogonal*".

4.3 Orthogonal trajectories in polar form

It is often convenient to express a given family of curves in terms of polar coordinates (r, θ). If ϕ denotes the angle between the tangent and the radius-vector at a point (r, θ) on a curve, we know from differential calculus that ϕ is given by

$$\tan\phi = r\frac{d\theta}{dr}. \qquad [4.4]$$

Then, it follows that any curve which cuts the given curve *orthogonally* will be given by replacing [4.4] by its negative reciprocal viz.,

$$-\frac{1}{r}\frac{dr}{d\theta}.$$

Thus, if the differential equation of the given family of curves in the polar form is

$$\frac{dr}{d\theta} = F(r, \theta) \qquad [4.5]$$

then the differential equation for the family of the *orthogonal trajectories* is

$$\frac{dr}{d\theta} = -\frac{r^2}{F(r, \theta)}. \qquad [4.6]$$

Example 4.5 Obtain the orthogonal trajectories of the family of circles

$$x^2 + y^2 = 2cx \qquad [1]$$

in the *polar form.*

Solution 4.5 Substituting $x = r\cos\theta, y = r\sin\theta$ and $x^2 + y^2 = r^2$ in [1], the polar form of [1] becomes

$$r = 2c\cos\theta. \qquad [2]$$

Differentiating [2] with respect to θ, we have

$$\frac{dr}{d\theta} = -2c\sin\theta. \qquad [3]$$

Eliminating c from [2] and [3] by dividing [2] by [3], we get

$$r\frac{d\theta}{dr} = -\frac{\cos\theta}{\sin\theta}$$

as the differential equation of the family of circles [1] in the polar form.

Accordingly, the differential equation of the orthogonal trajectories of [1] is given by

$$r\frac{d\theta}{dr} = \frac{\sin\theta}{\cos\theta} \quad \text{or} \quad \frac{dr}{r} = \frac{\cos\theta}{\sin\theta}\,d\theta \text{ (by separating the variables).}$$

On integration, we get $\log r = \log\sin\theta + \log(2a)$ (where $2a$ is an arbitrary constant)

or $r = 2a\sin\theta$ [4]

as the equation to the orthogonal trajectories of [1]. In fact, [4] represents another family of circles in the polar form.

Example 4.6 Find the orthogonal trajectories of the family of *cardioids*.

$$r = a(1 + \cos\theta) \qquad [1]$$

where a is the parameter.

Solution 4.6 Differentiating [1] with respect to θ, we get

$$\frac{dr}{d\theta} = -a\sin\theta. \qquad [2]$$

Dividing [2] by [3] and eliminating a, we get

$$\frac{r}{dr/d\theta} = \frac{1+\cos\theta}{-\sin\theta} \qquad [3]$$

which is the differential equation governing the family of cardioids [1].

Now, replacing $(dr/d\theta)$ by $(-r^2 d\theta/dr)$ in [3], we get the differential equation of the orthogonal trajectories as

$$\frac{r}{-r^2 d\theta/dr} = \frac{1+\cos\theta}{-\sin\theta}$$

or $$\frac{1}{r}\frac{dr}{d\theta} = \frac{1+\cos\theta}{\sin\theta}.$$

By separating the variables, we have $dr/r = [(1+\cos\theta)/\sin\theta] = (\text{cosec}\theta + \cot\theta)d\theta$.

Integrating, we get $\log r = \log\tan(\theta/2) + \log\sin\theta + \log c = \log(c\tan(\theta/2)\sin\theta)$

or $$r = c\tan\frac{\theta}{2}\cdot\left(2\sin\frac{\theta}{2}\cos\frac{\theta}{2}\right) = 2c\sin^2\frac{\theta}{2}$$

or $r = c\,(1-\cos\theta)$. [4]

Equation [4] represents another family of *cardioids*.

While the given family of cardioids [1] lies entirely to the *right* of the pole and symmetrical about the initial line ($\theta = 0$), the orthogonal trajectories [4] are the cardioids lying to the *left* of the pole and symmetrical about the line $\theta = \pi$, opposite to the initial line.

Exercise 4

Find the orthogonal trajectories of the following families of curves:

4.1 $xy = c$ **4.2** $y = cx^2$

4.3 $y = ce^x$ **4.4** $x - 4y = c$

4.5 $x^2 - y^2 = cx$ **4.6** $y^2 = cx^3$

4.7 $y^2 = \dfrac{x^3}{c - x}$ **4.8** $x + 1 = y + ce^{-y}$

4.9 $y = cx^n$ **4.10** $y = \tan x + c$

4.11 $y = c(\sec x + \tan x)$ **4.12** $ax^2 + by^2 = c$, a and b constants

4.13 $r = e^{k\theta}$ **4.14** $r(1 + \cos\theta) = 2a$

4.15 $r^n = a^n \cos n\theta$ **4.16** $r = \tan(\theta + \alpha)$

4.17 $r = a\theta$ **4.18** $r^n \sin n\theta = a^n$

CHAPTER FIVE

Higher Order Linear Differential Equations (With Constant Coefficients)

5.1 Higher order linear equations

In the preceding chapters we discussed the various methods of solving first order differential equations of different forms. In this chapter, we study the solutions of *linear* differential equations of *higher order* with constant coefficients.

An nth order linear differential equation with constant coefficients is of the form

$$y^{(n)} + a_1 y^{(n-1)} + a_2 y^{(n-2)} + \ldots + a_{n-1} y^{(1)} + a_n y = f(x) \qquad [5.1]$$

where $a_1, a_2, \ldots, a_n$ are given constants, $f(x)$ is a known function of the independent variable x and $y^{(n)} \equiv d^n y/dx^n, \ldots, y^{(1)} \equiv dy/dx$, the successive derivatives of the dependent variable y with respect to x.

We shall first consider the case $n = 2$ in [5.1] i.e. a *linear* differential equation of the *second order* with constant coefficients, and discuss the methods of solution. Then, these methods can be easily extended to the case of higher orders $(n > 2)$. We shall also consider some special cases which reduce to the form [5.1].

5.2 Homogeneous equations

In a linear differential equation of the form [5.1], if the right-hand side term $f(x)$, which is independent of y and its derivatives, is identically zero, then such a linear differential equation is called a *homogeneous* equation. It is important to note that the word "homogeneous" in this context, should not be confused with the same word used in a different sense in section 2.4. If $f(x) \not\equiv 0$ in [5.1], the equation is said to be *non-homogeneous. Now, taking* $n = 2$ and $f(x) = 0$ in [5.1], we have the second order homogeneous equation in the form

$$y'' + a_1 y' + a_2 y = 0. \qquad [5.2]$$

We shall construct the general solution of [5.2] using the following important theorems.

Theorem 5.1 If $\phi_1(x)$ and $\phi_2(x)$ are two solutions of [5.2] and c_1 and c_2 are arbitrary constants, then $\phi(x) = c_1\phi_1(x) + c_2\phi_2(x)$ is also a solution of [5.2].

Proof: Since ϕ_1 and ϕ_2 are solutions of [5.2], we have

$$\phi_1'' + a_1\phi_1' + a_2\phi_1 = 0 \tag{5.3}$$

$$\phi_2'' + a_1\phi_2' + a_2\phi_2 = 0. \tag{5.4}$$

From [5.3] and [5.4] we have

$$\phi'' + a_1\phi' + a_2\phi = (c_1\phi_1 + c_2\phi_2)'' + a_1(c_1\phi_1 + c_2\phi_2)' + a_2(c_1\phi_1 + c_2\phi_2)$$

$$= c_1(\phi_1'' + a_1\phi_1' + a_2\phi_1) + c_2(\phi_2'' + a_1\phi_2' + a_2\phi_2)$$

$$= c_1(0) + c_2(0) = 0$$

Thus, $\phi(x) = c_1\phi_1(x) + c_2\phi_2(x)$ satisfies [5.2] and hence a solution thereof.

Note: If ϕ_1 and ϕ_2 are such that $\phi_1 \neq k\phi_2$ (i.e. the ratio of ϕ_1 and ϕ_2 is not constant), for any constant k, then $\phi = c_1\phi_1 + c_2\phi_2$ involves *two* arbitrary constants. Then, $\phi(x)$ is a solution of the second order homogeneous equation [5.2] involving two arbitrary constants. Hence, $\phi(x)$ is the general solution of [5.2].

The solutions ϕ_1 and ϕ_2, with $\phi_1 \neq k\phi_2$ for any constant k, are called *linearly independent* solutions of [5.2] and ϕ is called a *linear combination* of ϕ_1 and ϕ_2.

Theorem 5.2: If m_1 and m_2 are two *distinct* roots of the equation

$m^2 + a_1m + a_2 = 0$ (called the *auxiliary equation* of [5.2])

then $\phi(x) = c_1e^{m_1x} + c_2e^{m_2x}$ (where c_1 and c_2 are arbitrary constants) is a solution of [5.2].

Proof: Let $\phi_1(x) = e^{m_1x}$. Now, ϕ_1 is a solution of [5.2] if

$$\phi_1'' + a_1\phi_1' + a_2\phi_1 = 0$$

or $(e^{m_1x})'' + a_1(e^{m_1x})' + a_2(e^{m_1x}) = 0$ or $e^{m_1x}\left[m_1^2 + a_1m_1 + a_2\right] = 0$

or $m_1^2 + a_1m_1 + a_2 = 0$ (since $e^{m_1x} \neq 0$)

which is true by hypothesis. Similarly,

$$\phi_2(x) = e^{m_2 x}$$

is a solution of [5.2]. Since ϕ_1 and ϕ_2 are solutions of [5.2], their linear combination

$$\phi(x) = c_1\phi_1(x) + c_2\phi_2(x) \quad \text{or} \quad \phi(x) = c_1 e^{m_1 x} + c_2 e^{m_2 x}$$

is also a solution of [5.2] from Theorem 5.1.

Note: If one of the roots of the auxiliary equation $m^2 + a_1 m + a_2 = 0$, say m_1, is a *complex number*, then we know from algebra that the other root, m_2, is also complex. In fact we have

$$m_2 = \bar{m}_1,$$

where m_1 is the complex conjugate of m_1. Therefore, if $m_1 = \alpha + i\beta$, then we have $m_2 = \alpha - i\beta$. Then, the solution of [5.2], $\phi(x) = c_1 e^{m_1 x} + c_2 e^{m_2 x}$, takes the form

$$\phi(x) = c_1 e^{\alpha x}\left[\cos\beta x + i\sin\beta x\right] + c_2 e^{\alpha x}\left[\cos\beta x - i\sin\beta x\right]$$

$$\text{or} \quad \phi(x) = e^{\alpha x}\left[A\cos\beta x + B\sin\beta x\right] \tag{5.5}$$

where $A = c_1 + c_2$ and $B = i(c_1 - c_2)$ are arbitrary constants.

Theorem 5.3 If m_1 is a *repeated* root of the auxiliary equation

$$m^2 + a_1 m + a_2 = 0$$

then $\phi(x) = (c_1 + c_2 x)e^{m_1 x}$ is a solution of [5.2] where c_1 and c_2 are arbitrary constants.

Proof: Since m_1 is a root of the quadratic equation

$$m^2 + a_1 m + a_2 = 0 \quad (\text{i.e.,} \quad m_1^2 + a_1 m_1 + a_2 = 0) \tag{5.6}$$

it follows, from Theorem 5.2, that $\phi_1(x) = e^{m_1 x}$ is a solution of [5.2].

Further, since m_1 is a repeated root, $m^2 + a_1 m + a_2 = (m - m_1)^2$ which gives

$$2m_1 + a_1 = 0. \tag{5.7}$$

Let $\phi_2(x) = xe^{m_1 x}$. Now, ϕ_2 is a solution of [5.2] if

$$(xe^{m_1 x})'' + a_1(xe^{m_1 x}) + a_2(xe^{m_1 x}) = 0$$

$$\text{or} \quad e^{m_1 x}\left[x(m_1^2 + a_1 m_1 + a_2) + (2m_1 + a_1)\right] = 0$$

$$\text{or} \quad x(m_1^2 + a_1 m_1 + a_2) + (2m_1 + a_1) = 0$$

which is true from [5.6] and [5.7]. Hence $\phi_1 = e^{m_1 x}$ and $\phi_2 = xe^{m_1 x}$ are solutions of [5.2] so that their linear combination

$$\phi(x) = (c_1 + c_2 x)e^{m_1 x}$$

is the general solution of [5.2].

Note: The above three theorems provide the methods to construct solutions of a second order homogeneous equation

$$y'' + a_1 y' + a_2 y = 0$$

using the roots of the quadratic equation $m^2 + a_1 m + a_2 = 0$

The quadratic equation is called the *auxiliary equation* or the *characteristic equation* of the given differential equation. The corresponding solution, which follows from any one of the theorems 5.1, 5.2 and 5.3, is the general solution of [5.2].

Example 5.1 Solve: $y'' + y' - 2y = 0$.

Solution 5.1 The characteristic equation is $m^2 + m - 2 = 0$ whose roots are 1 and -2 which are real and distinct. Therefore, the general solution is

$$y = c_1 e^x + c_2 e^{-2x}.$$

Example 5.2 Solve: $y'' - 6y' + 13y = 0$.

Solution 5.2 The auxiliary equation is $m^2 - 6m + 13 = 0$, and its roots are

$$m = \frac{6 \pm \sqrt{36 - 52}}{2} = 3 \pm 2i \equiv \alpha \pm \beta i.$$

Since the roots are complex, the general solution of the given equation is given by

$$y = e^{\alpha x}\left[A \cos \beta x + B \sin \beta x\right]$$

$$y = e^{3x}\left[A \cos 2x + B \sin 2x\right].$$

Example 5.3 Solve: $y'' - 2y' + y = 0.$

Solution 5.3 The auxiliary equation is $m^2 - 2m + 1 = 0$ or $(m-1)^2 = 0$ which has a *repeated root*, $m = 1$. Therefore, the general solution is

$$y = (c_1 + c_2 x)e^x.$$

Example 5.4 (*Simple pendulum problem*). A simple pendulum consists of a heavy particle of mass m supported by an inelastic string of length l and of negligible mass. If the string is always straight and the system is free to vibrate in a vertical plane, find the period of vibration of the pendulum.

Solution 5.4 In Fig. 5.1, AB denotes the inelastic string of the pendulum where A is the fixed end and B is the other end of the string at which a particle of mass m is attached. Let θ be the angle between the string and the vertical at an instant of time t. While the particle at B is in motion, two forces act on it– the tension τ along the string and the weight mg of the particle acting vertically downwards. Resolve the weight mg into two components, one along the path of motion i.e. along the tangent at B and the other perpendicular to it, i.e., along the normal.

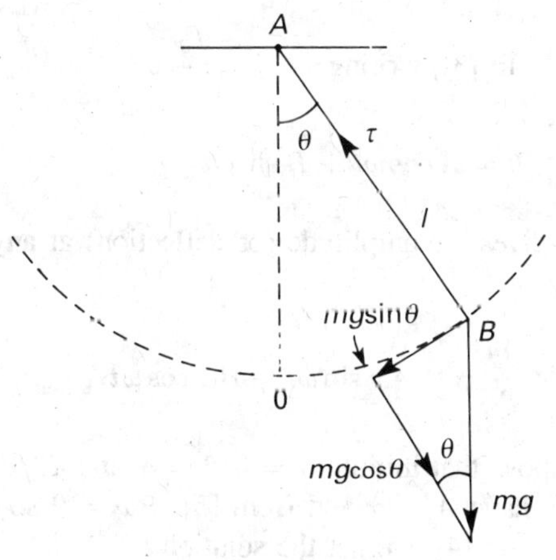

Fig. 5.1 *Simple pendulum*

The normal component $mg\cos\theta$ balances with the tension τ along the string. The force acting tangentially to the path is $mg\sin\theta$. Choose the sign for θ as positive or negative accordingly as the particle is to the right or to the left of the line AO of equilibrium. Then, the net force along the path is given by $(-mg\sin\theta)$ in magnitude and direction. Now, since the arc-length

$$\widehat{OB} = l\theta$$

we have $$m\frac{d^2s}{dt^2} = ml\frac{d^2\theta}{dt^2} = -mg\sin\theta \quad \text{(by Newton's law)}$$

or $$\frac{d^2\theta}{dt^2} = -\frac{g}{l}\sin\theta. \qquad [1]$$

It is important to note that [1] is *non-linear* and hence cannot be easily solved (non-linearity being caused by $\sin\theta$).

As a first approximation, we make [1] *linear* by assuming that the amplitude θ is small and hence $\sin\theta \approx \theta$; where θ is in radians. Then, within the small range of approximation for θ, [1] can be replaced by the *linear* equation

$$\frac{d^2\theta}{dt^2} + \frac{g}{l}\theta = 0 \qquad [2]$$

The roots of the auxiliary equation of [2] are $\pm i\sqrt{g/l}$ and its general solution is

$$\theta = A\sin(\sqrt{g/l})t + B\cos(\sqrt{g/l})t. \qquad [3]$$

From [3], it is clear that the period T is given by $T = \dfrac{2\pi}{\sqrt{g/l}}$ or $T = 2\pi\sqrt{\dfrac{l}{g}}$.

Note: In [3], writing $\omega = \sqrt{\dfrac{g}{l}}$ or $\omega^2 = \dfrac{g}{l}$, the general solution takes the form

$$\theta = A\cos\omega t + B\sin\omega t \qquad [4]$$

which gives the amplitude (or deflection) at any time t. Differentiating [4] with respect to t, we get

$$\frac{d\theta}{dt} = -A\omega\sin\omega t + B\omega\cos\omega t. \qquad [5]$$

Suppose that initially $t = 0, \theta = \alpha$ and $d\theta/dt = 0$. Using the initial conditions, from [4], we have $A = \alpha$ and from [5], $B\omega = 0$ so that $B = 0$. Substituting these values of A and B in [4], we get the solution

$$\theta = \alpha\cos\omega t \quad \text{or} \quad \theta = \alpha\cos\left[(\sqrt{g/l})t\right].$$

Exercise 5A

Solve the following equations:

5.1 $y'' - 9y = 0$

5.2 $y'' - y' = 0$

5.3 $y'' + y = 0$

5.4 $y'' + 4y' + 5y = 0$

5.5 $y'' - 9y' + 18y = 0$

5.6 $3y'' + 2y' - 2y = 0$

5.7 $3y'' + 19y' - 14y = 0$

5.8 $y'' - 6y' + 9y = 0$

5.9 $4y'' - 12y' + 9y = 0$

5.10 $y'' - 8y' + 25y = 0$

5.11 $y'' - y = 0; y(0) = 2, y'(0) = -3$

5.12 $y'' - 3y' + 2y = 0; y(0) = -1, y'(0) = 0$

5.13 $y'' - 4y = 0; y(0) = 4, y'(0) = 2$

5.14 $8y'' - 10y' + 3y = 0; y(0) = 1, y'(0) = 0$

5.15 $y'' + 36y = 0; y(0) = 5, y\left(\frac{\pi}{12}\right) = 7$

5.16 The charge q in an LRC circuit at any time t satisfies the equation

$$L\frac{d^2q}{dt^2} + R\frac{dq}{dt} + \frac{q}{C} = 0$$

where the inductance L, the resistance R and the capacitance C are constants.

Find the charge q at any time t given that

(i) $4L = 5CR^2$ (ii) initial charge is Q and (iii) initial current is zero.

5.2.1 Mechanical free oscillations

We consider a mechanical spring which resists compression as well as extension – in other words elastic – suspended vertically from a fixed point (Fig. 5.2). At the lower end of the spring a body of mass m is attached. Compared to m, the mass of the spring is neglected. When the body is pulled down by a certain distance and then released, it undergoes a certain type of motion. Let us assume that the body moves strictly vertically. We shall determine the motion of this mechanical system.

A study of the forces which act on the system leads to a differential equation. The solution of the resulting differential equation describes the motion of the system providing an expression for the displacement as a function of time.

For convenience, we choose the downward direction as positive and consider the downward forces as positive and the upward ones as negative. The following are the forces that come into play once the mechanical system is set into motion:

(i) The weight of the body of mass m given by

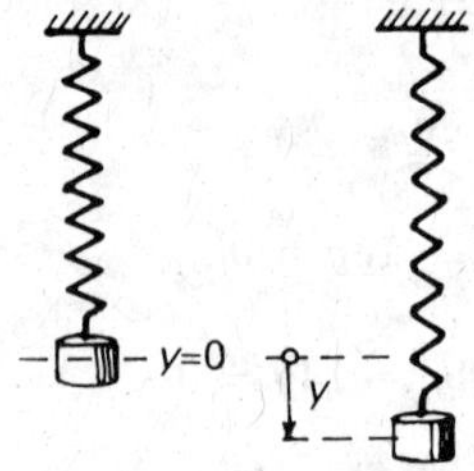

Fig.5.2 *An oscillating body*

$$F_1 = mg \qquad [1]$$

where g is the constant acceleration due to gravity.

(ii) The *spring force* exerted by the spring when it is stretched. We know by *Hooke's law* that this force called *stress* is proportional to the *strain* (or stretch), i.e.

$$F = ks$$

where s is the *stretch* of the spring, and k is the constant of proportionality called *spring modulus*. When the body is at rest, that position of the body is referred to as *static* equilibrium. In this position, the spring is stretched by a constant amount of length say s_0 such that the resultant spring force cancels with the weight [1]. That is

$$ks_0 = mg. \qquad [2]$$

Let $y = y(t)$ be the displacement of the body from the position of static equilibrium, the downward direction being taken as positive. From Hooke's law, it follows that the spring force corresponding to a displacement y is

$$F_2 = -k(s_0 + y) = -ks_0 - ky \qquad [3]$$

where $-ks_0$ is the spring force when the body is in static equilibrium and the additional spring force $-ky$ is caused by the displacement y.

From [1] and [3], the resultant of the weight and the spring force, acting vertically in opposite directions, is given by

$$F_1 + F_2 = mg - ks_o - ky.$$

From [2], this resultant force becomes

$$F_1 + F_2 = -ky. \qquad [4]$$

Since there are no *external* forces, in addition to the above, the system is said to be "*free*".

(a) Undamped system If the *damping* of the system is negligibly small, then it can be ignored. In that case [4] gives all the forces acting on the system. From Newton's second law of motion, we get

$$m\ddot{y} = -ky$$

where $\ddot{y} \equiv d^2y/dt^2$ is the acceleration. Therefore, the motion of an *undamped* system is governed by

$$m\ddot{y} + ky = 0 \tag{5}$$

which is a second order homogeneous linear differential equation with constant coefficients. The general solution of [5] is

$$y(t) = A\cos\omega_0 t + B\sin\omega_0 t (\text{where } \omega_0 = \sqrt{k/m}). \tag{6a}$$

The motion represented by [6a] is called *harmonic oscillation.* Equation [6a] can also be expressed in the form

$$y(t) = C\cos(\omega_0 t - \delta) \tag{6b}$$

$$\text{where} \quad C = \sqrt{A^2 + B^2} \quad \text{and} \quad \tan\delta = B/A.$$

Since the period of the sine and cosine functions in [6] is $2\pi/\omega_0$, the body executes $\omega_0/2\pi$ cycles per second. The quantity $\omega_0/2\pi$ is called the *frequency* of the oscillation and is measured in cycles/sec (or *hertz*, Hz).

(b) Damped system If the body of mass m, attached to the spring in the mechanical system, is connected to a dashpot (a hydraulic device to absorb vibrations), the dashpot causes a *damping* force opposing the motion. Assuming that the damping force is proportional to the velocity ($\dot{y} \equiv dy/dt$) of the body, it is given by $F_3 = -c\dot{y}$ where $c(> 0)$ is called *damping constant.* Now, the total force acting on the body is

$$F_1 + F_2 + F_3 = -ky - c\dot{y}$$

using [4]. Hence, by Newton's second law,

$$m\ddot{y} = -ky - c\dot{y}$$

$$\text{or} \quad m\ddot{y} + c\dot{y} + ky = 0 \tag{7}$$

which is again a homogeneous, linear differential equation of the second order with constant coefficients. The auxiliary (or characteristic) equation of [7] is

$$m\lambda^2 + c\lambda + k = 0 \quad \text{or} \quad \lambda^2 + \frac{c}{m}\lambda + \frac{k}{m} = 0$$

whose roots are $\lambda_1 = -\dfrac{c}{2m} + \dfrac{1}{2m}\sqrt{c^2 - 4mk}$ and $\lambda_2 = -\dfrac{c}{2m} - \dfrac{1}{2m}\sqrt{c^2 - 4mk}.$

The general solution of [7] depends on the nature of the roots. The following three cases arise:

case (i): $c^2 > 4mk$; λ_1, λ_2 real and distinct (over damping),
case (ii): $c^2 < 4mk$; λ_1, λ_2 complex roots (underdamping),
case (iii): $c^2 = 4mk$; $\lambda_1 = \lambda_2$, real (critical damping),

We shall discuss these different cases. Let

$$\alpha = \frac{c}{2m}; \beta = \frac{1}{2m}\sqrt{c^2 - 4mk}. \qquad [8]$$

Case(i): Overdamping When the damping constant is large such that $c^2 > 4mk$, the two roots λ_1 and λ_2, of the characteristic equation, are *real* and *distinct*. In that case, the general solution of the homogeneous equation [7], with $\lambda_1 = -\alpha + \beta$ and $\lambda_2 = -\alpha - \beta$, is given by

$$y(t) = c_1 e^{-(\alpha-\beta)t} + c_2 e^{-(\alpha+\beta)t} \qquad [9]$$

where $\alpha = -\frac{i}{2}(\lambda_1 + \lambda_2)$ and $\beta = \frac{1}{2}(\lambda_1 - \lambda_2)$.

In this case, the body *does not oscillate*. For $t > 0$, both the exponents in [9] are negative (since $\alpha > 0$, $\beta > 0$ and $\beta^2 = \alpha^2 - k/m < \alpha^2$) and hence both terms in [9] tend to zero as $t \to \infty$. This means that after a sufficiently long time the mass will be at rest in its static equilibrium state $(y = 0)$. Physically, this situation results from the fact that the *overdamping* takes away the energy of the system and there is no external force to keep the body in motion.

Case (ii) : Underdamping This situation is very significant and is caused when the damping constant c is small enough to have $c^2 < 4mk$. In that case, in [8], β is *purely imaginary*. Let

$$\beta = i\omega', \quad (\text{where } \omega' = \frac{1}{2m}\sqrt{4mk - c^2} \ (> 0)). \qquad [10]$$

The roots λ_1 and λ_2 are complex conjugate roots where $\lambda_1 = -\alpha + i\omega'$ and $\lambda_2 = -\alpha - i\omega'$. The general solution of [7] is

$$y(t) = e^{-\alpha t}(A\cos\omega' t + B\sin\omega' t)$$

or $y(t) = Ce^{-\alpha t}\cos(\omega' t - \delta)$ [11]

where $C^2 = A^2 + B^2$ and $\tan\delta = \dfrac{B}{A}$.

Solution [11] represents *damped oscillations*. Since $\cos(\omega' t - \delta)$ varies between -1 and 1, the integral curves represented by [11] lie between the curves

$$y = Ce^{-\alpha t} \quad \text{and} \quad y = -Ce^{-\alpha t}.$$

The integral curves [11] touch the two curves $y = \pm Ce^{-\alpha t}$ when $(\omega' t - \delta)$ is an integral multiple of π. The frequency of oscillations is $\omega'/2\pi$ cycles/second. The smaller the value of c is, from [10], the larger is ω' and hence the more rapid the oscillations are. As $c \to 0$, $\omega' \to \omega_0 = \sqrt{k/m}$ which corresponds to *harmonic oscillation* [6].

Case (iii): Critical damping If $c^2 = 4mk$, then $\beta = 0$ and $\lambda_1 = \lambda_2 = -\alpha$ in which case the general solution of [7] is

$$y(t) = (c_1 t + c_2)e^{-\alpha t}. \qquad [12]$$

Since $(c_1 t + c_2)$ can have at most one zero which is positive and the exponential function is always non-zero, it follows that the motion can have at most one passage through the equilibrium position ($y = 0$). If c_1 and c_2 have the same sign (i.e. $c_1 c_2 > 0$), from the initial conditions, there will be no such passage through the position of static equilibrium at all. This case of *critical damping* is similar to case (i).

5.3 Non-homogeneous equations (particular integrals by inverse operator)

A non-homogeneous linear differential equation of the second order is of the form

$$y'' + a_1 y' + a_2 y = f(x). \qquad [5.8]$$

We obtain the general solution of [5.8] in the form $y \equiv \phi(x) = \phi_h(x) + \phi_p(x)$ where ϕ_h is the *complementary function* (C.F.) and ϕ_p is called the *particular integral* (P.I.). Equation [5.8] can be written as

$$[L(D)]y = f(x)$$

where $L(D) \equiv (D^2 + a_1 D + a_2)$ is called a second order *linear differential operator* acting on the unknown function y. Here

$$D^2 \equiv \frac{d^2}{dx^2} \text{ and } D \equiv \frac{d}{dx}.$$

Correspondingly, $1/L(D)$ is called the *inverse differential operator* of the given differential operator $L(D)$. Now, before proceeding to find the particular integrals denoted by the term

$$\frac{1}{L(D)}[f(x)]$$

for different cases of $f(x)$, (in the right-hand side of [5.8]), we shall consider some useful results concerning the evaluation of $L(D)$ acting on different forms of $f(x)$.

5.3.1 Evaluation of $L(D)[f(x)]$

(i) $L(D)[e^{\lambda x}] = e^{\lambda x}L(\lambda)$

We have $L(D)[e^{\lambda x}] = (D^2 + a_1 D + a_2)[e^{\lambda x}]$

$$= D^2[e^{\lambda x}] + a_1 D[e^{\lambda x}] + a_2[e^{\lambda x}] = \lambda^2 e^{\lambda x} + a_1\lambda e^{\lambda x} + a_2 e^{\lambda x}$$

$$= e^{\lambda x}(\lambda^2 + a_1\lambda + a_2) \quad \text{or} \quad L(D)[e^{\lambda x}] = e^{\lambda x}L(\lambda)$$

where $L(\lambda)$ is obtained by replacing D by λ in $L(D)$. This result can be generalized for *higher order* differential operators and proved by induction.

(ii) $L(D)[e^{\lambda x}V(x)] = e^{\lambda x}L(D+\lambda)[V(x)]$

Here, $V(x)$ is some function of x. We have, for $V \equiv V(x)$,

$$L(D)[e^{\lambda x}V] = (D^2 + a_1 D + a_2)[e^{\lambda x}V]$$

$$= D^2[e^{\lambda x}V] + a_1 D[e^{\lambda x}V] + a_2 e^{\lambda x}V$$

$$= (\lambda^2 e^{\lambda x}V + 2\lambda e^{\lambda x}DV + e^{\lambda x}D^2 V) + a_1(\lambda e^{\lambda x}V + e^{\lambda x}DV) + a_2 e^{\lambda x}V$$

$$= e^{\lambda x}[(\lambda^2 + 2\lambda D + D^2)V + a_1(D+\lambda)V + a_2 V]$$

$$= e^{\lambda x}[(D+\lambda)^2 + a_1(D+\lambda) + a_2]V$$

or $L(D)[e^{\lambda x}V(x)] = e^{\lambda x}L(D+\lambda)V(x)$.

This result can be generalized for *higher orders* of $L(D)$ by induction.

(iii) $L(D^2)\left[\sin(ax)\right] = L(-a^2)\left[\sin(ax)\right]$

$$L(D^2)\left[\cos(ax)\right] = L(-a^2)\left[\cos(ax)\right]$$

Since $L(D) = D^2 + a_1 D + a_2,$ we have

$$L(D^2) = D^4 + a_1 D^2 + a_2.$$

Consider $L(D^2)\left[\sin(ax)\right] = (D^4 + a_1 D^2 + a_2)\left[\sin(ax)\right]$

$$= D^4\left[\sin(ax)\right] + a_1 D^2\left[\sin(ax)\right] + a_2\left[\sin(ax)\right]$$

$$= a^4 \sin\left[ax + 4 \cdot \frac{\pi}{2}\right] + a_1(a^2)\left[-\sin(ax)\right] + a_2\left[\sin(ax)\right]$$

$$= \left[a^4 + a_1(-a^2) + a_2\right]\left[\sin(ax)\right]$$

$$= \left[(-a^2)^2 + a_1(-a^2) + a_2\right]\left[\sin(ax)\right]$$

or $L(D^2)\left[\sin(ax)\right] = L(-a^2)\left[\sin(ax)\right].$

Similarly, it can be shown that $L(D^2)\left[\cos(ax)\right] = L(-a^2)\left[\cos(ax)\right].$

Note: The above result is true only in respect of *even order* derivatives like D^2, D^4 etc. in the linear differential operator $L(D)$.

5.3.2 Particular integral with $f(x) = e^{\lambda x}$

Consider an equation in the form $[L(D)]y = e^{\lambda x}$ where $L(D)$ is a linear operator. We have shown that $L(D)\left[e^{\lambda x}\right] = e^{\lambda x} L(\lambda)$ (see case (i) in section 5.3.1). Now, consider the following:

$$L(D)\left[\frac{1}{L(\lambda)} e^{\lambda x}\right] = \frac{1}{L(\lambda)} L(D)\left[e^{\lambda x}\right] = \frac{1}{L(\lambda)} e^{\lambda x} L(\lambda) = e^{\lambda x}.$$

Therefore, we have the particular integral

$$\frac{1}{L(D)}\left[e^{\lambda x}\right] = \frac{1}{L(\lambda)} e^{\lambda x} \qquad [5.9]$$

provided $L(\lambda) \neq 0$. In the case where λ is a root, simple or multiple, of the auxiliary equation we have the following result.

Theorem 5.4 If λ is a *multiple root* of order r of the auxiliary equation $L(m) = 0$, then $L(D) = (D - m)^r \phi(D)$; where $\phi(D)$ is some function of D such that $\phi(D) \neq 0$. In that case,

$$\frac{1}{L(D)}\left[e^{\lambda x}\right] = \frac{e^{\lambda x}}{\phi(D)}\left[\frac{x^r}{r!}\right]. \qquad [5.10]$$

Proof: Consider

$$L(D)\left[\frac{e^{\lambda x}}{\phi(\lambda)}\frac{x^r}{r!}\right] = (D-\lambda)^r \phi(D)\left[\frac{e^{\lambda x}}{\phi(\lambda)}\frac{x^r}{r!}\right]$$

$$= \phi(D)\left[(D-\lambda)^r\left\{\frac{e^{\lambda x}}{\phi(\lambda)}\frac{x^r}{r!}\right\}\right]$$

$$= \phi(D)\left[\frac{e^{\lambda x}}{\phi(\lambda)}D^r\left\{\frac{x^r}{r!}\right\}\right] \quad \left[\because (D+\lambda)-\lambda = D\right]$$

$$= \phi(D)\left[\frac{e^{\lambda x}}{\phi(\lambda)}\right] \quad \left[\because D^r\left\{\frac{x^r}{r!}\right\} = 1\right]$$

$$= \frac{1}{\phi(\lambda)}e^{\lambda x}\phi(\lambda) = e^{\lambda x}.$$

Therefore, we have the particular integral

$$\frac{1}{L(D)}\left[e^{\lambda x}\right] = \frac{e^{\lambda x}}{\phi(\lambda)}\frac{x^r}{r!}$$

where λ is a root of $L(m) = 0$ of multiplicity r and $\phi(\lambda)$ is the quotient obtained by dividing $L(D)$ by $(D - \lambda)^r$ and then replacing D by λ.

Note: As a special case, if λ is a *simple root* (i.e. $r = 1$) of the auxiliary equation $L(m) = 0$, then we have

$$\frac{1}{L(D)}\left[e^{\lambda x}\right] = \frac{e^{\lambda x}}{\phi(\lambda)}x \qquad [5.11]$$

where $\phi(\lambda)$ is the quotient obtained by dividing $L(D)$ by $(D - \lambda)$ and replacing D by λ.

Example 5.5 Solve: $(D^2 + 4D + 3)y = e^{-3x}$

Solution 5.5 The roots of the auxiliary equation $m^2 + 4m + 3 = 0$ are -1 and -3 so that the complementary function of the given equation is

$$\phi_h = c_1 e^{-x} + c_2 e^{-3x}. \qquad [1]$$

The particular integral is given by

$$\phi_p = \frac{1}{D^2 + 4D + 3}\left[e^{-3x}\right]$$

Since $D^2 + 4D + 3$ becomes 0 for $D = -3$ (i.e. value of λ) in this case, we write, factorizing that expression, with $(D + 3)$ as the factor, in the form

$$\phi_p = \frac{1}{(D+3)(D+1)}\left[e^{-3x}\right] = \frac{1}{(D+3)}\left[\frac{1}{(-3)+1}\right]\left[e^{-3x}\right] \qquad \text{(using [5.9])}$$

$$= \frac{1}{D+3}\left[-\frac{1}{2}e^{-3x}\right] = -\frac{x}{2}e^{-3x}. \qquad \text{(using [5.11])}$$

Therefore, the general solution of the given equation, using [1], is

$$y = \phi_n + \phi_h = c_1 e^{-x} + c_2 e^{-3x} - \frac{1}{2}xe^{-3x}.$$

5.3.3 Particular integral with $f(x) = {}^{\sin}_{\cos}\,(ax)$

We have seen in section 5.3.1 (iii) how to evaluate $L(D^2)[{}^{\sin}_{\cos}(ax)]$, i.e. when $L(D)$ contains *even* powers of D. In fact, if $L(D) = \phi(D^2)$ then we have

$$\frac{1}{\phi(D^2)}\left[\sin(ax)\right] = \frac{1}{\phi(-a^2)}\left[\sin ax\right]$$

$$\text{and} \quad \frac{1}{\phi(D^2)}\left[\cos(ax)\right] = \frac{1}{\phi(-a^2)}\left[\cos ax\right]$$

provided $\phi(-a^2) \neq 0$. Now, if $L(D)$ contains *odd* powers of D like D, D^3 etc. also, then we can write

$$L(D) = \phi_1(D^2) + D\phi_2(D^2)$$

by taking out D as a common factor from the terms containing only the odd powers of D. In that case also we can use the following *working rule* to obtain the particular integral:

$$\frac{1}{L(D)}[\sin(ax)] = \frac{1}{\phi_1(D^2) + D\phi_2(D^2)}[\sin(ax)]$$

$$= \frac{1}{\phi_1(-a^2) + D\phi_2(-a^2)}[\sin(ax)] = \frac{1}{p + Dq}[\sin(ax)]$$

[where $p = \phi_1(-a^2); q = \phi_2(-a^2)$]

$$= \frac{(p - Dq)}{(p^2 - D^2q^2)}[\sin(ax)] = \frac{(p - Dq)}{p^2 + a^2q^2}[\sin(ax)]$$

$$= \frac{1}{p^2 + a^2q^2}(p - Dq)[\sin(ax)] = \frac{1}{p^2 + a^2q^2}[p\sin(ax) - aq\cos(ax)]$$

This is the particular integral.

Example 5.6 Solve: $(D^2 + 4D + 3)y = 2\sin(3x)$.

Solution 5.6 The auxiliary equation of the given equation is $m^2 + 4m + 3 = 0$ whose roots are –1 and –3 so that the complementary function is

$$\phi_h = c_1e^{-x} + c_2e^{-3x} \qquad [1]$$

The particular integral of the given equation is given by

$$\phi_p = \frac{1}{D^2 + 4D + 3}[2\sin(3x)]$$

$$= \frac{1}{-9 + 4D + 3}[2\sin(3x)] = \frac{1}{4D - 6}[2\sin(3x)]$$

on replacing D^2 by $-(3)^2$ i.e., -9. Therefore

$$\phi_p = \frac{1}{4D - 6}[2\sin(3x)] = \frac{1}{2D - 3}[\sin(3x)]$$

$$= \frac{(2D + 3}{(4D^2 - 9)}[\sin(3x)] = \frac{(2D + 3)}{4(-9) - 9}[\sin(3x)]$$

$$= -\frac{1}{45}[2D\{\sin(3x)\} + 3\{\sin(3x)\}] = -\frac{1}{45}[6\cos(3x) + 3\sin(3x)]$$

or $$\phi_p = -\frac{1}{15}[2\cos(3x) + \sin(3x)]. \qquad [2]$$

Therefore, from [1] and [2], the general solution is $y = \phi_h + \phi_p$

or $y = c_1 e^{-x} + c_2 e^{-3x} - \frac{1}{15}\left[2\cos(3x) + \sin(3x)\right].$

Example 5.7 Solve: $(D^2 + 4)y = \cos 2x.$

Solution 5.7 The auxiliary equation of [1] is $m^2 + 1 = 0$ whose roots are $\pm 2i$ so that the complementary function of the given equation is

$$\phi_h = c_1 \cos 2x + c_2 \sin 2x. \qquad [1]$$

The particular integral is given by $\phi_p = \frac{1}{D^2 + 4}\left[\cos 2x\right].$

Here, $D^2 + 4$ becomes 0 when D^2 is replaced by $-(2)^2$, i.e. -4 and hence the method described at the beginning of this section fails. In that case we use the following procedure:

$$\phi_p = \text{ Real part of } \left[\frac{1}{D^2 + 4} e^{2ix}\right], \quad (\text{since } e^{2ix} = \cos 2x + i \sin 2x)$$

so that $\cos 2x =$ Real part of (e^{2ix}). Therefore,

$$\phi_p = \text{Real part of } \left[\frac{1}{(D - 2i)(D + 2i)} e^{2ix}\right]$$

$$= \text{Real part of } \left[\frac{1}{4i} \frac{1}{(D - 2i)} e^{2ix}\right] = \text{Real part of } \left[\frac{1}{4i}\, x e^{2ix}\right]$$

$$= \text{Real part of} \left[-\frac{i}{4} x(\cos 2x + i \sin 2x)\right]$$

$$= \text{Real part of} \left[\frac{1}{4} x \sin 2x - \frac{i}{4} x \cos 2x\right]$$

$$\text{or} \quad \phi_p = \frac{1}{4} x \sin 2x. \qquad [2]$$

Therefore, from [1] and [2], the general solution is

$$y = c_1 \cos 2x + c_2 \sin 2x + \frac{1}{4} x \sin 2x.$$

5.3.4 Particular integral with $f(x) = x^n$

In this case, the non-homogeneous equation takes the form

$$L(D)y = x^n$$

where n is a positive integer. The particular integral in this case is

$$\frac{1}{L(D)}[x^n].$$

In order to evaluate the particular integral, $1/L(D)$ is expanded in powers of D using the binomial theorem, and then allowing each term of the expression to operate on x^n. Since $(n+1)^{\text{th}}$ and higher derivatives of x^n are zero, it will be sufficient to expand $1/L(D)$ in ascending powers of D upto the term containing D^n.

Note: To expand $1/L(D)$ in powers of D, the following procedure may be followed:

(i) Take out the lowest degree term, along with its coefficient, in $L(D)$ as the common factor so that the first term becomes unity. Therefore, the binomial theorem can be applied to the remaining part which takes the form $[1+\phi(D)]$ or $[1-\phi(D)]$ for some polynomial $\phi(D)$ in D.

(ii) The factor $[1+\phi(D)]$ or $[1-\phi(D)]$ is taken to the numerator when its index becomes –1.

(iii) The above factor in the numerator is expanded in powers of D using the binomial theorem for the negative index. For this one may recall the results

$$(1+x)^{-1} = 1 - x + x^2 - x^3 + - \ldots \quad \text{and} \quad (1-x)^{-1} = 1 + x + x^2 + x^3 + \ldots$$

The expansion is carried out upto the term containing D^n since the higher order derivatives when operated on x^n vanish. Sometimes the index may be even -2 or -3 in which case the appropriate binomial expansion must be used.

Example 5.8 Solve: $(D^2 - 4)y = x^2$.

Solution 5.8 The auxiliary equation of the given equation is $m^2 - 4 = 0$ whose roots are ± 2 so that its complementary function is

$$\phi_h = c_1e^{2x} + c_2e^{-2x}. \qquad [1]$$

The particular integral of the given equation is

$$\phi_p = \frac{1}{D^2-4}(x^2) = \frac{1}{-4\left[1-D^2/4\right]}(x^2) = -\frac{1}{4}\left[1-\frac{D^2}{4}\right]^{-1}(x^2)$$

$$= -\frac{1}{4}\left[1+\frac{D^2}{4}+\frac{D^4}{16}+\ldots\right](x^2)$$

or $\phi_p = -\frac{1}{4}\left[x^2+\frac{1}{2}\right].$ [2]

Therefore the general solution, from [1] and [2], is

$$y = c_1e^{2x} + c_2e^{-2x} - \frac{1}{4}\left(x^2+\frac{1}{2}\right).$$

Example 5.9 Solve: $(D^4-2D^3+D^2)y = x^3$.

Solution 5.9 The auxiliary equation is

$$m^4 - 2m^3 + m^2 = 0$$

or $m^2(m^2-2m+1) = 0$ or $m^2(m-1)^2 = 0$

so that 0 and 1 are both *double* roots. The complementary function is

$$\phi_h = (c_1+c_2x)e^{(0)x} + (c_3+c_4x)e^{(1)x}$$

or $\phi_h = (c_1+c_2x) + (c_3+c_4x)e^x.$ [1]

The particular integral of the given equation is

$$\phi_p = \frac{1}{D^2(1-2D+D^2)}(x^3) = \frac{1}{D^2}(1-D)^{-2}(x^2)$$

$$= \frac{1}{D^2}[1+2D+3D^2+4D^3+...](x^3)$$

$$= \frac{1}{D^2}[x^3+2D(x^3)+3D^2(x^3)+4D^3(x^3)]$$

$$= \frac{1}{D^2}\left[x^3 + 2(3x^2) + 3(6x) + 4(6)\right] = \frac{1}{D^2}\left[x^3 + 6x^2 + 18x + 24\right]$$

$$= \frac{1}{D}\left[\frac{x^4}{4} + 6\left(\frac{x^3}{3}\right) + 18\left(\frac{x^2}{2}\right) + 24\right] = \frac{1}{D}\left[\frac{x^4}{4} + 2x^3 + 9x^2 + 24\right]$$

noting that $1/D$ stands for integration. Therefore, applying $1/D$ once again, we get

$$\phi_p = \left[\frac{x^5}{20} + \frac{x^4}{2} + 3x^3 + 24x\right]. \qquad [2]$$

The general solution,from [1] and [2], is given by

$$y = (c_1 + c_2 x) + (c_3 + c_4 x)e^x + \left[\frac{x^5}{20} + \frac{x^4}{2} + 3x^3 + 24x\right].$$

Example 5.10 Solve: $(D^3 + 3D^2 + 2D)y = x^2$.

Solution 5.10 The auxiliary equation is

$$m^3 + 3m^2 + 2m = 0 \quad \text{or} \quad m(m+1)(m+2) = 0$$

whose roots are $0, -1$ and -2. Therefore, the complementary function is

$$\phi_h = c_1 e^{(0)x} + c_2 e^{-x} + c_3 e^{-2x}$$

$$\text{or} \quad \phi_h = c_1 + c_2 e^{-x} + c_3 e^{-2x} \qquad [1]$$

The particular integral of [1] is

$$\phi_p = \frac{1}{D^3 + 3D^2 + 2D}(x^2) = \frac{1}{2D}\left[1 + \left(\frac{3D}{2} + \frac{D^2}{2}\right)\right]^{-1}(x^2)$$

$$= \frac{1}{2D}\left[1 - \left(\frac{3D}{2} + \frac{D^2}{2}\right) + \left(\frac{3D}{2} + \frac{D^2}{2}\right)^2\right](x^2)$$

$$= \frac{1}{2D}\left[1 - \frac{3D}{2} - \frac{D^2}{2} + \frac{9D^2}{4} + \ldots\right](x^2)$$

$$= \frac{1}{2D}\left[x^2 - \frac{3}{2}(2x) - \frac{1}{2}(2) + \frac{9}{4}(2)\right] = \frac{1}{2D}\left[x^2 - 3x + \frac{7}{2}\right]$$

$$= \frac{1}{2}\left[\frac{x^3}{3} - \frac{3x^2}{2} + \frac{7}{2}x\right]. \qquad [2]$$

Therefore, from [1] and [2], the general solution is

$$y = c_1 + c_2 e^{-x} + c_3 e^{-2x} + \frac{1}{12}[2x^3 - 9x^2 + 21x].$$

5.3.5 Particular integral with $f(x) = ke^{\lambda x}V(x)$

Suppose a non-homogeneous linear differential equation [5.8] is of the form

$$L(D)y = ke^{\lambda x}V(x) \qquad [5.12]$$

where $L(D) \equiv (D^2 + a_1 D + a_2)$, k and λ are constants and $V(x)$ is some given function. The particular integral

$$\phi_p = \frac{1}{L(D)}\left[ke^{\lambda x}V(x)\right]$$

is given by theorem 5.5.

Theorem 5.5 The particular integral of [5.12] is given by

$$\phi_p = \frac{1}{L(D)}\left[ke^{\lambda x}V(x)\right] = ke^{\lambda x}\left[\frac{1}{L(D+\lambda)}\right]V(x).$$

Proof: We have

$$L(D)\left[ke^{\lambda x}V(x)\right] = (D^2 + a_1 D + a_2)\left[ke^{\lambda x}V(x)\right]$$

$$= D^2\left[ke^{\lambda x}V(x)\right] + a_1 D\left[ke^{\lambda x}V(x)\right] + a_2\left[ke^{\lambda x}V(x)\right]$$

$$= ke^{\lambda x}\left[(D^2V + 2\lambda DV + \lambda^2 V) + a_1(DV + \lambda V) + a_2 V\right]$$

$$= ke^{\lambda x}\left[(D+\lambda)^2 V + a_1(D+\lambda) + a_2 V\right]$$

$$= ke^{\lambda x}\left[L(D+\lambda)\right]V(x).$$

Now, substituting $\left[\frac{1}{L(D+\lambda)}\right]V(x) = S(x)$ and working out as above, we get

$$L(D)\left[ke^{\lambda x}S(x)\right] = ke^{\lambda x}\left[L(D+\lambda)\right]S(x).$$

Substituting for $S(x)$, we have

$$L(D)\left[ke^{\lambda x}\left\{\frac{1}{L(D+\lambda)}V\right\}\right] = ke^{\lambda x}L(D+\lambda)\left\{\frac{1}{L(D+\lambda)}V\right\} = ke^{\lambda x}V,$$

or $$\frac{1}{L(D)}\left[ke^{\lambda x}V(x)\right] = ke^{\lambda x}\left[\frac{1}{L(D+\lambda)}\right]V(x) \qquad [5.13]$$

This completes the proof.

Method of evaluating $\left[\dfrac{1}{L(D+\lambda)}\right]V(x)$

Since $L(D+\lambda)$ is of the second order, let $L(D+\lambda) = (D-a)(D-b)$.

So $$\left[\frac{1}{L(D+\lambda)}\right]V = \left[\frac{1}{(D-a)(D-b)}\right]V = X, \text{ say. Then,}$$

$$(D-a)X = \frac{1}{(D-b)}V = Y, \text{ say.}$$

Therefore $X = \dfrac{1}{(D-a)}Y$ or $X = \dfrac{1}{(D-a)}\left[\dfrac{1}{(D-b)}V\right]$.

We shall now evaluate $(1/D-b)V$. We have

$$\frac{1}{(D-b)}V = Y \quad \text{or} \quad (D-b)Y = V$$

or $dY/dx - bY = V$ which is a first order linear equation. Its solution is

$$Y = e^{bx}\int Ve^{-bx}dx + C$$

where b is constant and C is an arbitrary constant. Since we require only a particular solution, we take $C = 0$. Therefore, we have

$$Y = \frac{1}{(D-b)}V = e^{bx}\int Ve^{-bx}dx \quad \text{and} \quad X = \frac{1}{(D-a)}Y = e^{ax}\int Ye^{-ax}dx$$

so that $$\phi_p = ke^{\lambda x}\frac{1}{(D-a)}\left[\frac{1}{(D-b)}V\right] = ke^{\lambda x}X$$

is a particular solution which can be simplified as explained above.

Note: If $\lambda = 0$, we have $f(x) = kV(x)$. In that case, the method of this section may be employed if $V(x)$ is not in anyone of the forms of $f(x)$ already discussed in the preceding sections.

Example 5.11 Solve: $y'' - 5y' + 6y = e^{4x}(x^2 + 9)$.

Solution 5.11 The auxiliary equation is $m^2 - 5m + 6 = 0$ whose roots are 2 and 3 so that the complementary function is

$$\phi_h = c_1 e^{2x} + c_2 e^{3x}. \quad [1]$$

Now, writing the equation in terms of the D- operator, we have

$$L(D)y = e^{4x}(x^2 + 9)$$

where $L(D) = D^2 - 5D + 6$. Therefore, the particular integral of the example is

$$\phi_p = \frac{1}{L(D)}\left[e^{4x}(x^2 + 9)\right] = e^{4x}\frac{1}{L(D+4)}[x^2 + 9] \quad \text{(using [5.13])}$$

$$= e^{4x}\left[\frac{1}{(D+4)^2 - 5(D+4) + 6}\right](x^2 + 9)$$

$$= e^{4x}\left[\frac{1}{(D+1)(D+2)}\right](x^2 + 9)$$

$$= e^{4x}\frac{1}{(D+1)}\left[\frac{1}{(D+2)}(x^2 + 9)\right]. \quad [2]$$

Now, $$\frac{1}{(D+2)}[x^2 + 9] = \frac{1}{2\left(1 + D/2\right)}[x^2 + 9] = \frac{1}{2}\left(1 + \frac{D}{2}\right)^{-1}[x^2 + 9]$$

$$= \frac{1}{2}\left[1 - \frac{D}{2} + \frac{D^2}{4} + \ldots\right][x^2 + 9]$$

$$= \frac{1}{2}\left[(x^2 + 9) - \frac{1}{2}(2x) + \frac{1}{4}(2)\right] = \frac{1}{2}\left[x^2 - x + \frac{19}{2}\right]$$

or $$\frac{1}{(D+2)}[x^2 + 9] = \frac{1}{4}[2x^2 - 2x + 19]. \quad [3]$$

The particular integral of the example, using [3] in [2], is

$$\phi_p = e^{4x}\frac{1}{(D+1)}\left[\frac{1}{4}\left\{2x^2 - 2x + 19\right\}\right] = \frac{1}{4}e^{4x}(1+D)^{-1}\left[2x^2 - 2x + 19\right]$$

$$= \frac{1}{4}e^{4x}(1 - D + D^2 - + ...)\left[2x^2 - 2x + 19\right]$$

$$= \frac{1}{4}e^{4x}\left[(2x^2 - 2x + 19) - (4x - 2) + (4)\right]$$

or $\phi_p = \dfrac{1}{4}e^{4x}\left[2x^2 - 6x + 25\right].$ [4]

Thus, from [1] and [4], the general solution is

$$y = c_1e^{2x} + c_2e^{3x} + \frac{1}{4}e^{4x}\left[2x^2 - 6x + 25\right].$$

Example 5.12 Solve: $y'' + 4y' + 29y = e^{-2x}\sin 5x$.

Solution 5.12 The auxiliary equation of the example is $m^2 + 4m + 29 = 0$ having the complex conjugate pair of roots $m = -2 \pm 5i$ so that the complementary function of the example is

$$\phi_h = e^{-2x}[A\cos 5x + B\sin 5x].$$ [1]

Writing the example in the D- form, we have

$$L(D)y = e^{-2x}\sin 5x$$

where $L(D) = D^2 + 4D + 29$.

The particular integral of the example, therefore, is

$$\phi_p = \frac{1}{L(D)}\left[e^{-2x}\sin 5x\right] = e^{-2x}\frac{1}{L(D-2)}\left[\sin 5x\right] \quad \text{(using [5.13])}$$

$$= e^{-2x}\frac{1}{(D-2)^2 + 4(D-2) + 29}\ \text{I.P}\left[e^{5ix}\right].$$

where $I.P. \equiv$ "imaginary part of", since $\sin 5x =$imaginary part of (e^{5ix}).

So $\phi_p = e^{-2x}\dfrac{1}{D^2+25}\,\text{I.P.}\left[e^{5ix}\right] = e^{-2x}\cdot\ \text{I.P.}\dfrac{1}{(D-5i)}\left[\dfrac{1}{(D+5i)}e^{5ix}\right]$

$$= e^{-2x}.\ \text{I.P.}\ \frac{1}{(D-5i)}\left[\frac{e^{5ix}}{10i}\right] = e^{-2x}.\ \text{I.P.}\ \left[\frac{e^{5ix}}{10i}x\right]$$

$$= e^{-2x}\left[-\frac{x\cos 5x}{10}\right]. \qquad [2]$$

Therefore, from [1] and [2], the general solution of the example is

$$y = e^{-2x}[A\cos 5x + B\sin 5x] - e^{-2x}\left[\frac{x\cos 5x}{10}\right]$$

or $y = e^{-2x}\left[\left(A - \dfrac{x}{10}\right)\cos 5x + B\sin 5x\right].$

5.3.6 When *f*(*x*) is a sum of functions

Suppose a second order non-homogeneous linear differential equation ([5.8]), with constant coefficients, is of the form

$$y'' + a_1y' + a_2y = f(x) = g(x) + h(x) \qquad [5.14]$$

where the right-hand side of [5.8], $f(x)$, is a sum of the functions $g(x)$ and $h(x)$. Let Ψ_1 and Ψ_2 respectively be particular solutions of the two non-homogeneous equations

$$y'' + a_1y' + a_2y = g(x) \quad \text{and} \quad y'' + a_1y' + a_2y = h(x).$$

We then have

$$\Psi_1'' + a_1\Psi_1' + a_2\Psi_1 = g(x) \qquad [5.15]$$

$$\Psi_2'' + a_1\Psi_2' + a_2\Psi_2 = h(x) \qquad [5.16]$$

where Ψ_1 and Ψ_2 are functions of x. Adding [5.15] and [5.16], we get

$$(\Psi_1 + \Psi_2)'' + a_1(\Psi_1 + \Psi_2)' + a_2(\Psi_1 + \Psi_2) = g(x) + h(x).$$

This means $(\Psi_1 + \Psi_2)$ is a particular integral of the given differential equation [5.14] in which $f(x)$ is the sum of the two functions $g(x)$ and $h(x)$. This result can be extended to a linear combination of more than two functions for $f(x)$.

Example 5.13 Find the general solution of

$$y'' - y = x^2 + 4e^x \qquad [1]$$

and hence the integral curve which touches the x-axis at the origin.

Solution 5.13 The auxiliary equation of [1] is $m^2 - 1 = 0$ whose roots are 1 and -1 so that the complementary function of [1] is

$$\phi_h = c_1 e^x + c_2 e^{-x}. \qquad [2]$$

Let Ψ_1 and Ψ_2 be the particular solutions corresponding to the terms x^2 and $4e^x$, respectively, taken separately in [1]. We have

$$\Psi_1 = \frac{1}{(D^2-1)}(x^2) = \frac{1}{-1(1-D^2)}(x^2) = -\left[(1-D^2)^{-1}\right](x^2)$$

$$= -\left[1 + D^2 + D^4 + ...\right](x^2) = -\left[x^2 + 2\right] \qquad [3]$$

$$\Psi_2 = \frac{1}{(D^2-1)}(4e^x) = \frac{4e^x}{(1+1)}x = 2xe^x. \qquad [4]$$

Therefore, the particular integral of [1] is

$$\phi_p = \Psi_1 + \Psi_2 = -(x^2+2) + 2xe^x \qquad [5]$$

using [3] and [4]. The general solution of [1], from [2] and [5], is

$$y = c_1 e^x + c_2 e^{-x} + 2xe^x - (x^2+2). \qquad [6]$$

Now, in order to find the required particular integral curve, it is given that the curve touches the x-axis at the origin. This means when $x = 0$, $y = 0$ and $dy/dx = 0$. Substituting $x = 0$ and $y = 0$ in [6], we get

$$c_1 + c_2 = 2. \qquad [7]$$

Differentiating [6] with respect to x, we have

$$\frac{dy}{dx} = c_1 e^x - c_2 e^{-x} + 2xe^x + 2e^x - 2x. \qquad [8]$$

Substituting $x = 0$ and $dy/dx = 0$ in [8], we get

$$c_1 - c_2 = -2. \qquad [9]$$

Solving [7] and [9], we get $c_1 = 0$ and $c_2 = 2$. Therefore, substituting these values in [6], the required particular integral curve is given by

$$y = 2e^{-x} + 2xe^{x} - (x^2 + 2).$$

Example 5.14 Solve: $(D^2 - 3D + 2)y = e^{2x}(3x^2 + 2\cos x)$.

Solution 5.14 The auxiliary equation is $m^2 - 3m + 2 = 0$ whose roots are 1 and 2 so that the complementary function is

$$\phi_h = c_1 e^{x} + c_2 e^{2x}. \qquad [1]$$

Let Ψ_1 and Ψ_2 be the particular integrals corresponding to the terms $3x^2e^{2x}$ and $2e^{2x}\cos x$ respectively, of the right-hand side of the given equation. We have

$$\Psi_1 = \frac{1}{L(D)}\left[3x^2 e^{2x}\right], \quad (\text{where} \quad L(D) = (D^2 - 3D + 2)).$$

Therefore, we get

$$\Psi_1 = 3e^{2x}\frac{1}{L(D+2)}(x^2) = 3e^{2x}\frac{1}{(D+2)^2 - 3(D+2) + 2}(x^2)$$

$$= 3e^{2x}\left[\frac{1}{D(D+1)}\right](x^2) = 3e^{2x}\frac{1}{D}(1+D)^{-1}(x^2)$$

$$= 3e^{2x}\frac{1}{D}\left[1 - D + D^2 - \ldots +\right](x^2)$$

$$= 3e^{2x}\frac{1}{D}\left[(x^2) - (2x) + (2)\right] = 3e^{2x}\left[\frac{x^3}{3} - x^2 + 2x\right]$$

or $\Psi_1 = (x^3 - 3x^2 + 6x)e^{2x}$ [2]

The other particular integral Ψ_2 corresponding to the term $2e^{2x}\cos x$ is given by

$$\Psi_2 = \frac{1}{L(D)}\left[2e^{2x}\cos x\right] = 2e^{2x}\frac{1}{L(D+2)}(\cos x)$$

$$= 2e^{2x}\frac{1}{D(D+1)}(\cos x) \qquad [3]$$

since $L(D+2) = D(D+1)$. Now

$$\frac{1}{D+1}(\cos x) = e^{-x}\int(\cos x)e^{x}\,dx = \frac{1}{2}(\cos x + \sin x). \qquad [4]$$

Therefore, using [4] in [3], we have

$$\Psi_2 = 2e^{2x}\frac{1}{D}\left[\frac{1}{2}(\cos x + \sin x)\right] = 2e^{2x}\left[\frac{1}{2}(\sin x - \cos x)\right]$$

or $\Psi_2 = e^{2x}(\sin x - \cos x)$. [5]

The general solution of the example, using [1], [2] and [5], is

$$y = c_1e^{x} + c_2e^{2x} + (x^3 - 3x^2 + 6x)e^{2x} + e^{2x}(\sin x - \cos x)$$

or $y = c_1e^{x} + (x^3 - 3x^2 + 6x + \sin x - \cos x + c_2)e^{2x}$.

Example 5.15 A battery of constant voltage E_0 is connected to an inductance L in series with a capacitance C so that the charge q in the circuit satisfies the equation

$$\frac{d^2q}{dt^2} + \omega^2 q = \frac{E_0}{L} \qquad [1]$$

where $\omega^2 = 1/LC$. Find the charge q and the current $i = dq/dt$ at any instant of time t if these are zero at $t = 0$. Show that $i = 0$ again when $t = \pi/\omega$ and that $q = 2E_0C$ at that time.

Solution 5.15 The auxiliary equation of [1] is $m^2 + \omega^2 = 0$ whose roots are $m = \pm i\omega$ which are complex. Therefore, the complementary function is

$$\phi_h = A\cos\omega t + B\sin\omega t \qquad [2]$$

where A and B are arbitrary constants. Now, the right-hand side of [1] is a constant, E_0/L. Therefore, we seek a particular integral of [1] in the form $\phi_p = \gamma$ where γ is a constant to be determined. By substituting $q = \phi_p = \gamma$ in [1], we get

$$\omega^2\gamma = \frac{E_0}{L} \quad \text{or} \quad \gamma = \frac{E_0}{\omega^2 L} = E_0 C$$

since $\omega^2 = 1/LC$. Therefore,

$$\phi_p = \gamma = E_0 C \qquad [3]$$

is a particular integral of [1]. Hence the general solution of [1], from [2] and [3], is

$$q = A\cos\omega t + B\sin\omega t + E_0 C. \qquad [4]$$

Differentiating [4] with respect to t, we get the current i given by

$$i = \frac{dq}{dt} = -A\omega\sin\omega t + B\omega\cos\omega t. \qquad [5]$$

It is given that $q = 0$ and $i = 0$ when $t = 0$. Using these in [4] and [5], we get $A + E_0C = 0$ and $B = 0$. Substituting the values for A and B in [4] and [5], we get the charge q and the current i given by

$$q = E_0C(1 - \cos\omega t) \quad \text{and} \quad i = E_0C\omega\sin\omega t.$$

At $t = \pi/\omega$, we get $i = 0$ again. At that instant, we have the charge given by

$$q = E_0C\left[1 - \cos\left(\omega\frac{\pi}{\omega}\right)\right] = E_0C\left[1 - \cos\pi\right] \quad \text{or} \quad q = 2E_0C.$$

Example 5.16 The displacement x of a particle moving along the x-axis is determined by the equation

$$\frac{d^2x}{dt^2} + 2\frac{dx}{dt} + 2x = \cos t. \qquad [1]$$

Show that, whatever be the initial conditions, the particle executes simple harmonic motion of period 2π after a large interval of time.

Solution 5.16 The auxiliary equation of [1] is $m^2 + 2m + 2 = 0$ whose roots are $-1 \pm i$. The complementary function is

$$\phi_h = e^{-t}(A\cos t + B\sin t). \qquad [2]$$

Now, we seek the particular integral of [1] in the form

$$\phi_p = \alpha \cos t + \beta \sin t. \qquad [3]$$

Substituting $x = \phi_p$ in the above relation and substituting it in [1], we get

$$(\alpha + 2\beta)\cos t + (\beta - 2\alpha)\sin t = \cos t.$$

By comparing the like terms, we have $\alpha = 1/5$ and $\beta = 2/5$. Therefore, the particular integral [3] becomes

$$\phi_p = \frac{1}{5}(\cos t + 2\sin t). \qquad [4]$$

The general solution of [1] from [2] and [4], therefore, is

$$x = e^{-t}(A\cos t + B\sin t) + \frac{1}{5}(\cos t + 2\sin t). \qquad [5]$$

Whatever be the values of the arbitrary constants A and B, (i.e. independent of the initial conditions), we notice that in [5], as $t \to \infty$, $e^{-t}(A\cos t + B\sin t) \to 0$, so that

$$x \to \frac{1}{5}(\cos t + 2\sin t)$$

Equation [5] gives the displacement of the particle at any instant t, when t is very large. Differentiating [5] with respect to t, twice in succession, the velocity and the acceleration of the particle at that instant are respectively given by

$$\frac{dx}{dt} = \frac{1}{5}(-\sin t + 2\cos t) \quad \text{and} \quad \frac{d^2x}{dt^2} = -\frac{1}{5}(\cos t + 2\sin t) = -x$$

or $$\frac{d^2x}{dt^2} + x = 0 \qquad [6]$$

using [5]. Equation [6] represents a simple harmonic motion (SHM). Since in [5], the displacement remains the same when t is replaced by $(2\pi + t)$, it follows that the period of the SHM is 2π.

Example 5.17 An inductor of 0.5 henry is connected in series with a resistor of 6 ohms, a capacitor of 0.02 farad, a generator having alternative voltage given by $24\sin(10t)$, $t \geqslant 0$ and a switch K.

(i) Set up a differential equation for the instantaneous charge on the capacitor.

(ii) Find the charge and the current at time t if the charge on the capacitor is zero when the switch K is closed at $t = 0$.

[**Note:** For details of electrical circuits see section 5.3.8].

Solution 5.17 When the switch K (Fig. 5.3.) is closed, an instantaneous current flows. If Q is the instantaneous charge on the capacitor C, then by Kirchhoff's law, we have

$$L\frac{dI}{dt} + RI + \frac{Q}{C} = E(t) \qquad [1]$$

where $E(t) = 24\sin(10t)$ is the electromotive force (e.m.f.) which, in this case, depends on time t. Here, the inductance L, the resistance R and the capacitance C are the given constants.

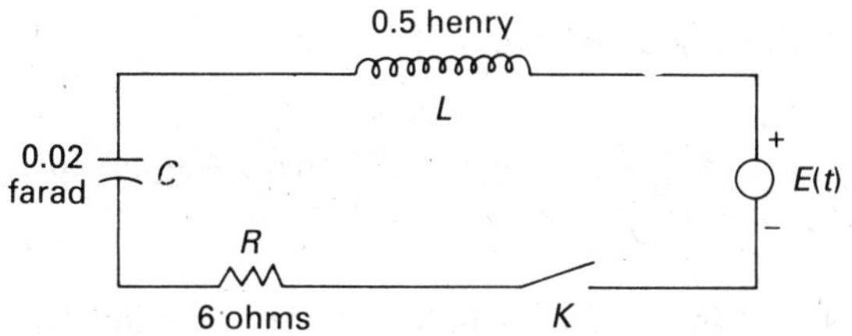

Fig. 5.3 *Electrical circuit in series with a generator.*

Now, the voltage drop across the resistor is given by $RI = 6I$, the voltage drop across the inductor is given by, $L(dI/dt) = (0.5)(dI/dt)$ and the voltage drop across the capacitor is given by

$$\frac{Q}{C} = \frac{Q}{0.02} = 50Q.$$

Hence, by Kirchhoff's law, we have

$$(0.5)\frac{dI}{dt} + 6I + 50Q = 24\sin(10t)$$

or $$(0.5)\frac{d^2Q}{dt^2} + 6\frac{dQ}{dt} + 50Q = 24\sin(10t)\left(\text{since } I = \frac{dQ}{dt}\right)$$

or $$\frac{d^2Q}{dt^2} + 12\frac{dQ}{dt} + 100Q = 48\sin(10t). \qquad [2]$$

Thus, equation [2] is the non-homogeneous second order differential equation governing the given electrical circuit. The initial conditions are $Q = 0$ and $I = dQ/dt = 0$ when $t = 0$. The complementary function of [2] can be verified to be

$$\phi_h = e^{-6t}(A\cos 8t + B\sin 8t). \qquad [3]$$

Now, assuming the particular solution of [1] in the form $\phi_p = \alpha\sin(10t) + \beta\cos(10t)$, we find that $\alpha = 0$ and $\beta = -2/5$ and hence the particular solution of [1] is

$$\phi_p = -\frac{2}{5}\cos(10t). \qquad [4]$$

The general solution of [1], from [3] and [4], is therefore

$$Q = e^{-6t}(A\cos 8t + B\sin 8t) - \frac{2}{5}\cos(10t). \qquad [5]$$

With the given initial conditions, namely $Q = 0$ and $dQ/dt = 0$ for $t = 0$, being used in [5] and its derivative we get

$$A = \frac{2}{5} \quad \text{and} \quad B = \frac{3}{10}.$$

Substituting these values of A and B in [5], the solution is

$$Q = \frac{1}{10}e^{-6t}(4\cos 8t + 3\sin 8t) - \frac{2}{5}\cos 10t.$$

Note: In the above solution, the first term involving e^{-6t} is called the *transient solution* which becomes negligible with time. The second term is called the *steady solution* which dominates the solution.

Exercise 5B

Solve the following equations

5.17 $y'' - 3y' + 2y = 2e^{-4x}$

5.18 $y'' + 4y' + 5y = 3e^{-2x}$; $y(0) = 4$, $y'(0) = -7$

5.19 $\dfrac{d^2x}{dt^2} - 2\dfrac{dx}{dt} + x = 2e^{-2t}$

5.20 $y'' - y' - 6y = 20e^{-2x}$

5.21 $y'' + 10y' + 25y = 14e^{-5x}$

5.22 $y'' + 4y = 3\sin x$

5.23 $y'' - 3y' + 2y = 14\sin 2x - 18\cos 2x$

5.24 $y'' + y = 2\cos x$

5.25 $y'' - 2y' + 5y = 25x^2 + 12$

5.26 $y'' - 2y' = 12x - 10$

5.27 The national income N in a certain economic model is governed by the equation

$$\frac{d^2N}{dt^2} + m\frac{dN}{dt} = n$$

where m and n are constants. How does the national income vary with time t if $N = 0$ and $dN/dt = 0$ when $t = 0$?

5.28 A particle moving along the x-axis has its equation of motion

$$\frac{d^2x}{dt^2} + 4\frac{dx}{dt} + 5x = 2a\sin t$$

where x is the displacement at time t. If the particle starts from rest at $x = a$, show that

$$x = \frac{1}{4}ae^{-2t}\left[9\sin t + 5\cos t\right] + \frac{1}{4}a(\sin t - \cos t).$$

5.29 Determine A and B such that the damped oscillation $y = (A\cos t + B\sin t)e^{-t}$ starts with initial displacment unity and initial velocity (i) -1, (ii) 0 and (iii) 1. Represent these functions in the form $y = ce^{-t}\cos(t - \delta)$.

5.30 Find the values of t at which $y = e^{-t}\cos t$ has maxima and minima.

5.3.7 Forced oscillations

In section 5.2.1, we discussed the problem of *free* oscillations of a body of mass m attached to a vertically suspended spring. The motion of the body is governed by the *homogeneous* equation

$$m\ddot{y} + c\dot{y} + ky = 0 \qquad [1]$$

where $\dot{y} = dy/dt$, $\ddot{y} = d^2y/dt^2$ and m is the mass of the body, c is the damping constant and k is the spring constant. We shall now extend this problem by including a *variable* external force $f(t)$ which acts on the body. This extended situation has great significance in the concept of *resonance*. The equation of motion of the body is now obtained by adding the force $f(t)$ in the right-hand side of [1] obeying Newton's second law. Thus, we have $m\ddot{y} + c\dot{y} + ky = f(t)$, where the force $f(t)$ is called the input or *driving force* and the corresponding solution is called the output or the *response* of the system to the driving force $f(t)$. The resulting motion of the body is called *forced motion*.

The forced motion is of particular interest when the inputs $f(t)$ are *periodic*. We shall consider for example $f(t) = F_0\cos\omega t; F_0 > 0, \omega > 0$. In that case the equation of motion becomes

$$m\ddot{y} + c\dot{y} + ky = F_0\cos\omega t. \qquad [2]$$

Equation [2] is *non-homogeneous* and its complementary function is given by

$$y_c(t) = a\cos\omega t + b\sin\omega t.$$

We shall find the particular solution of [2] by the method of *undetermined coefficients.* By differentiating [3] twice successively with respect to t, we have

$$\dot{y}_c = -a\omega \sin \omega t + b\omega \cos \omega t \quad \text{and} \quad \ddot{y}_c = -a\omega^2 \cos \omega t - b\omega^2 \sin \omega t.$$

Substituting $\dot{y} = \dot{y}_c$ and $\ddot{y} = \ddot{y}_c$ in [2] and regrouping the terms, we get

$$\left[(k - m\omega^2)a + bc\omega\right] \cos \omega t + \left[-ca\omega + (k - m\omega^2)b\right] \sin \omega t = F_0 \cos \omega t.$$

By equating the coefficients of the sine and cosine terms on the two sides, we have

$$\left.\begin{aligned} (k - m\omega^2)a + bc\omega &= F_0 \\ -ca\omega + (k - m\omega^2)b &= 0 \end{aligned}\right\} \qquad [4]$$

Equations [4] form a system of two linear algebraic equations in two unknowns *a* and *b*. Solving the system [4], we get

$$a = F_0 \frac{k - m\omega^2}{(k - m\omega^2)^2 + c^2\omega^2} \quad \text{and} \quad b = F_0 \frac{c\omega}{(k - m\omega^2)^2 + c^2\omega^2}$$

provided the denominator is not zero. Now, if we set $\omega_0 = \sqrt{k/m}(> 0)$ as in section 5.2.1, we get

$$a = F_0 \frac{m(\omega_0^2 - \omega^2)}{m^2(\omega_0^2 - \omega^2)^2 + c^2\omega^2} \quad \text{and} \quad b = F_0 \frac{c\omega}{m^2(\omega_0^2 - \omega^2)^2 + c^2\omega^2}. \qquad [5]$$

Thus, the general solution of the non-homogeneous equation [2] is given by

$$y(t) = y_c(t) + y_p(t) \qquad [6]$$

where $y_c(t)$, given by [3], is the general solution of the *homogeneous* equation [1] and $y_p(t)$ is the particular solution given by [3] with coefficients a and b as in [5].

The behaviour of the *forced* oscillations is discussed separately when the system is *damped* $(c > 0)$ and *undamped* $(c = 0)$.

Case (i) : Undamped forced oscillations We shall study the behaviour of the mechanical system governed by equation [2] with $c = 0$ and $F_0 \neq 0$ which is the case of *undamped* and *forced* oscillations. Suppose $\omega^2 \neq \omega_0^2$. Then, from [3] and [5], we have

$$y_p(t) = \frac{F_0}{m(\omega_0^2 - \omega^2)} \cos \omega t = \frac{F_0}{k\left[1 - (\omega/\omega_0)^2\right]} \cos \omega t. \qquad [7]$$

From [6b] of section 5.2.1 and [7] above, we have the general solution for this case as

$$y_t = C\cos(\omega_0 t - \delta) + \frac{F_0}{m(\omega_0^2 - \omega^2)}\cos\omega t. \qquad [8]$$

The *output* [8] is clearly a superposition of two harmonic oscillations. The frequencies of these two oscillations are the *natural frequency* $\omega_0/2\pi$ cycles per second of the system (i.e. frequency of the free undamped motion) and the frequency $\omega/2\pi$ of the *input.*

From [7], we see that the maximum *amplitude* of y_p is

$$a_0 = \frac{F_0}{k[1-(\omega/\omega_0)^2]}$$

or $$a_0 = \frac{F_0}{k}\rho \text{ (where } \rho = \frac{1}{[1-(\omega/\omega_0)^2]}) \qquad [9]$$

As $\omega \to \omega_0$, ρ, and hence the maximum amplitude a_0, both tend to ∞. This phenomenon of obtaining large oscillations by matching the input and natural frequencies, i.e. $\omega = \omega_0$ is called *resonance* which is important in the study of vibrating systems. The factor ρ in [9] is called the *resonance factor* (Fig. 5.4). From [9], ρ/k is the ratio of the amplitudes of y_p and the input. In the case of *resonance*, [2] becomes

$$\ddot{y} + \omega_0^2 y = \frac{F_0}{m}\cos\ \omega_0 t. \qquad [10]$$

A particular solution of [10] is of the form $y_p(t) = t(a\cos\ \omega_0 t + b\sin\omega_0 t)$. By substituting this in [10], we get $a = 0, b = F_0/2m\omega_0$ and

$$y_p(t) = \frac{F_0}{2m\omega_0}\ t\sin\omega_0 t. \qquad [11]$$

Another very important oscillation is obtained when ω is *close to* (but not equal to) ω_0. Consider, for example, the particular solution (see [8])

math $$y(t) = \frac{F_0}{m(\omega_0^2 - \omega^2)}(\cos\omega t - \cos\omega_0 t) \qquad [12]$$

with $\omega \neq \omega_0$ and the initial conditions $y(0) = 0$ and $\dot{y}(0) = 0$. Using trigonometry, [12] can be simplified and expressed in the form

$$y(t) = \frac{2F_0}{m(\omega_0^2 - \omega^2)}\sin\left[\frac{\omega_0 + \omega}{2}t\right]\sin\left[\frac{\omega_0 - \omega}{2}t\right].$$

Since ω is close to ω_0, the difference $(\omega_0 - \omega)$ is very small so that the period of

$$\sin\left[\frac{\omega_0 - \omega}{2}t\right]$$

is very large. This yields an oscillation of the type shown in Fig. 5.6.

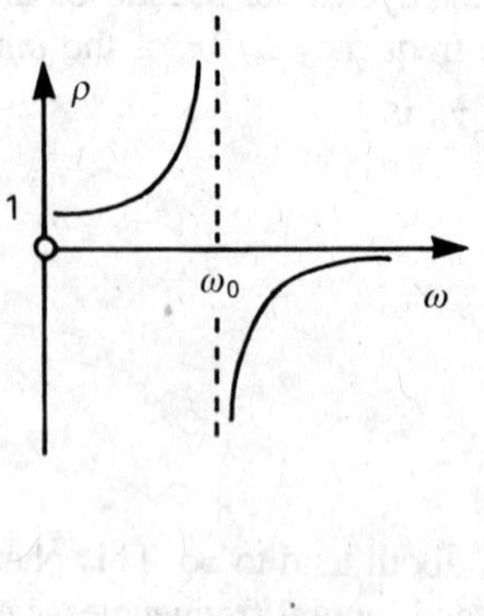

Fig. 5.4

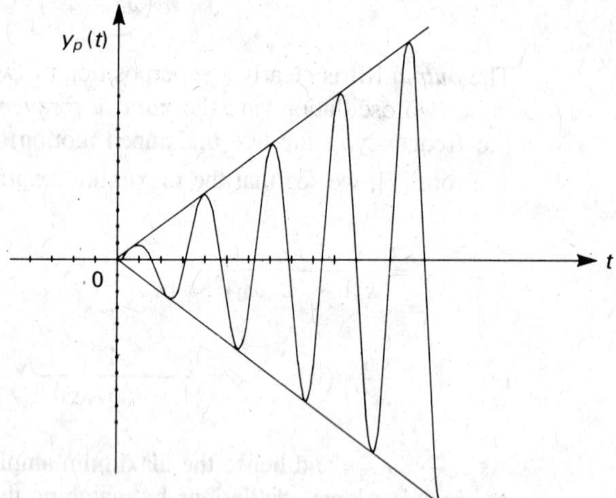

Fig. 5.5

Fig. 5.4 *Resonance factor ρ*
Fig. 5.5 *Particular solution for resonance*

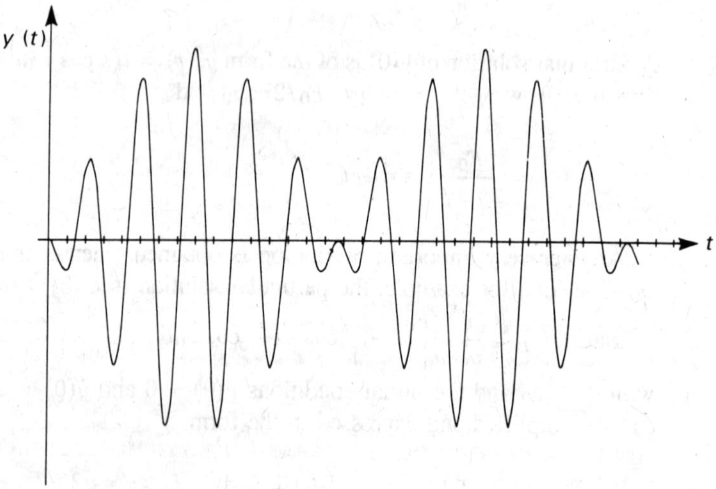

Fig. 5.6 *Forced undamped oscillations,* $(\omega_0 - \omega)$ *small*

Case (ii): Damped forced oscillations If there is *damping*, then the damping constant $c > 0$. In that case, the general solution y_c of [1] approaches zero as $t \to \infty$, i.e., t becomes very large (see section 5.2.1). That is, the general solution [6] of equation [2] now represents the *transient solution* and tends to the *steady-state solution* y_p. In other words, *after a sufficiently long time, the output of a purely sinusoidal input will* *approximate a harmonic oscillation whose frequency is that of the input. This is quite realistic since no physical system is completely undamped.*

In the case of an undamped system, we saw that the amplitude of y_p approaches infinity as $\omega \to \omega_0$. But, in the present case this situation does not arise. The amplitude is always finite but may attain a maximum for some value of ω depending on c. This is a situation of *practical resonance*. This shows that some input may excite oscillations with such large amplitudes that the system may collapse. In many practical situations — for example, in the construction of vehicles, aeroplanes etc — it is very difficult to avoid altogether undesired resonance effects.

Now, to study the amplitude of y_p as a function of ω, the particular solution [3] can be written as

$$y_p(t) = C_1 \cos(\omega t - \eta) \tag{13}$$

where, from [5], we have

$$C_1(\omega) = \sqrt{a^2 + b^2} = \frac{F_0}{\sqrt{m^2(\omega_0^2 - \omega^2)^2 + c^2\omega^2}} \tag{14}$$

and $\tan \eta = \dfrac{b}{a} = \dfrac{c\omega}{m(\omega_0^2 - \omega^2)}$.

From the first equation in [14], we shall find the *maximum* of $C_1(\omega)$. By differentiating C_1 with respect to ω and equating $dC_1/d\omega$ to zero, we get $\left[-2m^2(\omega_0^2 - \omega^2) + c^2\right]\omega = 0$. The coefficient of ω vanishes when

$$c^2 = 2m^2(\omega_0^2 - \omega^2). \tag{15}$$

When the damping is sufficiently large, i.e., $c^2 > 2m^2\omega_0^2 = 2mk$, equation [15] has no real roots; C_1 decreases monotonically with respect to ω (Fig. 5.7).

If $c^2 \leq 2mk$, then equation [15] has a real solution, $\omega = \omega_{\max}$ which increases as c decreases and approaches ω_0 as $c \to 0$. The amplitude $C_1(\omega)$ has a maximum at $\omega = \omega_{\max}$. By substituting $\omega = \omega_{\max}$ in [14], we get

$$C_1(\omega_{\max}) = \frac{2mF_0}{c\sqrt{4m^2\omega_0^2 - c^2}} \tag{16}$$

When $c > 0$, $C_1(\omega_{\max})$ is finite. Since

$$\frac{dC_1(\omega_{\max})}{dc} < 0$$

when $c^2 < 2mk$, the value of $C_1(\omega_{\max})$ increases as $c(\leq \sqrt{2mk})$ decreases, and approaches infinity as $c \to 0$. This agrees with the result in case (i).

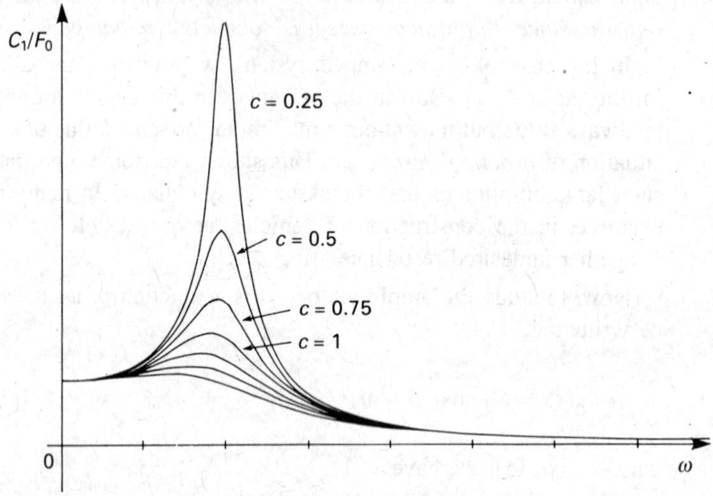

Fig. 5.7 *Amplification C_1/F_0 against ω*

In Fig. 5.7, the variation of the *amplification* C_1/F_0, the ratio of the amplitude of the output to the amplitude of the input, with respect to the frequency ω for $m = 1$, $k = 1$ and different values of the damping constants is plotted.

The angle η in [14] is called the *phase angle* or *phase lag* (Fig.5.8) since it measures the "lag" of the output with respect to the input. For $\omega < \omega_0$, we have $\eta < \pi/2$; when $\omega = \omega_0$, $\eta = \pi/2$ and for $\omega > \omega_0$, we have $\eta > \pi/2$.

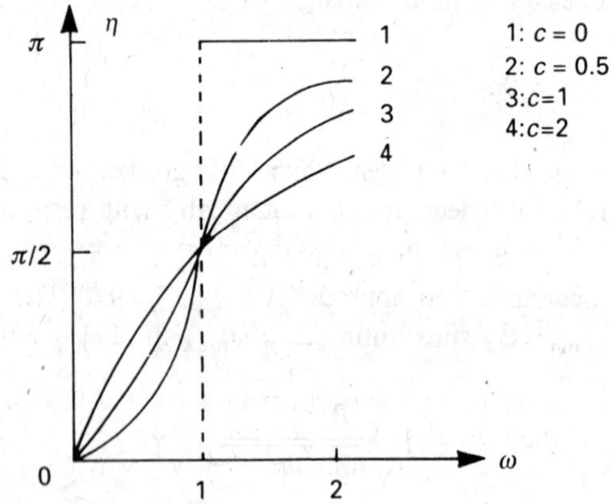

Fig. 5.8 *Variation of phase lag against frequency*

5.3.8 Electrical circuits

We shall study in this section the formation and solution of the differential equation which governs the flow of electric current in a simple circuit shown in Fig. 5.9.

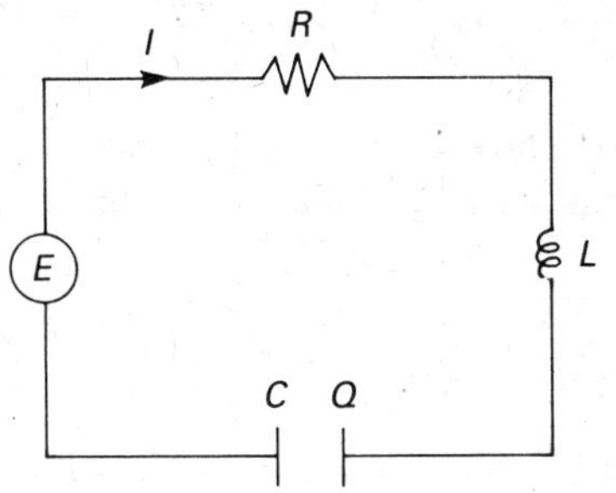

Fig. 5.9 *Electrical circuit*

The circuit consists of the following four elements:

(i) A source of *electromotive force* (e.m.f.) E, which could be a battery or a generator, which drives electric charge and produces a current I. The e.m.f. E may be a constant or a function of time t.

(ii) A resistor of *resistance* R which opposes the current by producing a drop in the e.m.f. of magnitude $E_R = IR$ according to *Ohm's law*, named after the German physicist, Georg Simon Ohm (1787–1854).

(iii) An inductor of *inductance* L which opposes any change in the current by producing a drop in the e.m.f. of magnitude $E_L = LdI/dt$.

(iv) A capacitor (or condenser) of *capacitance* C which stores the charge Q. The charge accumulated by the capacitor resists the inflow of additional charge. The drop in the e.m.f. arising due to the capacitor is $E_c = Q(1/C)$.

Since the *current* is the rate of flow of charge and hence the rate at which charge builds up on the capacitor, we have

$$I = \frac{dQ}{dt}.$$

The elements of the electrical circuit act according to Kirchhoff's law which states that the algebraic sum of the electromotive forces around a closed circuit is zero. (The law is named after the German scientist, Gustav Robert Kirchhoff [1824–1887]).

Accordingly, we have

$$E - E_R - E_L - E_C = 0 \text{ or } E - IR - L\frac{dI}{dt} - \frac{1}{C}Q = 0$$

$$\text{or} \quad L\frac{dI}{dt} + RI + \frac{1}{C}Q = E \qquad [1]$$

Now, in equation [1], either I or Q may be treated as the dependent variable (t being the independent variable). If I is to be treated as the dependent variable, then Q is eliminated by differentiating [1] with respect to t and then replacing dQ/dt by I. We get

$$L\frac{d^2I}{dt^2} + R\frac{dI}{dt} + \frac{1}{C}I = \frac{dE}{dt}. \qquad [2]$$

On the other hand, if Q be treated as the dependent variable, by eliminating I, we simply replace I by dQ/dt in [1]. In that case, we get

$$L\frac{d^2Q}{dt^2} + R\frac{dQ}{dt} + \frac{1}{C}Q = E. \qquad [3]$$

Equation [1], without capacitor C, will be in the form

$$L\frac{dI}{dt} + RI = E \qquad [4]$$

and has been dealt with in Chapter 2 as a *first order* linear differential equation.

Now, suppose a *periodic* electromotive force, $E = E_0 \cos \omega t$ acts in the given simple circuit (Fig. 5.9). Then the charge Q in the capacitor is governed by the differential equation [3] in the modified form

$$L\frac{d^2Q}{dt^2} + R\frac{dQ}{dt} + \frac{1}{C}Q = E_0 \cos \omega t. \qquad [5]$$

It is remarkable to note that equation [5] is similar to equation [2] of section 5.3.7 governing the *periodic* force oscillations of a mechanical system.

The similarity of the two equations provides for a mathematically identical treatment for the study of their solutions. In other words, the mathematical discussion of the equation for the periodic forced oscillation of a mechanical system is equally valid for the equation governing the electrical circuit, namely [5] above. It is interesting to note the analogy between the elements of the mechanical and the electrical systems:

Mechanical		*Electrical*
Mass, m	:	Inductance, L
Damping const.,c	:	Resistance, R
Spring modulus, k	:	Reciprocal of capacitance, $1/C$
Displacement, y	:	Charge Q of the capacitor.

In the electrical circuit, by analogy, we have a *critical* resistance below which the free behaviour of the circuit will be vibratory with a certain natural frequency, a forced steady-state vibration of the charge Q, and the *resonance* under favourable conditions.

5.4 Non-homogeneous equation of higher orders

The methods discussed rather elaborately in section 5.3, in the case of second order equations, for finding the general solutions of non-homogeneous linear equations with constant coefficients, can be applied to those of *higher orders* also.

Example 5.18 Solve: $(D^4 - 2D^3 + D^2)y = x^3$.

Solution 5.18 Here, the auxiliary equation is

$$m^4 - 2m^3 + m^2 = 0 \quad \text{or} \quad m^2(m^2 - 2m + 1) = 0, \text{ or } m^2(m-1)^2 = 0$$

which has both 0 and 1 as *double roots*. The complementary function is

$$\phi_h = (c_1 + c_2 x)e^{(0)x} + (c_3 + c_4 x)e^{(1)x}$$

or $\phi_h = (c_1 + c_2 x) + (c_3 + c_4 x)e^x$. [1]

The particular integral of the example is given by

$$\phi_p = \frac{1}{D^2(1 - 2D + D^2)}(x^3) = \frac{1}{D^2}(1 - D)^{-2}(x^3)$$

$$= \frac{1}{D^2}\left[1 + 2D + 3D^2 + 4D^3 + \ldots\right](x^3)$$

$$= \frac{1}{D^2}\left[(x^3) + 2(3x^2) + 3(6x) + 4(6)\right]$$

$$= \frac{1}{D^2}\left[x^3 + 6x^2 + 18x + 24\right] = \frac{1}{D}\left[\frac{x^4}{4} + 2x^3 + 9x^2 + 24x\right]$$

or $\phi_p = \left[\frac{x^5}{20} + \frac{x^4}{2} + 3x^3 + 12x^2\right]$. [3]

Therefore, from [2] and [3], the general solution of [1] is given by

$$y = (c_1 + c_2 x) + (c_3 + c_4 x)e^x + \frac{1}{20}\left[x^5 + 10x^4 + 60x^3 + 240x^2\right].$$

Example 5.19 Solve: $(D^3 + 3D^2 + 2D)y = x^2$.

Solution 5.19 The auxiliary equation is

$$m^3 + 3m^2 + 2m = 0 \quad \text{or} \quad m(m^2 + 3m + 2) = 0 \quad \text{or} \quad m(m+1)(m+2) = 0$$

whose roots are 0, –1 and –2. The complementary function is

$$\phi_h = c_1 e^{(0)x} + c_2 e^{-x} + c_3 e^{-2x}$$

or $\phi_h = c_1 + c_2 e^{-x} + c_3 e^{-2x}$. [1]

The particular integral of the example is

$$\phi_p = \frac{1}{D^3 + 3D^2 + 2D}(x^2) = \frac{1}{2D\left[1 + \frac{3}{2}D + D^2/2\right]}(x^2)$$

$$= \frac{1}{2D}\left[1 + \left(\frac{3}{2}D + \frac{D^2}{2}\right)\right]^{-1}(x^2)$$

$$= \frac{1}{2D}\left[1 - \left(\frac{3}{2}D + \frac{D^2}{2}\right) + \left(\frac{3}{2}D + \frac{D^2}{2}\right)^2 - \ldots\right](x^2)$$

$$= \frac{1}{2D}\left[1 - \frac{3}{2}D - \frac{D^2}{2} + \frac{9}{4}D^2 + \ldots\right](x^2)$$

$$= \frac{1}{2D}\left[x^2 - 3x + \frac{7}{2}\right] = \frac{1}{2}\left[\frac{x^3}{3} - \frac{3x^2}{2} + \frac{7}{2}x\right]$$

or $\phi_p = \frac{1}{12}\left[2x^3 - 9x^2 + 21x\right]$. [2]

Therefore, from [1] and [2], the general solution, is

$$y = c_1 + c_2 e^{-x} + c_3 e^{-2x} + \frac{1}{12}\left[2x^3 - 9x^2 + 21x\right].$$

Example 5.20 Solve: $(D-1)^2(D^2+1)^2 y = \sin^2\left(\frac{x}{2}\right) + e^x + x$.

Solution 5.20 The auxiliary equation is $(m-1)^2(m^2+1)^2 = 0$ whose roots are 1, $\pm i$ — all *double roots.* Correspondingly, the complementary function is

$$\phi_h = (c_1 + c_2 x)e^x + (c_3 + c_4 x)\cos x + (c_5 + c_6 x)\sin x. \quad [1]$$

The particular integral is given by

$$\phi_p = \frac{1}{(D-1)^2(D^2+1)^2}\left[\sin^2\left(\frac{x}{2}\right) + e^x + x\right]$$

or $$\phi_p = \frac{1}{(D-1)^2(D^2+1)^2}\left[\frac{1}{2}(1 - \cos x) + e^x + x\right]$$

We consider ϕ_p as the sum of the particular integrals Ψ_1, Ψ_2, Ψ_3 and Ψ_4 corresponding to the terms $1/2, -1/2\cos x, e^x$ and x. We have

(i) $$\Psi_1 = \frac{1}{(D-1)^2(D^2+1)^2}\left[\frac{1}{2}e^{(0)x}\right]$$

$$= \frac{1}{2}\frac{1}{(D-1)^2(D^2+1)^2}\left[e^{(0)x}\right] = \frac{1}{2}\frac{1}{(0-1)^2(0^2+1)^2}e^{(0)x}$$

or $\Psi_1 = \frac{1}{2}$. [2]

(ii) $$\Psi_2 = -\frac{1}{2}\frac{1}{(D-1)^2(D^2+1)^2}[\cos x]$$

$$= -\frac{1}{2}\frac{1}{(D^2+1)^2(D^2-2D+1)}[\cos x]$$

$$= -\frac{1}{2}\frac{1}{(D^2+1)^2\left\{-(1)^2-2D+1\right\}}[\cos x]$$

$$= \frac{1}{4}\frac{1}{(D^2+1)^2}\frac{1}{D}[\cos x] = \frac{1}{4}\frac{1}{(D^2+1)^2}[\sin x]$$

$$= \frac{1}{4}\text{ I.P}\left[\frac{1}{(D^2+1)^2}e^{ix}\right] \quad (\text{since} \quad \sin x = \text{imaginary part of } e^{ix}).$$

Therefore

$$\Psi_2 = \frac{1}{4}\text{ I.P.}\left[\frac{e^{ix}1}{\{(D+i)^2+1\}^2}(1)\right] = \frac{1}{4}\text{ I.P.}\left[e^{ix}\frac{1}{\{D^2+2iD\}^2}(1)\right]$$

$$= \frac{1}{4}\text{ I.P.}\left[e^{ix}\frac{1}{(2iD)^2\{1+(D/2i)\}^2}(1)\right]$$

$$= -\frac{1}{16}\text{ I.P.}\left[e^{ix}\cdot\frac{1}{D^2}\frac{1}{(1+0)^2}e^{(0)x}\right]$$

$$= -\frac{1}{16}\text{ I.P.}\left[e^{ix}\cdot\frac{1}{D^2}(1)\right]\left[\because \quad \frac{1}{D^2}(1) = \frac{1}{D}(x) = \frac{x^2}{2}\right]$$

$$= -\frac{1}{16}\text{ I.P.}\left[e^{ix}\left(\frac{x^2}{2}\right)\right] = -\frac{1}{32}\text{ I.P.}\left[x^2\cos x + ix^2\sin x\right]$$

$$\text{or } \Psi_2 = -\frac{1}{32}x^2 \sin x. \qquad [3]$$

(iii) $$\Psi_3 = \frac{1}{(D-1)^2(D^2+1)^2}e^x = \frac{1}{(D-1)^2[(1)^2+1]^2}e^x$$

$$= \frac{1}{4}\left(\frac{1}{(D-1)^2}e^x\right) = \frac{1}{4}\left(\frac{x^2}{2!}e^x\right)$$

$$\text{or } \Psi_3 = \frac{1}{8}x^2 e^x. \qquad [4]$$

(iv) $$\Psi_4 = \frac{1}{(D-1)^2(D^2+1)^2}(x) = [(1-D)^{-2}(1+D^2)^{-2}](x)$$

$$= [(1+2D+\ldots)(1-+\ldots)](x) = [1+2D+\ldots](x)$$

$$\text{or } \Psi_4 = x + 2. \qquad [5]$$

Now from [1], [2], [3], [4] and [5], the general solution is

$$y = (c_1 + c_2 x)e^x + (c_3 + c_4 x)\cos x + (c_5 + c_6 x)\sin x$$

$$+\frac{1}{2} - \frac{1}{32}x^2 \sin x + \frac{1}{8}x^2 e^x + x + 2.$$

Example 5.21 Solve: $(D^5 - D)y = 12e^x + 8\sin x - 2x$.

Solution 5.21 The auxiliary equation is $m^5 - m = 0$ or

$$m(m^2-1)(m^2+1) = 0$$

so that the roots are $0, 1, -1, \pm i$. Therefore, the complementary function is

$$\phi_h = c_1 + c_2 e^x + c_3 e^{-x} + c_4 \cos x + c_5 \sin x. \qquad [1]$$

The particular integral of the example is given by

$$\phi_p = \frac{1}{D^5 - D}\left[12e^x + 8\sin x - 2x\right].$$

We consider ϕ_p as the sum of the particular integrals corresponding to the terms $12e^x, 8\sin x$ and $-2x$ of the right-hand side of the example. We have

(i) $$\Psi_1 = \frac{1}{(D^5 - D)}(12e^x) = \frac{1}{(D-1)D(D+1)(D^2+1)}(12e^x)$$

$$= 12\frac{1}{(D-1)}\frac{1}{(1)(1+1)(1^2+1)}e^x = 3\frac{1}{(D-1)}e^x = 3 \cdot \frac{x}{1!}e^x$$

or $\Psi_1 = 3xe^x$. [2]

(ii) $$\Psi_2 = \frac{1}{(D^5 - D)}(8\sin x) = 8\frac{1}{(D^2+1)D(D^2-1)}\sin x$$

$$= 8\frac{1}{(D^2+1)D(-1^2-1)}(\sin x)$$

$$= -4\frac{1}{(D^2+1)}\left[\frac{1}{D}(\sin x)\right] = -4\frac{1}{(D^2+1)}(-\cos x)$$

or $\Psi_2 = 4\frac{x}{2(1)}\sin x = 2x\sin x.$ [3]

since $\frac{1}{D^2+k^2}\cos(kx) = \frac{x}{2k}\sin(kx).$

(iii) $$\Psi_3 = \frac{1}{(D^5 - D)}(-2x) = -2\frac{1}{D(D^2-1)(D^2+1)}(x)$$

$$= 2\frac{1}{D(1-D^2)(1+D^2)}(x) = 2\frac{1}{D}\left[(1-D^2)^{-1}(1+D^2)^{-1}\right](x)$$

$$= 2\frac{1}{D}\left[(1+D^2+\ldots)(1-D^2+\ldots)\right](x) = 2\frac{1}{D}[x] = 2\left[\frac{x^2}{2}\right]$$

or $\Psi_3 = x^2$. [4]

Now from [1], [2], [3] and [4], the general solution is

$$y = c_1 + c_2e^x + c_3e^{-x} + c_4\cos x + c_5\sin x + 3xe^x + 2x\sin x + x^2$$

or $y = c_1 + x^2 + (c_2 + 3x)e^x + c_3e^{-x} + c_4\cos x + (c_5 + 2x)\sin x.$

Exercise 5C

Solve the following equations

5.31 $(D^2 + 3D + 2)y = e^{2x} \sin x$

5.32 $(D^2 - D + 1)y = e^{2x} + e^x$

5.33 $y'' - 2y' + y = (1 + e^{-x})^2$

5.34 $(D^2 + D + 1)y = (1 + \sin x)^2$

5.35 $(D^2 - 2D + 5)y = e^{-x} \sin 2x + e^x \sin 2x$

5.36 $2y'' - 3y' - 5y = 2 \sin 2x + 3 \cos 2x$

5.37 $(D^2 - 4)y = x^2$

5.38 $(D^2 - 5D + 6)y = x$

5.39 $(D^2 + 4)y = x^2 + 3$

5.40 $(D^3 + 8)y = x^4 + 2x + 1$

5.41 $(D^2 - 2D + 1)y = x - 1$

5.42 $(D^3 - 2D^2 + D)y = e^{2x} + x^2 + x$

5.43 $(D^3 - 3D^2 + 3D - 1)y = xe^x + e^x$

5.44 $(D^2 - 1)y = (1 + x^2)e^x$

5.45 $(D^3 + 2D^2)y = 5x^2 - 7$

5.46 $(D^3 - 12D + 16)y = 4$

5.47 $(D^3 - D^2 + 3D + 5)y = e^x \cos 2x$

5.48 $(D^3 - 7D - 6)y = (1 + x)e^{2x}$

5.50 $(D^3 - D^2 - 4D + 4)y = e^x$

5.49 $(D^4 + 3D^3 + 3D^2 + D)y = 20x^2 + 2x - 7$

CHAPTER SIX

Cauchy–Euler Differential Equations

In this chapter we consider a special type of *linear* differential equations with *variable coefficients* of the form

$$a_0 x^n \frac{d^n y}{dx^n} + a_1 x^{n-1} \frac{d^{n-1}y}{dx^{n-1}} + \ldots + a_n y = f(x) \qquad [6.1]$$

where $a_0, a_1, \ldots, a_n$ are constants. An equation of the type [6.1] is called a *Cauchy–Euler* differential equation, named after the famous mathematicians, Cauchy (1789–1857) of France and Euler (1707–1783) of Switzerland. However, solution of differential equations with variable coefficients in their general form will be discussed in Chapter 8. Equation [6.1] can be reduced to one with *constant coefficients* by the transformation

$$x = e^z, \qquad \text{i.e., } z = \log x. \qquad [6.2]$$

We have $\dfrac{dy}{dx} = \dfrac{dy}{dz} \cdot \dfrac{dz}{dx} = \dfrac{dy}{dz} \cdot \dfrac{1}{x} = \dfrac{1}{x}\dfrac{dy}{dz}$

$$\frac{d^2y}{dx^2} = \frac{d}{dx}\left[\frac{1}{x}\frac{dy}{dz}\right] = \frac{1}{x}\frac{d}{dx}\left(\frac{dy}{dz}\right) + \frac{dy}{dz}\frac{d}{dx}\left(\frac{1}{x}\right) = \frac{1}{x}\left[\frac{d^2y}{dz^2}\frac{dz}{dx}\right] + \frac{dy}{dz}\left(-\frac{1}{x^2}\right)$$

$$= \frac{1}{x^2}\left[\frac{d^2y}{dz^2} - \frac{dy}{dz}\right]\left[\text{since } \frac{dz}{dx} = \frac{1}{x}\right].$$

Similarly, $\dfrac{d^3y}{dx^3} = \dfrac{1}{x^3}\left[\dfrac{d^3y}{dz^3} - 3\dfrac{d^2y}{dz^2} + 2\dfrac{dy}{dz}\right]$ and so on.

Now, denoting d/dz by D, d^2/dz^2 by D^2 and so on, the above successive derivatives can be expressed as

$$x\frac{dy}{dx} = \frac{dy}{dz} \equiv Dy$$

$$x^2\frac{d^2y}{dx^2} = \frac{d^2y}{dz^2} - \frac{dy}{dz} = D^2y - Dy \equiv D(D-1)y$$

$$x^3\frac{d^3y}{dx^3} = (D^3 - 3D^2 + 2D)y \equiv D(D-1)(D-2)y$$

and so on. In fact, by induction, we have

$$x^n\frac{d^ny}{dx^n} = \big[D(D-1)(D-2)\dots(D-n+1)\big]y.$$

Substituting the above expressions in [1], the Cauchy–Euler equation becomes

$$\Bigg[\Big\{a_0D(D-1)\dots(D-n+1)\Big\} + \Big\{a_1D(D-1)\dots(D-n+2)\Big\} + \dots$$

$$+\Big\{a_{n-1}D\Big\} + a_n\Bigg]y = f(e^z) \qquad [6.3]$$

Equation [6.3] is a *linear* non-homogeneous equation with *constant coefficients* which can be readily solved by the methods discussed in Chapter 5. For $n = 2$, the Cauchy–Euler equation [6.1] takes the form

$$a_0x^2\frac{d^2y}{dx^2} + a_1x\frac{dy}{dx} + a_2y = f(x) \qquad [6.4]$$

which is of the *second order*. Similarly, a *third order* Cauchy–Euler equation is of the following form

$$a_0x^3\frac{d^3y}{dx^3} + a_1x^2\frac{d^2y}{dx^2} + a_2x\frac{dy}{dx} + a_3y = f(x). \qquad [6.5]$$

Example 6.1 Solve: $x^2y'' + 5xy' + 13y = \log x + x^2$.

Solution 6.1 The above equation is a second order Cauchy–Euler equation of the form [6.4]. This equation is converted into an equation with constant coefficients by the transformation

$$x = e^z \quad \text{or} \quad z = \log x. \qquad [1]$$

Accordingly, we have $x^2y'' \equiv D(D-1)y$ and $xy' \equiv Dy$ so that the example becomes

$$D(D-1)y + 5Dy + 13y = z + e^{2z} \quad \text{or} \quad \big[D(D-1) + 5D + 13\big]y = z + e^{2z}$$

or $\big[D^2 + 4D + 13\big]y = z + e^{2z}$ [2]

where $Dy \equiv dy/dz$ and $D^2y \equiv d^2y/dz^2$.

Now, [2] is a linear non-homogeneous equation with constant coefficients. The auxiliary equation of [2] is $m^2 + 4m + 13 = 0$, which has the pair of complex roots $m = -2 \pm 3i$. Therefore, the complementary function of [2] is

$$\phi_h = e^{-2z}[A\cos 3z + B\sin 3z]. \qquad [3]$$

It can be verified that the particular integral of [2] is given by

$$\phi_p = \frac{1}{25}e^{2z} + \frac{1}{13}z - \frac{4}{169}. \qquad [4]$$

Therefore, from [3] and [4], the complete solution of [2] is

$$y = e^{-2z}[A\cos 3z + B\sin 2z] + \frac{e^{2z}}{25} + \frac{z}{13} - \frac{4}{169}.$$

Substituting for z and e^z from [1], the complete solution is

$$y = \frac{1}{x^2}[A\cos(3\log x) + B\sin(3\log x)] + \frac{x^2}{25} + \frac{\log x}{13} - \frac{4}{169}.$$

Example 6.2 Solve : $(2x+1)^2y'' - 2(2x+1)y' - 12y = 6x$.

Solution 6.2 The example is converted into an equation with constant coefficients by the transformation

$$(2x+1) = e^z \quad (\text{and hence } z = \log(2x+1)). \qquad [1]$$

We have $\dfrac{dy}{dx} = \dfrac{dy}{dz}.\dfrac{dz}{dx} = \dfrac{2}{2x+1}\dfrac{dy}{dz}$ so that

$$(2x+1)\frac{dy}{dx} = 2\frac{dy}{dz} \equiv 2Dy \ (\text{where } D \equiv d/dz). \qquad [2]$$

Also $\dfrac{d^2y}{dx^2} = \dfrac{d}{dx}\left(\dfrac{dy}{dx}\right) = \dfrac{d}{dz}\left(\dfrac{dy}{dx}\right)\cdot\dfrac{dz}{dx} = \dfrac{d}{dz}\left[\dfrac{2}{2x+1}\dfrac{dy}{dz}\right]\dfrac{dz}{dx}$

$$= 2\left[\frac{1}{2x+1}\frac{d^2y}{dz^2} + \left\{\frac{(-2)}{(2x+1)^2}\frac{dx}{dz}\right\}\frac{dy}{dz}\right]\frac{dz}{dx} = \frac{4}{(2x+1)^2}\left[\frac{d^2y}{dz^2} - \frac{dy}{dz}\right]$$

on substituting $dz/dx = 2/(2x+1)$. Therefore, we have

$$(2x+1)^2\frac{d^2y}{dx^2} \equiv 4D(D-1)y. \qquad [3]$$

Using [2] and [3] in the given differential equation, we have

$$4D(D-1)y - 2(2D)y - 12y = 6\left[\frac{1}{2}(e^z - 1)\right]$$

or $4\big[D(D-1) - D - 3\big]y = 3(e^z - 1)$

or $\big[D^2 - 2D - 3\big]y = \frac{3}{4}(e^z - 1).$ [4]

Equation [4] is a second order linear non-homogeneous equation with *constant coefficients.* The auxiliary equation of [4] has the roots -1 and 3 and is given by

$$m^2 - 2m - 3 = 0 \quad \text{or} \quad (m+1)(m-3) = 0.$$

Therefore, the complementary function of [4] is $\phi_h = c_1e^{-z} + c_2e^{3z}$ or

$$\phi_h = \frac{c_1}{2x+1} + c_2(2x+1)^3. \qquad [5]$$

The particular integral of [4] is

$$\phi_p = \frac{3}{4}\frac{1}{(D-3)(D+1)}\left[e^z - e^{(0)z}\right]$$

$$= \frac{3}{4}\left[\frac{1}{(1-3)(1+1)}e^z - \frac{1}{(0-3)(0+1)}\right] = -\frac{3}{16}e^z + \frac{1}{4}$$

or $\phi_p = -\frac{3}{16}(2x+1) + \frac{1}{4}.$ [6]

From [5] and [6], the complete solution of [4] (and hence of the given equation) is

$$y = \frac{c_1}{2x+1} + c_2(2x+1)^3 - \frac{3}{16}(2x+1) + \frac{1}{4}$$

or $y = \frac{c_1}{2x+1} + c_2(2x+1)^3 - \frac{1}{16}(6x-1).$

Example 6.3 Solve: $x^3\frac{d^3y}{dx^3} - 3x\frac{dy}{dx} + 3y = 4x$ given that

$$y = \frac{dy}{dx} = \frac{d^2y}{dx^2} = 0 \text{ at } x = 1.$$

Solution 6.3 Substituting $x = e^z$, i.e. $z = \log x$, and hence

$$x\frac{dy}{dx} \equiv Dy,\ x^2\frac{d^2y}{dx^2} \equiv D(D-1)y,\ x^3\frac{d^3y}{dx^3} \equiv D(D-1)(D-2)y$$

The given equation becomes

$$\big[D(D-1)(D-2) - 3D(D-1) + 3\big]y = 4e^z$$

or $\big[D^3 - 3D^2 - D + 3\big]y = 4e^z (\text{where} \quad D \equiv d/dz).$ [1]

The auxiliary equation of [1] is

$$m^3 - 3m^2 - m + 3 = 0 \quad \text{or} \quad m^2(m-3) - 1(m-3) = 0$$

or $(m-3)(m^2-1) = 0$

whose roots are 3 and $\pm$ 1. Therefore, the complementary function of [1] is

$$\phi_h = c_1e^{3z} + c_2e^z + c_3e^{-z}.$$ [2]

The particular integral of the example is given by

$$\phi_p = \frac{1}{(D+1)(D-1)(D-3)}[4e^z] = \frac{1}{2}\left[\frac{1}{D+1} - \frac{2}{D-1} + \frac{1}{D-3}\right]e^z$$

or $\phi_p = -ze^z.$ [3]

From [2] and [3], the complete solution of [1] is

$$y = c_1e^{3z} + c_2e^z + c_3e^{-z} - ze^z.$$

Substituting $e^z = x$ or $z = \log x$ the complete solution of the given differential equation is given by

$$y = c_1x^3 + c_2x + \frac{c_3}{x} - x\log x.$$ [4]

We use the given initial conditions to determine the arbitrary constants c_1, c_2, c_3. Differentiating [4] with respect to x, we get

$$\frac{dy}{dx} = 3c_1x^2 + c_2 - \frac{c_3}{x^2} - 1 - \log x.$$ [5]

Differentiating [5] with respect to x,

$$\frac{d^2y}{dx^2} = 6c_1x + \frac{2c_3}{x^3} - \frac{1}{x}. \qquad [6]$$

Substituting $x = 1, y = 0, dy/dx = 0$ and $d^2y/dx^2 = 0$ in [4], [5] and [6], we get

$$c_1 + c_2 + c_3 = 0, 3c_1 + c_2 - c_3 = 1 \quad \text{and} \quad 6c_1 + 2c_3 = 1.$$

Solving these three simultaneous algebraic equations, we get $c_1 = 1/4, c_2 = 0$ and $c_3 = -1/4$. Substituting these values in [4], the required solution is

$$y = \frac{x^3}{4} - \frac{1}{4x} - x\log x.$$

Example 6.4 Solve: $(x^3d^3y/dx^3) + (2x^2d^2y/dx^2) + 2y = 10(x + 1/x)$.

Solution 6.4 We use the transformation $x = e^z$ or $z = \log x$ so that the given equation takes the form

$$[D(D-1)(D-2) + 2D(D-1) + 2]y = 10(e^z + e^{-z})$$

or $[D^3 - D^2 + 2]y = 10(e^z + e^{-z})$. [1]

The auxiliary equation of [1] is

$$m^3 - m^2 + 2 = 0 \qquad [2]$$

By inspection, we see that $m = -1$ is a root of [2]. To find the other roots, [2] is divided by the factor $(m + 1)$ using the "*synthetic division*":

		1	−1	0	2
m =	−1	0	−1	2	−2
		1	−2	2	0

The resulting quotient is $m^2 - 2m + 2 = 0$, whose roots are $1 \pm i$. Thus, the roots of [2] are -1 and $1 \pm i$ so that the complementary function of [1] is

$$\phi_h = c_1e^{-z} + e^z(c_2\cos z + c_3\sin z). \qquad [3]$$

The particular integral of [1] is given by

$$\phi_p = \frac{1}{D^3 - D^2 + 2}[10(e^z + e^{-z})]$$

$$= 10\left[\frac{1}{D^3 - D^2 + 2}\right](e^z) + 10\left[\frac{1}{D^3 - D^2 + 2}\right](e^{-z})$$

$$= 10\left[\frac{1}{(1)^3 - (1)^2 + 2}\right](e^{-z}) + 10\left[\frac{1}{(D+1)(D^2 - 2D + 2)}\right](e^{-z})$$

$$= 5e^z + \frac{10}{D+1}\left[\frac{1}{(-1)^2 - 2(-1) + 2}\right](e^{-z})$$

or $\phi_p = 5e^z + 2ze^{-z}$ [4]

From [3] and [4], the general solution of [1] and hence of the given equation is

$$y = c_1 e^{-z} + e^z(c_2 \cos z + c_3 \sin z) + 5e^z + 2ze^{-z}$$

or $y = \dfrac{c_1}{x} + x[c_2 \cos(\log x) + c_3 \sin(\log x)] + 5x + \dfrac{2\log x}{x}$

or $y = \dfrac{(c_1 + 2\log x)}{x} + x[c_2 \cos(\log x) + c_3 \sin(\log x) + 5].$

Exercise 6

Solve the following equations:

6.1 $x^2y'' + 3xy' + 10y = 0$

6.2 $x^2y'' - 3xy' + 4y = 0$

6.3 $x^2y'' + xy' - 4y = x^3$

6.4 $x^2y'' + 3xy' + 2y = x^2 + 2x$

6.5 $3x^2y'' + xy' - 16y = x^3(2\log x + 3)$

6.6 $x^2y'' - 2xy' + 2y = 2x^3$ given that $y = 0$ when $x = \pm 1$.

6.7 $x^3y''' + x^2y'' - 6xy' + 6y = x + \dfrac{3}{x^2}$

6.8 $x^3y''' + xy' - y = 3x^4$ given that $y = y' = y'' = 0$ when $x = 2$.

6.9 $(x+2)^2y'' + 7(x+2)y' - 12y = 3x^2 + 2x + 7$

6.10 $(2x+1)^2y'' - 2(2x+1)y' - 12y = 6x$

6.11 $(1+x)^2y'' + (1+x)y' + y = 4\cos\log(1+x)$

6.12 $(2x-1)^3y''' + (2x-1)y' - 2y = 0$

6.13 $(x^2D^2+1)y = 3x^2$

6.14 $(x^2D^2 - 3xD + 5)y = x^2 \sin \log x$

6.15 $(3x+2)^2y'' + 3(3x+2)y' - 36y = 3x^2 + 4x + 1$

CHAPTER SEVEN

Simultaneous Differential Equations (In Two Variables With Constant Coefficients)

7.1 Linear system of equations

A *system* of two first order equations in two variables (or unknown functions) x and y, in terms of an independent variable t, is of the form

$$\frac{dx}{dt} = F(t, x, y), \quad \frac{dy}{dt} = G(t, x, y). \tag{7.1}$$

Further, the system is said to be one of *linear* equations with *constant coefficients* if F and G functions in [7.1] are of the form

$$\left.\begin{aligned} F(t, x, y) &= a_1 x + b_1 y + f(t) \\ G(t, x, y) &= a_2 x + b_2 y + g(t) \end{aligned}\right\} \tag{7.2}$$

where a_1, b_1, a_2 and b_2 are given constants and $f(t)$ and $g(t)$ are functions of t only.

It is possible to find the solution of such a system by reducing it to a single equation of a higher order when the latter equation is linear with constant coefficients.

7.2 Euler's method for homogeneous linear system

A system of linear equations [7.2], with constant coefficients, is said to be homogeneous if $f(t) = 0$ and $g(t) = 0$. In other words, a *homogeneous* system of *linear* differential equations with constant coefficients is of the form

$$\left.\begin{aligned} dx/dt &= a_1 x + b_1 y \\ dy/dt &= a_2 x + b_2 y \end{aligned}\right\} \tag{7.3}$$

where a_1, b_1, a_2 and b_2 are constants. We seek a solution of the system [7.3] in the form given by [7.4].

$$x = \alpha e^{\lambda t}, \quad y = \beta e^{\lambda t} \tag{7.4}$$

where α, β and λ are constants to be determined. Substituting [7.4] in [7.3], we get

$$\frac{d}{dt}(\alpha e^{\lambda t}) = a_1(\alpha e^{\lambda t}) + b_1(\beta e^{\lambda t}), \quad \frac{d}{dt}(\beta e^{\lambda t}) = a_2(\alpha e^{\lambda t}) + b_2(\beta e^{\lambda t})$$

or on simplification $\alpha\lambda = a_1\alpha + b_1\beta$ and $\beta\lambda = a_2\alpha + b_2\beta$ or

$$\left.\begin{aligned} (a_1 - \lambda)\alpha + b_1\beta = 0 \\ a_2\alpha + (b_2 - \lambda)\beta = 0 \end{aligned}\right\} \tag{7.5}$$

For the system [7.5] of homogeneous linear algebraic equations in the two unknowns α and β to have a non-trivial solution, a necessary and sufficient condition is that the determinant of the coefficient matrix of [7.5] should vanish. That is

$$\begin{vmatrix} a_1 - \lambda & b_1 \\ a_2 & b_2 - \lambda \end{vmatrix} = 0. \tag{7.6}$$

The *characteristic equation* [7.6] is a quadratic equation in λ in this case. Solving the quadratic equation, we get the two roots, say λ_1 and λ_2.

If the roots λ_1 and λ_2 are distinct, then substituting their values in [7.5], we obtain the corresponding non-trivial solution α and β and consequently find the solutions of the system of differential equations [7.3] from [7.4] in the form

$$x = \alpha_1 e^{\lambda_1 t}, \quad y = \beta_1 e^{\lambda_1 t} \quad \text{and} \quad x = \alpha_2 e^{\lambda_2 t}, \quad y = \beta_2 e^{\lambda_2 t}.$$

Example 7.1 Solve the system of equations

$$\frac{dx}{dt} = -x + 2y; \quad \frac{dy}{dt} = 2x - y. \tag{1}$$

Solution 7.1 We seek the solution of system [1] in the form

$$x = \alpha e^{\lambda t}, \quad y = \beta e^{\lambda t} \tag{2}$$

Correspondingly, the characteristic equation is

$$\begin{vmatrix} -1 - \lambda & 2 \\ 2 & -1 - \lambda \end{vmatrix} = 0$$

or $\lambda^2 + 2\lambda - 3 = 0$ which has the roots $\lambda_1 = 1$ and $\lambda_2 = -3$. On substituting [2] in [1], we get the algebraic system of equations

$$\left.\begin{aligned} (-1-\lambda)\alpha + 2\beta = 0 \\ 2\alpha + (-1-\lambda)\beta = 0 \end{aligned}\right\} \qquad [3]$$

Now, substituting $\lambda = 1$ in [3] and taking the corresponding values of α and β as α_1 and β_1, we get

$$-2\alpha_1 + 2\beta_1 = 0, \quad \text{and} \quad 2\alpha_1 - 2\beta_1 = 0$$

so that $\alpha_1 = \beta_1$. Hence, substituting $\lambda = 1$ in [2], we have

$$x = \alpha_1 e^t, \quad y = \alpha_1 e^t. \qquad [4]$$

Similarly, setting $\lambda = -3$, and taking the corresponding values of α and β as α_2 and β_2, we have, from [3]:

$$2\alpha_2 + 2\beta_2 = 0 \quad \text{and} \quad 2\alpha_2 + 2\beta_2 = 0$$

so that $\alpha_2 = -\beta_2$. Hence

$$x = \alpha_2 e^{-3t}, \quad y = -\alpha_2 e^{-3t} \qquad [5]$$

From [4] and [5], the general solution of the given system [1] consists of

$$x(t) = c_1(\alpha_1 e^t) + c_2(\alpha_2 e^{-3t}) \quad \text{and} \quad y(t) = c_1(\alpha_1 e^t) + c_2(-\alpha_2 e^{-3t})$$

the linear combinations of sets of solutions [4] and [5] or

$$x(t) = C_1 e^t + C_2 e^{-3t} \quad \text{and} \quad y(t) = C_1 e^t - C_2 e^{-3t}$$

where $C_1 = c_1\alpha_1$ and $C_2 = c_2\alpha_2$.

7.3 Linear independence of solutions

In order to make sure that the two solutions obtained (corresponding to the two values of λ) are *linearly independent* and hence yielding the general solution, we have to examine in detail the different possibilities for λ_1 and λ_2, the two values of λ.

Case (i) λ_1 and λ_2 real and distinct: If the roots of the characteristic equation [7.6] are real and distinct, then the two sets of solutions

$$\begin{bmatrix} x \\ y \end{bmatrix} = \begin{bmatrix} \alpha_1 e^{\lambda_1 t} \\ \beta_1 e^{\lambda_1 t} \end{bmatrix} \quad \text{and} \quad \begin{bmatrix} x \\ y \end{bmatrix} = \begin{bmatrix} \alpha_2 e^{\lambda_2 t} \\ \beta_2 e^{\lambda_2 t} \end{bmatrix}$$

are linearly independent and hence the general solution of the system [7.3] is a linear combination of the set of solutions viz.,

$$x = c_1\alpha_1 e^{\lambda_1 t} + c_2\alpha_2 e^{\lambda_2 t}; \quad y = c_1\beta_1 e^{\lambda_1 t} + c_2\beta_2 e^{\lambda_2 t}.$$

Case (ii) λ_1 and λ_2 are complex conjugate roots: If the roots of the characteristic equation [7.6] are a pair of complex conjugate numbers of the form $(a \pm ib)$, with $b \neq 0$, then $\alpha_1, \beta_1, \alpha_2$ and β_2 are all complex and we get two linearly independent solutions

$$\begin{bmatrix} x \\ y \end{bmatrix} = \begin{bmatrix} \alpha_1' & e^{(a+ib)t} \\ \beta_1' & e^{(a+ib)t} \end{bmatrix}$$

$$\begin{bmatrix} x \\ y \end{bmatrix} = \begin{bmatrix} \alpha_2' & e^{(a-ib)t} \\ \beta_2' & e^{(a-ib)t} \end{bmatrix} \qquad [7.7]$$

Since these solutions are complex-valued, we proceed as follows to obtain the real solutions therefrom. The complex numbers α_1' and β_1' are expressed in the standard form $\alpha_1' = \alpha_1 + i\alpha_2$ and $\beta_1' = \beta_1 + i\beta_2$. Now, using the well-known Euler's formula, $e^{i\theta} = \cos\theta + i\sin\theta$, the first of the two complex solutions [7.7] can be written as

$$x = (\alpha_1 + i\alpha_2)e^{at}(\cos bt + i\sin bt); \quad y = (\beta_1 + i\beta_2)e^{at}(\cos bt + i\sin bt)$$

or $x = e^{at}[(\alpha_1\cos bt - \alpha_2\sin bt) + i(\alpha_1\sin bt + \alpha_2\cos bt)]$;

$$y = e^{at}[(\beta_1\cos bt - \beta_2\sin bt) + i(\beta_1\sin bt + \beta_2\cos bt)].$$

The above complex values of x and y yield the *real* solutions

$$x = e^{at}[\alpha_1\cos bt - \alpha_2\sin bt] \equiv x_1(\text{say});$$

$$y = e^{at}[\beta_1\cos bt - \beta_2\sin bt] \equiv y_1(\text{say}).$$

Also $x = e^{at}[\alpha_1\sin bt + \alpha_2\cos bt] \equiv x_2(\text{say})$;

$$y = e^{at}[\beta_1\sin bt + \beta_2\cos bt] \equiv y_2(\text{say}).$$

Then, the general solution of the given system [7.3] of linear homogeneous differential equations, with constant coefficients, is given by the linear combinations

$$x = c_1x_1 + c_2x_2, \quad y = c_1y_1 + c_2y_2.$$

Case (iii) λ_1 and λ_2 are real and equal: Suppose the characteristic equation [7.6] of the system [7.3] has real and equal roots $\lambda_1 = \lambda_2 = \lambda$, a real number. In that case, the two solutions

$$\begin{bmatrix} \alpha_1 e^{\lambda t} \\ \beta_1 e^{\lambda t} \end{bmatrix} \quad \text{and} \quad \begin{bmatrix} \alpha_2 e^{\lambda t} \\ \beta_2 e^{\lambda t} \end{bmatrix}$$

are *not* linearly independent and hence we get basically only one solution $x = \alpha e^{\lambda t}$ and $y = \beta e^{\lambda t}$. However, we can look for a second solution of the form $x = (\alpha_1 + \alpha_2 t)e^{\lambda t}$ and $y = (\beta_1 + \beta_2 t)e^{\lambda t}$. The constants $\alpha_1, \alpha_2, \beta_1$ and β_2 are actually determined by substituting the above expressions for x and y in the given system of differential equations [7.3]. Then, the general solution of the system [7.3] is given by

$$x = c_1\alpha e^{\lambda t} + c_2(\alpha_1 + \alpha_2 t)e^{\lambda t}$$

$$y = c_1\beta e^{\lambda t} + c_2(\beta_1 + \beta_2 t)e^{\lambda t}.$$

Note: In practice, a system of differential equations can more conveniently be solved, in many cases, by the "*method of elimination*" using the operator $D \equiv d/dt$ and transforming a system of two first order equations into a single second order equation. The method is demonstrated in the following example.

Example 7.2 Solve the system of equations

$$\frac{dx}{dt} = x + y, \qquad \frac{dy}{dt} = 4x - 2y. \tag{1}$$

Solution 7.2 The equations of the system [1] can be written as

$$\frac{dx}{dt} - x = y, \qquad \frac{dy}{dt} + 2y = 4x$$

or $(D - 1)x = y,$ [2]

$$(D + 2)y = 4x. \tag{3}$$

Now, multiplying [2] by 4 and [3] by $(D - 1)$, we get

$$4(D - 1)x = 4y, \tag{4}$$

$$4(D - 1)x = (D - 1)(D + 2)y. \tag{5}$$

Subtracting [4] from [5], we get $(D-1)(D+2)y-4y=0$, or

$$(D^2+D-6)y=0 \qquad [6]$$

which is a *linear* and *homogeneous* equation with constant coefficients.

The auxiliary equation of [6] is $m^2+m-6=0$ whose roots are -3 and 2. Therefore, the solution of [6] is

$$y=c_1e^{-3t}+c_2e^{2t}. \qquad [7]$$

Substituting [7] in [2], we get $(D-1)x=c_1e^{-3t}+c_2e^{2t}$ so that

$$x=\frac{1}{D-1}\left[c_1e^{-3t}+c_2e^{2t}\right]=\frac{1}{(-3)-1}(c_1e^{-3t})+\frac{1}{(2)-1}(c_2e^{2t})$$

or $$x=-\frac{1}{4}c_1e^{-3t}+c_2e^{2t}. \qquad [8]$$

Thus, [7] and [8] together constitute the general solution of [1]:

$$x=-\frac{1}{4}c_1e^{-3t}+c_2e^{2t};\ y=c_1e^{-3t}+c_2e^{2t}.$$

Example 7.3 Solve the system of equations

$$\frac{dx}{dt}=3x-4y, \qquad \frac{dy}{dt}=x-y. \qquad [1]$$

Solution 7.3 In the D-operator form, equations [1] can be written as

$$(D-3)x+4y=0, \qquad [2]$$

$$x-(D+1)y=0. \qquad [3]$$

Multiplying [2] by $(D+1)$ and [3] by 4, we have

$$(D+1)(D-3)x+4(D+1)y=0, \qquad [4]$$

$$4x-4(D+1)y=0. \qquad [5]$$

Adding [4] and [5], and eliminating y, we get

$$[(D+1)(D-3)+4]x=0$$

or $[D^2 - 2D + 1]x = 0.$ [6]

The auxiliary equation of the homogeneous equation [6] is $m^2 - 2m + 1 = 0$ or $(m-1)^2 = 0$ which has 1 as a *double root.* Therefore, the solution of [6] is

$$x = (c_1 + c_2 t)e^t. \quad [7]$$

Substituting [7] in [2], we have $(D-3)[c_1 + c_2 t]e^t + 4y = 0$ so that

$$y = -\frac{1}{4}[(D-3)\{c_1 e^t + c_2 t e^t\}]$$

or $y = -\frac{1}{4}[c_1 e^t + c_2(te^t + e^t) - 3(c_1 e^t + c_2 t e^t)]$

or $y = -\frac{1}{4}[-2c_1 + c_2(1-2t)]e^t.$ [8]

Thus, the general solution of the given system of equations [1] consists of [7] and [8].

Note: The examples considered above are systems of linear equations of the *homogeneous type.* However, the method of elimination of one of the variables using the D-operator form can be effectively employed even in the case of *non-homogeneous* linear equations with constant coefficients, as demonstrated in the following example.

Example 7.4 Solve the system

$$\frac{dx}{dt} - 3x - 6y = t^2; \quad \frac{dy}{dt} + \frac{dx}{dt} - 3y = e^t. \quad [1]$$

Solution 7.4 Equations [1] can be written in the D-operator form as

$$(D-3)x - 6y = t^2 \quad [2]$$

$$Dx + (D-3)y = e^t. \quad [3]$$

Now, multiplying [2] by D and [3] by $(D-3)$, we get

$$D(D-3)x - 6Dy = D(t^2) = 2t \quad [4]$$

$$D(D-3)x + (D-3)^2 y = (D-3)e^t = -2e^t. \quad [5]$$

Subtracting [4] from [5] and simplifying, we get

$$(D^2+9)y = -2e^t - 2t. \qquad [6]$$

The general solution of [6] is

$$y = c_1 \cos 3t + c_2 \sin 3t - \frac{1}{5}e^t - \frac{2}{9}t. \qquad [7]$$

(Obtaining [7] is left as an exercise to students).
Next, multiplying [2] by $[D-3]$ and [3] by 6, we have

$$(D-3)^2x - 6(D-3)y = (D-3)t^2 = 2t - 3t^2 \qquad [8]$$

$$6Dx + 6(D-3)y = 6e^t. \qquad [9]$$

Adding [8] and [9], we get

$$(D^2+9)x = 2t - 3t^2 + 6e^t. \qquad [10]$$

The general solution of [10] can be verified (left as an exercise) to be

$$x = c_3 \cos 3t + c_4 \sin 3t + \frac{3}{5}e^t - \frac{1}{3}t^2 + \frac{2}{9}t + \frac{2}{27}. \qquad [11]$$

From [7] and [11] there appears to be more arbitrary constants than we need. However, these constants are related, and not independent, as we can see in the following discussion.

Substituting for x and y from [7] and [11] into one of the equations [2] and [3], say for example [2], we find $(\sin t)[-3c_3 - 3c_4 - 6c_2] + (\cos t)[3c_4 - 3c_3 - 6c_1] = 0$ since t^2 cancels out. Since this must be an identity, we must have $-3c_3 - 3c_4 - 6c_2 = 0$ and $3c_4 - 3c_3 - 6c_1 = 0$. These equations yield $c_3 = -(c_1 + c_2)$ and $c_4 = (c_1 - c_2)$. Substituting these in [11], and with [7], the general solution of the non-homogeneous system [1] consists of

$$x = (c_1 - c_2)\sin 3t - (c_1 + c_2)\cos 3t + \frac{3}{5}e^t - \frac{1}{3}t^2 + \frac{2}{9}t + \frac{2}{27}$$

$$y = c_1 \cos 3t + c_2 \sin 3t - \frac{1}{5}e^t - \frac{2}{9}t$$

where c_1 and c_2 are two arbitrary constants.

Example 7.5 Solve the system

$$2\frac{dx}{dt} + \frac{dy}{dt} - 4x - y = e^t \qquad [1]$$

$$\frac{dx}{dt} + 3x + y = 0. \qquad [2]$$

Solution 7.5 Differentiating [2], we get

$$\frac{d^2x}{dt^2} + 3\frac{dx}{dt} + \frac{dy}{dt} = 0. \qquad [3]$$

Now multiplying [1] by -1, [2] by -1 and [3] by 1 and adding, we obtain

$$\frac{d^2x}{dt^2} + x = -e^t$$

which is independent of y and its derivatives, and can be easily solved.

Thus $x = c_1 \cos t + c_2 \sin t - \dfrac{1}{D^2+1}e^t = c_1 \cos t + c_2 \sin t - \dfrac{1}{2}e^t.$

Proceeding in the same way by differentiating [1], we get

$$y = -\frac{dx}{dt} - 3x = (c_1 - 3c_2) \sin t - (3c_1 + c_2) \cos t + 2e^t.$$

So, $x = c_1 \cos t + c_2 \sin t - \dfrac{1}{2}e^t; y = (c_1 - 3c_2) \sin t - (3c_1 + c_2) \cos t + 2e^t.$

Exercise 7

Solve the following systems of equations:

7.1 $\dfrac{dy}{dt} = x, \quad \dfrac{dx}{dt} = -y; \quad y(0) = 0, x(0) = 1$

7.2 $\dfrac{dx}{dt} = -3x + 4y; \quad \dfrac{dy}{dt} = -2x + 3y$

7.3 $\dfrac{dx}{dt} = 4x - 2y; \quad \dfrac{dy}{dt} = 5x + 2y$

7.4 $\dfrac{dx}{dt} = 5x + 4y; \quad \dfrac{dy}{dt} = -x + y$

7.5 $\dfrac{dx}{dt} = 4x - 3y; \quad \dfrac{dy}{dt} = 8x - 6y$

7.6 $\dfrac{dx}{dt}+5x-2y=t;\quad \dfrac{dy}{dt}+2x+y=0$

7.7 $\dfrac{dx}{dt}+2x+3y=0;\quad \dfrac{dy}{dt}+3x+2y=2e^{2t}$

7.8 $(5D+4)x+(2D+1)y=e^{-t};\quad (D+8)x-3y=5e^{-t}$ where $D\equiv d/dt$.

7.9 $D^2x-3x-y=e^t;\quad Dy-2x=0$

7.10 $\dfrac{d^2x}{dt^2}-3x-4y=0;\quad \dfrac{d^2y}{dt^2}+x+y=0$

7.11 $\dfrac{dx}{dt}+\dfrac{dy}{dt}-2y=2\cos t-7\sin t;$

$$\frac{dx}{dt}-\frac{dy}{dt}+2x=4\cos t-3\sin t$$

7.12 $\dfrac{dx}{dt}=y+t,\quad \dfrac{dy}{dt}+x=1$

7.13 $\dfrac{dx}{dt}=3x+2y;\quad \dfrac{dy}{dt}=5x+3y$

7.14 $(D-3)x+2(D+2)y=2\sin t;$

$$2(D+1)x+(D-1)y=\cos t \quad \text{where} \quad D\equiv\frac{d}{dt}$$

7.15 $Dx-(D+1)y=-e^t; x+(D-1)y=e^{2t}$ where $D\equiv\dfrac{d}{dt}$

7.16 $\dfrac{dx}{dt}+y=\sin t;\quad \dfrac{dy}{dt}+x=\cos t$

given that $x=2$ and $y=0$ when $t=0$

7.17 $(5D+4)y-(2D+1)z=e^{-x};$

$$(D+8)y-3z=5e^{-x} \quad \text{where} \quad D\equiv\frac{d}{dx}$$

7.18 $\dfrac{dx}{dt}+5x+y=e^t;\quad \dfrac{dy}{dt}-x+3y=e^{2t}$

7.19 $2\dfrac{d^2y}{dx^2}-\dfrac{dz}{dx}-4y-2x=0; 2\dfrac{dy}{dx}+4\dfrac{dz}{dx}-3z=0$

7.20 A mechanical system with two degrees of freedom satisfies the equations

$$2\frac{d^2x}{dt^2} + 3\frac{dy}{dt} = 4; \quad 2\frac{d^2y}{dx^2} - 3\frac{dx}{dt} = 0.$$

Obtain expressions for x and y in terms of t, given $x, y, dx/dt, dy/dt$ vanish at $t = 0$.

CHAPTER EIGHT

Second Order Differential Equations With Variable Coefficients

In Chapter 5, we discussed at length the solution of second order linear differential equations with *constant coefficients.* In Chapter 6, we considered the Cauchy–Euler equation where the coefficients are variable but the given differential equation can be reduced to one with constant coefficients. In this chapter, we shall consider an equation with variable coefficients where the equation cannot be reduced to the Cauchy–Euler form. An ordinary second-order *linear* differential equation with *variable coefficients* is of the form

$$a_0 y'' + a_1 y' + a_2 y = f(x) \quad [8.1]$$

where $a_0(\neq 0), a_1, a_2$ and $f(x)$ are given functions of x, and $y(x)$ is an unknown function of x, the independent variable. Dividing [8.1] throughout by $a_0(\neq 0)$, we can write the equation in the form

$$y'' + Py' + Qy = R \quad [8.2]$$

where $P = a_1/a_0, Q = a_2/a_0$ and $R = f(x)/a_0$ all being functions of x and hence variables in general. In [8.2], if $R = 0$, the equation is said to be *homogeneous* and if $R \neq 0$, it is *non-homogeneous.* As in the case of an equation with constant coefficients, here also the *general* (or *complete) solution* of [8.2] is of the form

$$y = \phi_h + \phi_p \quad [8.3]$$

where ϕ_h is the general solution of the homogeneous equation, [8.2] with $R = 0$, and ϕ_p is a particular solution satisfying [8.2]. The function ϕ_h is called the *complementary function* and ϕ_p a *particular integral* (or *solution)* of [8.2]. When P and Q are *constants,* ϕ_h and ϕ_p and hence the general solution [8.3] can be obtained by the methods discussed in Chapter 5. But when P and Q are functions of x, of no particular form, there are no general methods for finding ϕ_h and ϕ_p. We shall discuss the following five special methods to obtain the complete solution of equation [8.1] or [8.2]:

(i) When a part of the complementary function is known.
(ii) Method of variation of parameters.
(iii) By changing the independent variable.
(iv) By changing the dependent variable.
(v) When the equation is *exact.*

8.1 When a part of the complementary function is known

Consider the equation in the form [8.2]: $y'' + Py' + Qy = R$. Suppose a *part* of the complementary function ϕ_h, say $u = u(x)$, is known. This means $y = u(x)$, satisfies the *homogeneous* equation

$$y'' + Py' + Qy = 0. \tag{8.4}$$

We now seek a complete solution of [8.2] in the form

$$y = uv \tag{8.5}$$

where $v = v(x)$ is a function to be determined. Differentiating [8.5] twice successively, we get

$$y' = uv' + u'v \tag{8.6}$$

and $y'' = uv'' + 2u'v' + u''v$. Substituting [8.5] and [8.6] in [8.2], we get

$$uv'' + (2u' + Pu)v' + (u'' + Pu' + Qu)v = R.$$

But $y = u$ is a solution of [8.4] so that $u'' + Pu' + Qu = 0$. Therefore [8.2] now reduces to the form

$$uv'' + (2u' + Pu)v' = R \tag{8.7a}$$

$$\text{or} \quad v'' + \left[P + 2\frac{u'}{u}\right]v' = \frac{R}{u}. \tag{8.7b}$$

On substituting $\omega = v'$ in [8.7], the equation becomes

$$u\omega' + (2u' + Pu)\omega = R. \tag{8.8}$$

Equation [8.8] is a *first order linear* differential equation in ω which can be solved by the usual method (see section 2.6) by finding the *integrating factor*. Thus ω is determined. Since $\omega = v'$, integration of ω with respect to x yields v. Thus, when v is determined, $y = uv$ is the general solution of [8.2]. Note that the two arbitrary constants are contained in v since determination of v involves integration twice.

Then the question arises: how to *know* a part $u = u(x)$ of the complementary function ϕ_h? The following results help us to know a factor u, of the complementary function of [8.2], if there is one.

Theorem 8.1 A necessary and sufficient condition that $y = e^{mx}$ may be a solution of the homogeneous equation [8.4] is that

$$m^2 + mP + Q = 0. \tag{8.9}$$

Proof: Let $y = e^{mx}$ so that $y' = me^{mx}$ and $y'' = m^2e^{mx}$. Hence

$$y'' + Py' + Qy = 0$$

holds if and only if

$$(m^2 + mP + Q)e^{mx} = 0 \quad \text{or} \quad m^2 + mP + Q = 0$$

since $e^{mx} \neq 0$.

Note: As particular cases of Theorem 8.1, we have
(i) $y = e^x$ is a solution of [8.4] if and only if (with $m = 1$)

$$1 + P + Q = 0;$$

(ii) $y = e^{-x}$ is a solution of [8.4] if and only if (with $m = -1$)

$$1 - P + Q = 0.$$

Theorem 8.2 A necessary and sufficient condition that $y = x^m$ may be a solution of [8.4] is that

$$m(m-1) + mPx + Qx^2 = 0. \tag{8.10}$$

Proof: Let $y = x^m$ so that we have $y' = mx^{m-1}$ and $y'' = m(m-1)x^{m-2}$. Now, equation [8.4] holds if and only if $m(m-1)x^{m-2} + mPx^{m-1} + Qx^m = 0$, or, dividing by x^{m-2}, we have $m(m-1) + mPx + Qx^2 = 0$.

Note: As particular cases of [8.10], we have
(i) $y = x$ is a solution of [8.4], with $m = 1$, if and only if $P + Qx = 0$;
(ii) $y = x^2$ is a solution of [8.4], with $m = 2$, if and only if $2 + 2Px + Qx^2 = 0$;
(iii) $y = 1/x$ is a solution of [8.4], with $m = -1$, if and only if $2 - Px + Qx^2 = 0$.

Example 8.1 Solve: $y'' + y = \text{cosec } x$.

Solution 8.1 The auxiliary equation is $m^2 + 1 = 0$ whose roots are $\pm i$ so that the complementary function is $\phi_h = A\cos x + B\sin x$. Now, $\cos x$ and $\sin x$ are the two parts of the complementary function. Let us choose one of them, say $\sin x$, as the part *known*. That is, let

$$u = \sin x. \tag{1}$$

Now, we seek the general solution of the given equation in the form

$$y = uv \tag{2}$$

so that on substituting for y and its derivatives in the given equation we get

$$uv'' + (2u' + Pu)v' = R$$

$$\text{or} \quad v'' + \left[P + 2\frac{u'}{u}\right]v' = \frac{R}{u}. \tag{3}$$

In the given equation we have $P = 0, Q = 1$ and $R = \text{cosec}\ x$. Substituting the above relationships and [1] in [3], we get

$$v'' + \left[2\frac{\cos x}{\sin x}\right]v' = \frac{\text{cosec}\ x}{\sin x}$$

$$\text{or} \quad \omega' + (2\cot x)\omega = \text{cosec}^2 x \tag{4}$$

where $\omega = v'$. Equation [4] is a first order linear equation and hence its solution is

$$\omega e^{\int 2\cot x dx} = \int (\text{cosec}^2 x)\left[e^{\int 2\cot x dx}\right]dx + c_1$$

$$\text{or} \quad \omega \sin^2 x = x + c_1 \quad \text{or} \quad \omega = (x + c_1)\text{cosec}^2 x.$$

Since $\omega = v'$, we have $v' = (x + c_1)\text{cosec}^2 x$, so that

$$v = \int (x + c_1)\text{cosec}^2 x dx = (x + c_1)(-\cot x) - \int (1)(-\cot x)dx$$

$$\text{or} \quad v = -(x + c_1)\cot x + \log|\sin x| + C_2 \tag{5}$$

where c_1 and c_2 are two arbitrary constants. Using [1] and [5] in [2], the complete solution is $y = uv$, that is,

$$y = (\sin x)\left[\log|\sin x| - (x + c_1)\cot x + c_2\right]$$

$$\text{or} \quad y = c_2 \sin x + \{\log|\sin x|\}\sin x - (x + c_1)\cos x.$$

Note: In Example 8.1, although the coefficients are *constant,* we have applied this method for illustration.

Example 8.2 Solve $x^2y''+xy'-y = 2x^2$ by first finding a part of the complementary function.

Solution 8.2 Dividing the equation by x^2, we get

$$y'' + \frac{1}{x}y' - \frac{1}{x^2}y = 2. \quad [1]$$

In [1], we have $P = 1/x, Q = -1/x^2$ and $R = 2$. Since $P + Qx = 0$ from [8.10], **Note (i),** $y = x$ is a part of the complementary function. Therefore, we take

$$u = x. \quad [2]$$

We seek the solution of [1] in the form

$$y = uv \quad [3]$$

where $v = v(x)$ satisfies the equation [8.7b], i.e.,

$$v'' + \left[P + 2\frac{u'}{u}\right]v' = \frac{R}{u} \quad \text{or} \quad v'' + \left[\frac{1}{x} + 2\frac{1}{x}\right]v' = \frac{2}{x} \quad \text{or} \quad v'' + \left(\frac{3}{x}\right)v' = \frac{2}{x}.$$

Let $\omega = v'$, so that the equation becomes

$$\omega' + \left(\frac{3}{x}\right)\omega = \frac{2}{x}. \quad [4]$$

The *integrating factor* of the first order linear equation [4] is

$$e^{\int(3/x)dx} = e^{3\log x} = x^3$$

so that the general solution of [4] is

$$\omega(x^3) = \int\left(\frac{2}{x}\right)x^3dx \quad \text{or} \quad \omega x^3 = \frac{2x^3}{3} + c_1 \quad \text{or} \quad \omega = \frac{2}{3} + c_1x^{-3}.$$

Since $\omega = v'$, the above equation becomes $v' = 2/3 + c_1x^{-3}$. On integration, we get

$$v = \frac{2}{3}x + \frac{c_1x^{-2}}{-2} + c_2$$

or $\quad v = \dfrac{2}{3}x - \dfrac{c_1}{2x^2} + c_2.$ [5]

Using [2] and [5] in [3], the general solution of [1] is

$$y = x\left[\frac{2}{3}x - \frac{c_1}{2x^2} + c_2\right] \quad \text{or} \quad y = \frac{2}{3}x^2 - \frac{c_1}{2x} + c_2 x.$$

Note: The general solution of [1] can also be obtained by taking $u = 1/x$ since $2 - Px + Qx^2 = 0$. This is left as an exercise to students.

Example 8.3 Solve: $xy'' + (1-x)y' - y = e^x$.

Solution 8.3 Dividing the equation throughout by x, it can be written as

$$y'' + \left[\frac{1}{x} - 1\right]y' - \frac{1}{x}y = \frac{e^x}{x}. \qquad [1]$$

Here, $P = 1/x - 1, Q = -1/x, R = e^x/x$. Since $1 + P + Q = 0$,

$$u = e^x \qquad [2]$$

is a part of the complementary function (from [8.9], **Note**(i)). We seek the solution of [1] in the form

$$y = uv \qquad [3]$$

where v satisfies the equation [8.7b]:

$$v'' + \left[P + \frac{2u'}{u}\right]v' = \frac{R}{u} \quad \text{or} \quad v'' + \left[\left(\frac{1}{x} - 1\right) + 2\right]v' = \frac{1}{x}$$

or $\quad \omega' + \left(\dfrac{1}{x} + 1\right)\omega = \dfrac{1}{x}.$ [4]

where $\omega = v'$.The integrating factor of [4] is

$$e^{\int((1/x)+1)dx} = e^{\log x + x} = e^{\log x}.e^x = xe^x.$$

Therefore, the general solution of [4] is

$$\omega(xe^x) = \int\left(\frac{1}{x}\right)xe^x dx \quad \text{or} \quad \omega x e^x = e^x + c_1$$

or $\omega = \dfrac{1}{x} + \dfrac{c_1 e^{-x}}{x}$.

Since $\omega = v'$, we have $v' = 1/x + c_1 x^{-1} e^{-x}$. On integration, we get

$$v = \log x + c_1 \int x^{-1} e^{-x} dx + c_2. \qquad [5]$$

Using [2] and [5] in [3], the general solution of [1] is

$$y = e^x \left[\log x + c_1 \int x^{-1} e^{-x} dx + c_2 \right].$$

Note: In the general solution obtained, $\int x^{-1} e^{-x} dx$ cannot be expressed in terms of elementary functions. However, for fixed limits of integration, numerical methods can be employed for integration.

Example 8.4 Solve $xy'' - 2(x+1)y' + (x+2)y = (x-2)e^{2x}$; $(x \neq 0)$ given that $y = e^2$ and $y' = 2e^2$ when $x = 1$.

Solution 8.4 Dividing the equation by $x (\neq 0)$, the given equation becomes

$$y'' - 2\left(1 + \frac{1}{x}\right)y' + \left(1 + \frac{2}{x}\right)y = \left(1 - \frac{2}{x}\right)e^{2x} \qquad [1]$$

Here $P = -2\left(1 + \dfrac{1}{x}\right)$, $Q = 1 + \dfrac{2}{x}$ and $R = \left(1 - \dfrac{2}{x}\right)e^{2x}$.

We readily see that $1 + P + Q = 0$ and hence (from [8.9], **Note** (i)), we have

$$u = e^x \qquad [2]$$

as a part of the complementary function of [1]. We seek the solution of [1] in the form

$$y = uv \qquad [3]$$

where v satisfies the equation

$$v'' + \left[P + 2\frac{u'}{u}\right]v' = \frac{R}{u} \quad \text{or} \quad v'' + \left[\left(-2 - \frac{2}{x}\right) + 2\right]v' = \left(1 - \frac{2}{x}\right)e^x$$

$$\text{or} \quad \omega' + \left(\frac{-2}{x}\right)\omega = \left(1 - \frac{2}{x}\right)e^x \qquad [4]$$

where $\omega = v'$. The integrating factor of [4] is

$$e^{\int((-2/x))dx} = e^{-2\log x} = e^{\log(1/x^2)} = \frac{1}{x^2}$$

so that the general solution of [4] is

$$\omega\left(\frac{1}{x^2}\right) = \int\left[\left(1-\frac{2}{x}\right)e^x\right]\frac{1}{x^2}dx + c_1 = \int\left[\left(\frac{1}{x^2}\right) - \frac{2}{x^3}\right]e^x dx + c_1$$

$$= \frac{1}{x^2}e^x + c_1.$$

Note: With $f(x) = 1/x^2$ and $f'(x) = -2/x^3$, we have

$$\int\left[\frac{1}{x^2} - \frac{2}{x^3}\right]e^x dx = \int[f(x) + f'(x)]e^x dx = f(x).e^x = \frac{1}{x^2}e^x.$$

Therefore, the general solution of [4] is

$$\omega = e^x + c_1x^2 \quad \text{or} \quad v' = e^x + c_1x^2.$$

On integration, we get

$$v = e^x + \frac{c_1x^3}{3} + c_2. \qquad [5]$$

Using [2] and [5] in [3], the complete solution of [1] is

$$y = e^x\left[e^x + \frac{c_1x^3}{3} + c_2\right] \qquad [6]$$

Now, we have to evaluate c_1 and c_2 using the given initial conditions $y = e^2$ and $y' = 2e^2$ when $x = 1$. Substituting $x = 1$ and $y = e^2$ in [6], we have

$$e^2 = e\left[e + \frac{c_1}{3} + c_2\right]$$

or $\frac{1}{3}c_1 + c_2 = 0.$ [7]

Differentiating [6] with respect to x, we get

$$y' = e^x\left[\left\{e^x + c_1x^2\right\} + \left\{e^x + \frac{c_1x^3}{3} + c_2\right\}\right]$$

$$\text{or} \quad y' = e^x \left[2e^x + \frac{c_1 x^3}{3} + c_1 x^2 + c_2 \right].$$

Substituting $y' = 2e^2$ and $x = 1$ in this equation, we get $2e^2 = e[2e + c_1/3 + c_1 + c_2]$

$$\text{or} \quad \frac{4c_1}{3} + c_2 = 0. \qquad [8]$$

Solving the simultaneous equations [7] and [8] in c_1 and c_2 we get $c_1 = 0$ and $c_2 = 0$. Substituting these values in [6], the required solution of the given initial-value problem is given by

$$y = e^{2x}.$$

Example 8.5 Solve: $y'' + (1 - \cot x)y' - y \cot x = \sin^2 x.$ [1]

Solution 8.5 Here $P = 1 - \cot x$, $Q = -\cot x$ and $R = \sin^2 x$. Since $1 - P + Q = 0$, (from [8.9] **Note** (ii)),

$$u = e^{-x} \qquad [2]$$

is a part of the complementary function of the example. We seek the general solution of it in the form

$$y = uv \qquad [3]$$

so that v satisfies the equation

$$v'' + \left[P + \frac{2u'}{u} \right] v' = \frac{R}{u} \quad \text{or} \quad v'' + \left[(1 - \cot x) + 2\{-e^{-x}/e^{-x}\} \right] v' = e^x \sin^2 x$$

$$\text{or} \quad \omega' - (1 + \cot x)\omega = e^x \sin^2 x \qquad [4]$$

where $\omega = v'$. The linear equation [4] has its integrating factor

$$e^{\int -(1+\cot x)dx} = e^{-x - \log \sin x} = e^{-x}.e^{\log \operatorname{cosec} x} = e^{-x} \operatorname{cosec} x$$

so that the general solution of [4] is

$$\omega(e^{-x} \operatorname{cosec} x) = \int (e^x \sin^2 x)(e^{-x} \operatorname{cosec} x) dx = \int \sin x \, dx$$

$$\text{or} \quad \omega(e^{-x} \operatorname{cosec} x) = -\cos x + c_1$$

or $\omega = -e^x \sin x \cos x + c_1 e^x \sin x.$

Since $\omega = v'$, on integration, we get

$$v = -\frac{1}{2}\int e^x \sin 2x dx + c_1 \int e^x \sin x dx$$

or $v \equiv -\frac{1}{2} I_1 + c_1 I_2 + c_2$ [5]

where we have

$$I_1 = \int e^x \sin 2x dx = (\sin 2x) e^x - \int (2\cos 2x) e^x dx$$

$$= e^x \sin 2x - 2\left[(e^x)\cos 2x - \int (-2\sin 2x) e^x dx\right]$$

$$= e^x \sin 2x - 2e^x \cos 2x - 4I_1 \quad \text{or} \quad 5I_1 = e^x[\sin 2x - 2\cos 2x]$$

or $I_1 = \frac{1}{5} e^x[\sin 2x - 2\cos 2x].$ [6]

Next, we have

$$I_2 = \int e^x \sin x dx = (\sin x) e^x - \int (\cos x) e^x dx$$

$$= e^x \sin x - \left[(\cos x) e^x - \int (-\sin x) e^x dx\right] = e^x \sin x - e^x \cos x - I_2$$

or $I_2 = \frac{1}{2} e^x[\sin x - \cos x].$ [7]

Finally, substituting [6] and [7] in [5], we have

$$v = -\frac{1}{10} e^x[\sin 2x - 2\cos 2x] + \frac{c_1}{2} e^x[\sin x - \cos x] + c_2. \quad [8]$$

Using [2] and [8] in [3], the general solution of [1] is

$$y = \frac{c_1}{2}[\sin x - \cos x] - \frac{1}{10}[\sin 2x - 2\cos 2x] + c_2 e^{-x}.$$

Exercise 8A

8.1 Solve the following equations by finding a part of the complementary function:

(i) $y'' + y = \sec x$ (ii) $(3 - x)y'' - (9 - 4x)y' + (6 - 3x)y = 0$

(iii) $xy'' - (2x + 1)y' + (x + 1)y = (x^2 + x - 1)e^{2x}$

(iv) $x^2y'' - 2x(x + 1)y' + 2(x + 1)y = x^3$

(v) $(x + 2)y'' - (2x + 5)y' + 2y = (x + 1)e^x$

8.2 Solve: $x^2y'' - x(x + 2)y' + (x + 2)y = x^3e^x$ given that x is a part of the complementary function.

8.3 Solve: $xy' - y = (x - 1)(y'' - x + 1)$ given that x is a part of the complementary function.

8.4 Solve: $y'' + y' \cot x - y \operatorname{cosec}^2 x = 0$ given that $\cot x$ is a solution.

8.5 Solve: $y'' + 2y' \tan x = y$ given that $\sin x$ is a solution.

8.6 Solve: $xy'' + (1 - 2x)y' - (1 - x)y = 0, x > 0$ given that e^x is a solution.

8.7 Solve: $x^2y'' + xy' - y = 0, x > 0$ given that $x + \frac{1}{x}$ is a solution.

8.8 Solve: $(x^2 + 1)y'' - 2xy' + 2y = 0$ given that x is a solution and $y = 1$ and $y' = 0$ at $x = 0$.

8.9 Solve: $x^2y'' + xy' - y = 2x^2, x > 0$ given that $1/x$ is a part of the complementary function and $y(1) = y'(1) = 0$.

8.10 Solve: $y'' - y' \cot x - (1 - \cot x)y = e^x \sin x$ given that e^x is a part of the complementary function and that $y(0) = y'(0) = -1/2$.

8.11 Verify that $y = e^{2x}$ is a solution of the equation $(x + 2)y'' - (2x + 5)y' + 2y = 0$. Hence find the general solution of $(x + 2)y'' - (2x + 5)y' + 2y = (x + 1)e^x$.

8.12 Given that the equation $4x(1 - x)y'' + 2(3 - 4x)y' - y = 0$ has a solution of the form x^k, determine k and hence solve the equation completely.

8.2 Variation of parameters

Suppose we know the *full* complementary function of the differential equation [8.2]. Let it be

$$y_h = c_1y_1(x) + c_2y_2(x). \qquad [8.11]$$

We seek the general solution of [8.2] in the form

$$y = v_1 y_1 + v_2 y_2 \tag{8.12}$$

where v_1 and v_2 are functions of x to be appropriately determined.

Differentiating [8.12] with respect to x, we have

$$y' = (v_1 y_1' + v_2 y_2') + (v_1' y_1 + v_2' y_2). \tag{8.13a}$$

If we set

$$v_1' y_1 + v_2' y_2 = 0 \tag{8.13b}$$

then [8.13a] becomes on using [8.13b]:

$$y' = v_1 y_1' + v_2 y_2' \tag{8.14}$$

so that, on differentiation again, we get

$$y'' = v_1 y_1'' + v_1' y_1' + v_2 y_2'' + v_2' y_2'. \tag{8.15}$$

On substituting [8.12], [8.14] and [8.15] into [8.2] and rearranging, we get

$$v_1(y_1'' + P y_1' + Q y_1) + v_2(y_2'' + P y_2' + Q y_2) + v_1' y_1' + v_2' y_2' = R \tag{8.16}$$

Since y_1 and y_2 are both solutions of the *homogeneous* counterpart [8.4] of the given non-homogeneous equation [8.2], we have

$$y_1'' + P y_1' + Q y_1 = 0 \quad \text{and} \quad y_2'' + P y_2' + Q y_2 = 0$$

and hence [8.16] becomes

$$v_1' y_1' + v_2' y_2' = R. \tag{8.17}$$

Now, considering [8.13b] and [8.17] together, we have the following two simultaneous linear algebraic equations in the two unknowns v_1' and v_2' :

$$v_1' y_1 + v_2' y_2 = 0; \quad v_1' y_1' + v_2' y_2' = R.$$

Solving these equations, we get

$$v_1' = \frac{-y_2 R}{W}; \quad v_2' = \frac{y_1 R}{W} \tag{8.18}$$

where W is the *wronskian* of y_1 and y_2 defined by the determinant

$$W(y_1, y_2) = \begin{vmatrix} y_1 & y_2 \\ y_1' & y_2' \end{vmatrix}$$

introduced by J. M. Höne (1778–1853) who changed his name to *Wrönski*.
On integrating [8.18], we get

$$v_1 = \int \frac{-y_2 R}{W} dx; \quad v_2 = \int \frac{y_1 R}{W} dx \quad [8.19]$$

provided the wronskian $W \neq 0$. Now, substituting [8.19] in [8.12], we get the general solution of the non-homogeneous differential equation [8.12] in the form

$$y = v_1 y_1 + v_2 y_2 \quad \text{or} \quad y = y_1 \int \frac{-y_2 R}{W} dx + y_2 \int \frac{y_1 R}{W} dx.$$

Note: Since the solution [8.12] is sought by *varying* the *parameters* c_1 and c_2 in [8.11] by replacing them by $v_1(x)$ and $v_2(x)$, the method is called *variation of parameters*. Some authors retain the parameters c_1 and c_2 as $c_1(x)$ and $c_2(x)$ instead of using a different notation viz., $v_1(x)$ and $v_2(x)$. This method is applicable only when $W \neq 0$.

Example 8.6 Solve $y'' + y = \operatorname{cosec} x$ by the method of variation of parameters.
Solution 8.6 Although the equation can be solved by the method discussed in Chapter5, since the coefficients are constant, we apply the method of variation of parameters for illustration. Here, $P = 0, Q = 1$ and $R = \operatorname{cosec} x$. The auxiliary equation is $m^2 + 1 = 0$ which has the roots $\pm i$ so that the complementary function is

$$y_h = c_1 \sin x + c_2 \cos x$$

Taking

$$y_1 = \sin x; \quad y_2 = \cos x \quad [1]$$

the Wronskian $W(y_1, y_2)$ is given by

$$W(y_1, y_2) = \begin{vmatrix} y_1 & y_2 \\ y_1' & y_2' \end{vmatrix} = \begin{vmatrix} \sin x & \cos x \\ \cos x & -\sin x \end{vmatrix}$$

$$= -\sin^2 x - \cos^2 x \text{ or } W = -1 \neq 0.$$

We seek the solution of the example in the form

$$y = v_1 y_1 + v_2 y_2 \quad [2]$$

where $v_1(x)$ and $v_2(x)$ are given by $v_1 = \int(-y_2 R/W)dx$ and $v_2 = \int(y_1 R/W)dx$.

Therefore, we have

$$\left.\begin{aligned} v_1 &= -\int \frac{\cos x \ \text{cosec}\ x}{-1} dx = \log|\sin x| + c_1 \\ v_2 &= \int \frac{\sin x \ \text{cosec}\ x}{-1} dx = -x + c_2 \end{aligned}\right\} \qquad [3]$$

Substituting [1] and [3] in [2], we get the general solution as

$$y = (\sin x)[\log|\sin x| + c_1] + (\cos x)[c_2 - x].$$

Example 8.7 Solve: $x^2y'' - 4xy' + 6y = x^4 \sin x$.

Solution 8.7 Dividing the equation by $x^2(\neq 0)$, we have

$$y'' - \frac{4}{x}y' + \frac{6}{x^2}y = x^2 \sin x. \qquad [1]$$

Here $P = -4/x, Q = 6/x^2$ and $R = x^2 \sin x$. Since there is a power of x as a factor in the right-hand side of [1], we seek the parts of the complementary function of [1] in the form $y = x^m$. This is a solution of the *homogeneous* counterpart of [1] viz.,

$$y'' - 4y'/x + 6y/x^2 = 0 \text{ if and only if } m(m-1) + mPx + Qx^2 = 0$$

from [8.10]. Substituting the values of P and Q, we get

$$(m^2 - m) + m(-4/x)x + (6/x^2)x^2 = 0 \quad \text{or} \quad m^2 - 5m + 6 = 0$$

whose roots are $m_1 = 2$ and $m_2 = 3$. This means

$$y_1 = x^{m_1} = x^2; \quad y_2 = x^{m_2} = x^3 \qquad [2]$$

are the parts of the complementary function so that the complementary function is $y_h = c_1y_1 + c_2y_2$. We seek the general solution of [1] in the form

$$y = v_1y_1 + v_2y_2 \qquad [3]$$

where v_1 and v_2 are given by

$$v_1 = -\int \frac{y_2R}{W}dx = -\int x^3\frac{(x^2 \sin x)}{W}dx = -\int \frac{x^5 \sin x}{W}dx.$$

and $$v_2 = \int \frac{y_1 R}{W} dx = \int \frac{x^2(x^2 \sin x)}{W} dx = \int \frac{x^4 \sin x}{W} dx.$$

Here, the Wronskian W is given by

$$W = \begin{vmatrix} y_1 & y_2 \\ y_1' & y_2' \end{vmatrix} = \begin{vmatrix} x^2 & x^3 \\ 2x & 3x^2 \end{vmatrix} = x^4 (\neq 0).$$

So $$v_1 = -\int \frac{x^5 \sin x}{(x^4)} dx = -\int x \sin x \, dx = -\left[x(-\cos x) - \int (-\cos x) dx \right]$$

or $$v_1 = x \cos x - \sin x + c_1; \qquad [4]$$

$$v_2 = \int \frac{x^4 \sin x}{(x^4)} dx = \int \sin x \, dx = -\cos x + c_2. \qquad [5]$$

Using [2], [4] and [5] in [3], the general solution of [1] is

$$y = x^2(x \cos x - \sin x + c_1) + x^3(c_2 - \cos x).$$

Example 8.8 Obtain the solution of $y'' + k^2 y = f(x)$ in the form

$$y = \frac{1}{k} \int_0^x f(t) \sin k(x - t) \, dt \quad (k \neq 0)$$

given the initial conditions $y = 0$ and $y' = 0$ when $x = 0$.

Solution 8.8 The auxiliary equation of the given equation is $m^2 + k^2 = 0$ whose roots are $\pm\, ik$ so that the complementary function is $y_h = c_1 \cos kx + c_2 \sin kx$.

Let $y_1 = \cos kx$ and $y_2 = \sin kx$ so that the Wronskian W is given by

$$W = \begin{vmatrix} y_1 & y_2 \\ y_1' & y_2' \end{vmatrix} = \begin{vmatrix} \cos kx & \sin kx \\ -k \sin kx & k \cos kx \end{vmatrix} = k.$$

We seek the solution of the given equation in the form

$$y = v_1 y_1 + v_2 y_2 \qquad [1]$$

where $$v_1 = -\int \frac{y_2 R}{W} dx = -\int \frac{(\sin kx) f(x)}{k} dx \qquad [2]$$

$$v_2 = \int \frac{y_1 R}{W} dx = \int \frac{(\cos kx) f(x)}{k} dx. \qquad [3]$$

The general solution [1] takes the form

$$y = v_1 \cos kx + v_2 \sin kx. \qquad [4]$$

Since initially for $x = 0$, we have $y = 0 = y'$, from [4] we get $v_1 = 0$ and $v_2 = 0$ for $x = 0$. To meet this requirement, we take v_1 and v_2, in view of [2] and [3], in the form

$$v_1 = -\frac{1}{k}\int_0^x f(t) \sin ktdt \quad \text{and} \quad v_2 = \frac{1}{k}\int_0^x f(t) \cos ktdt.$$

Here, t is used as the variable of integration to avoid possible confusion. Then the general solution [4] becomes

$$y = \frac{1}{k}\left[-(\cos kx)\int_0^x f(t) \sin ktdt + (\sin kx)\int_0^x f(t) \cos ktdt\right]$$

or $$y = \frac{1}{k}\int_0^x f(t)[\sin kx \cos kt - \cos kx \sin kt]$$

or $$y = \frac{1}{k}\int_0^x f(t) \sin k(x-t)dt.$$

Example 8.9 Solve: $(1-x)y'' + xy' - y = (1-x)^2, x \neq 1.$

Solution 8.9 Dividing the given equation by $(1-x) \neq 0$, the equation becomes

$$y'' + \frac{x}{1-x}y' - \frac{1}{1-x}y = 1 - x. \qquad [1]$$

Here $P = \dfrac{x}{1-x}$ and $Q = -\dfrac{1}{1-x}, R = 1 - x.$

Since $1 + P + Q = 0$ from [8.9], **Note (i)**, $y_1 = e^x$ is a part of the complementary function of [1]. Also, since $P + Qx = 0$ from [8.10], **Note (i)**, $y_2 = x$ is also a part of the complementary function. since [1] is of second order, y_1 and y_2 are the only two parts of the complementary function. Therefore, we have $y_h = c_1e^x + c_2x$. We seek the solution of [1] in the form

$$y = v_1y_1 + v_2y_2 \qquad [2]$$

with $y_1 = e^x; \quad y_2 = x.$ [3]

The Wronskian W is given by

$$W = \begin{vmatrix} y_1 & y_2 \\ y_1' & y_2' \end{vmatrix} = \begin{vmatrix} e^x & x \\ e^x & 1 \end{vmatrix} = e^x(1-x) \neq 0.$$

The unknown functions v_1 and v_2 are given by

$$v_1 = -\int \frac{y_2 R}{W} dx = -\int \frac{x(1-x)}{e^x(1-x)} dx = -\int x e^{-x} dx$$

$$= -[x(-e^{-x}) - \int (1)(-e^{-x} dx] = xe^{-x} + e^{-x} + c_1$$

or $v_1 = (x+1)e^{-x} + c_1$ [4]

$$v_2 = \int \frac{y_1 R}{W} dx = \int \frac{e^x(1-x)}{e^x(1-x)} dx = x + c_2. \qquad [5]$$

Using [3], [4] and [5] in [2], the general solution of [1] is obtained as

$$y = e^x[(x+1)e^{-x} + c_1] + x[x + c_2] = (x^2 + x + 1) + c_1 e^x + c_2 x$$

or $y = x^2 + (c_2 + 1)x + c_1 e^x + 1.$

Exercise 8B

Solve the following by the method of variation of parameters.

8.13 $y'' + y = \cot x$

8.14 $y'' + a^2 y = \sec(ax)$

8.15 $y'' + 4y = \operatorname{cosec}(2x)$

8.16 $y'' + 3y' + 2y = 3e^{-2x} + x$

8.17 $y'' - 6y' + 9y = \dfrac{e^{3x}}{x^2}$

8.18 $x^2 y'' - 2xy' + 2y = \dfrac{1}{x^2}$

8.19 $(1 - x^2)y'' - 4xy' - (1 + x^2)y = x$

8.20 $(1 - x)y'' + xy' - y = (1 - x^2), x \neq 1$

8.21 Solve $(x^2+1)y'' - 2xy' + 2y = 6(x^2+1)^2$ given that $y_1 = x$ and $y_2 = x^2 - 1$ are the linearly independent solutions of the corresponding homogeneous equation and that $y = -2$ when $x = -2$ and $y = 4$ when $x = 2$.

8.22 $y'' - \dfrac{2}{x}y' + \dfrac{2}{x^2}y = x\cos x$ given that $y = y' = \dfrac{\pi}{2}$ when $x = \dfrac{\pi}{2}$.

8.23 $x^2y'' + xy' - y = x^2e^x$

8.3 Changing the independent variable

Consider equation [8.2] viz.,

$$\frac{d^2y}{dx^2} + P(x)\frac{dy}{dx} + Q(x)y = R(x).$$

Suppose we change the independent variable x to another variable z, related to x in an appropriately chosen manner. Then, we have

$$\frac{dy}{dx} = \frac{dy}{dz}\cdot\frac{dz}{dx} \qquad [8.20]$$

and differentiating [8.20] again with respect to x,we have

$$\frac{d^2y}{dx^2} = \frac{d}{dx}\left[\frac{dy}{dx}\right] = \frac{d}{dx}\left[\frac{dy}{dz}\cdot\frac{dz}{dx}\right] = \frac{dy}{dz}\frac{d^2z}{dx^2} + \frac{dz}{dx}\cdot\frac{d^2y}{dz^2}\cdot\frac{dz}{dx}$$

or $$\frac{d^2y}{dx^2} = \frac{dy}{dz}\frac{d^2z}{dx^2} + \left[\frac{dz}{dx}\right]^2\frac{d^2y}{dz^2}. \qquad [8.21]$$

Using [8.20] and [8.21] in [8.2], we get

$$\left[\frac{dz}{dx}\right]^2\frac{d^2y}{dz^2} + \frac{dy}{dz}\frac{d^2z}{dx^2} + P\frac{dy}{dz}\frac{dz}{dx} + Qy = R$$

or $$\frac{d^2y}{dz^2} + \frac{[(d^2z/dx^2) + P(dz/dx)]}{(dz/dx)^2}\frac{dy}{dz} + \frac{Qy}{(dz/dx)^2} = \frac{R}{(dz/dx)^2}$$

or $$\frac{d^2y}{dz^2} + P_1\frac{dy}{dz} + Q_1y = R_1 \qquad [8.22]$$

where $P_1 = \dfrac{(d^2z/dx^2) + P(dz/dx)}{(dz/dx)^2}$, $Q_1 = \dfrac{Q}{(dz/dx)^2}$ and $R_1 = \dfrac{R}{(dz/dx)^2}$.

We notice that [8.22] is an equation similar to [8.2] in form, in which z is the independent variable. To solve [8.32], we choose z in *one* of the following ways usually:

(i) We choose z such that

$$\left[\frac{dz}{dx}\right]^2 = kQ$$

for some appropriate constant k. For this choice of z, if P_1 and Q_1 turn out to be constants, then [8.22] can be solved by the usual procedure of Chapter 5.

(ii) We choose z such that $P_1 = 0$. For this choice of z, if Q_1 turns out to be a constant, then again [8.22] can be solved by the procedure for constant coefficients discussed in Chapter 5.

Note: For some particular equations, the choice of z other than the two mentioned above may be more convenient. Such cases are regarded as exceptional.

Example 8.10 Solve the following equation by changing the independent variable:

$$\frac{d^2y}{dx^2} + \frac{2}{x}\frac{dy}{dx} + \frac{a^2}{x^4}y = 0. \qquad [1]$$

Solution 8.10 Here, we have $P = 2/x, Q = a^2/x^4$ and $R = 0$. We choose the value of z such that

$$\left[\frac{dz}{dx}\right]^2 = Q = \frac{a^2}{x^4} \quad \text{or} \quad \frac{dz}{dx} = \pm\frac{a}{x^2}$$

or $$\frac{d^2z}{dx^2} = \mp\frac{2a}{x^3}; \quad z = \mp\frac{a}{x}. \qquad [2]$$

Therefore, we have, on substitution

$$P_1 = \frac{(d^2z/dx^2) + P(dz/dx)}{(dz/dx)^2} = 0 \quad \text{and} \quad Q_1 = \frac{Q}{(dz/dx)^2} = 1.$$

Hence the change of independent variable from x to z reduces [1] to

$$\frac{d^2y}{dz^2} + P_1\frac{dy}{dz} + Q_1y = R_1$$

or $$\frac{d^2y}{dz^2} + y = 0 \qquad [3]$$

since $R_1 = R/(dz/dx)^2 = 0$. The general solution of [3] is $y = c_1 \cos z + c_2 \sin z$. Substituting for z from [2], we get the general solution of [1] as

$$y = c_1 \cos\frac{a}{x} + c_2 \sin\frac{a}{x}.$$

Note: When alternatives like $\pm$ occur, as in [2], students are advised to choose one of them and proceed consistently with that choice of the sign.

Example 8.11 Solve: $(\cos x)y'' + (\sin x)y' - 2y\cos^3 x = 2\cos^5 x$.

Solution 8.11 Dividing the equation by $\cos x (\neq 0)$, it can be written as

$$y'' + (\tan x)y' - (2\cos^2 x)y = 2\cos^4 x. \qquad [1]$$

Here, we have $P = \tan x$, $Q = -2\cos^2 x$ and $R = 2\cos^4 x$. Let us choose z such that $(dz/dx)^2 = -Q = 2\cos^2 x$ or $dz/dx = \sqrt{2}\cos x$ so that we have

$$\frac{d^2z}{dx^2} = -\sqrt{2}\sin x; \quad z = \sqrt{2}\sin x. \qquad [2]$$

Then $$P_1 = \frac{Pz' + z''}{(z')^2} = \frac{(\tan x)(\sqrt{2}\cos x) - \sqrt{2}\sin x}{2\cos^2 x} = 0;$$

$$Q_1 = \frac{Q}{(z')^2} = -1 \quad \text{and} \quad R_1 = \frac{R}{(z')^2} = \frac{2\cos^4 x}{2}\cos^2 x = \cos^2 x = 1 - \frac{1}{2}z^2$$

using [2]. By the change of the independent variable, from x to z, [1] reduces to

$$\frac{d^2y}{dz^2} + P_1\frac{dy}{dz} + Q_1 y = R_1$$

or $$\frac{d^2y}{dz^2} - y = 1 - \frac{1}{2}z^2. \qquad [3]$$

The auxiliary equation of [3] is $m^2 - 1 = 0$ whose roots are ± 1 so that the complementary function of [3] is

$$y_h = c_1 e^z + c_2 e^{-z}. \qquad [4]$$

The particular integral of [3] is given by

$$y_p = \frac{1}{D^2 - 1}\left[1 - \frac{1}{2}z^2\right] = (1 - D^2)^{-1}\left[\frac{1}{2}z^2 - 1\right]$$

$$= (1 + D^2 + D^4 + \ldots)\left[\frac{1}{2}z^2 - 1\right]$$

or $$y_p = \left(\frac{1}{2}z^2 - 1\right) + 1 = \frac{1}{2}z^2. \qquad [5]$$

Therefore, the complete solution of [3], from [4] and [5] is

$$y = y_h + y_p = c_1 e^z + c_2 e^{-z} + (1/2)z^2.$$

Using [2], the general solution of [1] is

$$y = c_1 e^{\sqrt{2}\sin x} + c_2 e^{-\sqrt{2}\sin x} + \sin^2 x.$$

Example 8.12 Solve $xy'' + (2x^2 - 1)y' - 24x^3 y = 4x^3 \sin(x^2)$ using the transformation $z = x^2$ given that $y = -7/50$ when $x = \sqrt{\pi/2}$ and $y = 1/50$ when $x = \sqrt{\pi}$.

Solution 8.12 Dividing the equation of the given *boundary value problem* by x throughout, it becomes

$$y'' + \frac{2x^2 - 1}{x} y' - 24x^2 y = 4x^2 \sin(x^2). \qquad [1]$$

Here, $P = (2x^2 - 1)/x, Q = -24x^2$ and $R = 4x^2 \sin(x^2)$. Using the *given* transformation $z = x^2$, [1] reduces to the form

$$\frac{d^2y}{dz^2} + P_1 \frac{dy}{dz} + Q_1 y = R_1$$

where $P_1 = \dfrac{z'' + Pz'}{(z')^2} = \dfrac{2 + (2x^2 - 1/x)(2x)}{4x^2} = 1;\quad Q_1 = \dfrac{Q}{(z')^2} = \dfrac{-24x^2}{4x^2} = -6$

and $R_1 = \dfrac{R}{(z')^2} = \dfrac{4x^2 \sin(x^2)}{4x^2} = \sin(x^2) = \sin z.$

Therefore, equation [1] becomes

$$\frac{d^2y}{dz^2} + \frac{dy}{dz} - 6y = \sin z. \qquad [2]$$

It can be verified that the general solution of [2], by the methods of Chapter 5, is

$$y = c_1 e^{2z} + c_2 e^{-3z} - \frac{1}{50}[\cos z + 7\sin z].$$

Substituting $z = x^2$, the general solution of [1] becomes

$$y = c_1 e^{2x^2} + c_2 e^{-3x^2} - \frac{1}{50}[\cos(x^2) + 7\sin(x^2)]. \qquad [3]$$

Now, substituting $x = \sqrt{\pi/2}$ and $y = -7/50$ in [3], we get

$$c_1 e^{\pi} + c_2 e^{-3\pi/2} = 0. \qquad [4]$$

Substituting $x = \sqrt{\pi}$ and $y = 1/50$ in [3], we get

$$c_1 e^{2\pi} + c_2 e^{-3\pi/2} = 0. \qquad [5]$$

Solving [4] and [5], we get $c_1 = 0$ and $c_2 = 0$. Substituting these values in [3], the solution of the given boundary value problem is

$$y = -\frac{1}{50}[\cos(x^2) + 7\sin(x^2)].$$

Example 8.13 Solve: $y'' - (1+4e^x)y' + 3e^{2x}y = e^{2(x+e^x)}$.

Solution 8.13 Here we have

$$P = -(1+4e^x), Q = 3e^{2x},$$

and $R = e^{2(x+e^x)}$.

We shall choose the independent variable z such that

$$\left[\frac{dz}{dx}\right]^2 = \frac{1}{3}Q = e^{2x},$$

so $z' \equiv \dfrac{dz}{dx} = e^x, z'' \equiv \dfrac{d^2z}{dx^2} = e^x, \quad z = e^x.$ [1]

Then, the given equation takes the form

$$\frac{d^2y}{dz^2} + P_1\frac{dy}{dz} + Q_1y = R_1$$

where $P_1 = \dfrac{Pz' + z''}{(z')^2} = \dfrac{-(1+4e^x)e^x + e^x}{e^{2x}} = -4; \quad Q_1 = \dfrac{Q}{(z')^2} = \dfrac{3e^{2x}}{e^{2x}} = 3$

and $R_1 = \dfrac{R}{(z')^2} = \dfrac{e^{2(x+e^x)}}{e^{2x}} = \dfrac{e^{2x}\exp(2e^x)}{e^{2x}}$ or $R_1 = \exp(2e^x) = e^{2z}$

since $z = e^x$. Therefore, the transformed equation becomes

$$\frac{d^2y}{dz^2} - 4\frac{dy}{dz} + 3y = e^{2z}.$$ [2]

The auxiliary equation of [2] is $m^2 - 4m + 3 = 0$ whose roots are 1 and 3 so that the complementary function of [2] is $y_h = c_1e^z + c_2e^{3z}$. The particular integral of [2] is given by

$$y_p = \left[\frac{1}{D^2 - 4D + 3}\right]e^{2z}$$

where $D \equiv d/dz$. Therefore, we have

$$y_p = \frac{1}{(2)^2 - 4(2) + 3} e^{2z} = -e^{2z}.$$

The general solution of the given equation, with $z = e^x$, is

$$y = c_1 e^z + c_2 e^{3z} - e^{2z}.$$

Exercise 8C

I. Solve the following equations by changing the independent variable:

8.24 $y'' + (\cot x)y' + 4y \operatorname{cosec}^2 x = 0$

8.25 $xy'' - y' + 4x^3 y = x^5$

8.26 $x^6 y'' + 3x^5 y' + a^2 y = \dfrac{1}{x^2}$

8.27 $(\sin^2 x)y'' + (\sin x \cos x)y' + 4y = 0$

8.28 $y'' + (2\cos x + \tan x)y' + y\cos^2 x = \cos^4 x$

8.29 In Example 8.11, find the solution if $y = e$ when $x = 0$ and $y = 3/2$ when $x = \pi/4$.

II. Solve the following equations by changing the independent variable to z by the transformation given:

8.30 $xy'' - 2y' + x^5 y = 2x^5, x > 0$, taking $z = x^3$

8.31 $(1 + x^2)^2 y'' + 2x(1 + x^2)y' + y = 0$, taking $z = \tan^{-1} x$

8.32 $x^6 y'' + 2x^5 y' + x^2 y = 2x^2 + 1, x > 0$, taking $z = -\dfrac{1}{x}$

8.33 $y'' - y' \cot x - y \sin^2 x = \cos x \sin^2 x; z = -\cos x$ given that $y = 0$ when $x = \pi/2$ and $y = 2$ when $x = \pi$.

8.34 $(1 - x^2)y'' - xy' + 4y = 2x\sqrt{1 - x^2}; z = \sin^{-1} x$

8.35 $xy'' - y' - 4x^3 y = 8x^3 \sin(x^2); z = x^2$

8.4 Changing the dependent variable
[Or reduction to the *normal form* or removal of the first derivative]

Let us consider the differential equation [8.2] viz.,

$$y'' + Py' + Qy = R$$

where P, Q and R are functions of x . We seek the solution of [8.2] in the form

$$y = uv \tag{8.23}$$

where u and v are functions of x. Note that we adopted a similar technique in section 8.1; but in that case there was the need to know a part u of the complementary function. In this case, we know neither u nor v and both are to be determined. Differentiating [8.23] with respect to x, we get $y' = uv' + u'v$ and $y'' = uv'' + 2u'v' + u''v$. Substituting [8.23] and its above successive derivatives in [8.22], we obtain

$$uv'' + (2u' + Pu)v' + (u'' + Pu' + Qu)v = R. \tag{8.24}$$

Now, let us choose the function $u = u(x)$ in [8.23] such that the coefficient of v' in [8.24] is zero, i.e.,

$$2u' + Pu = 0. \tag{8.25}$$

Since by the choice [8.25], the *first derivative* term in [8.24] vanishes, the method is referred to as *removal of the first derivative* . Equation [8.25] gives

$$u' = -\frac{1}{2}Pu \quad \text{and} \quad u'' = -\frac{1}{2}[Pu' + uP']$$

$$\text{or} \quad u'' = -\frac{1}{2}\left[P\left(-\frac{1}{2}Pu\right) + uP'\right] \quad \text{or} \quad u'' = -\frac{1}{2}\left[\ \frac{1}{2}P^2u + uP'\right]$$

where the primes denote derivatives with respect to x. From [8.25], we have $2u'/u = -P$. On integration, this gives $2\log u = -P$ or

$$u = e^{-\frac{1}{2}\int P dx}. \tag{8.26}$$

Now, substituting the above values of u, u' and u'' in [8.24], we get the equation in the form given by

$$v'' + Iv = \frac{R}{u} \tag{8.27}$$

$$\text{(where)} \quad I = Q - \frac{P^2}{4} - \frac{P'}{2}. \tag{8.28}$$

Equation [8.27] is called the *normal form* of [8.2] in which the term containing the first derivative is eliminated. Further, equation [8.2] is transformed into the normal form [8.27] by *changing the dependent variable* from y to v by using [8.23] and [8.26].

Now, equation [8.27] can be solved if I is in one of the following forms:

(i) I is a *constant* in which case [8.27] will be a linear equation with *constant coefficients*
(ii) I is of the form κ/x^2, where κ is a constant. In that case [8.27] can be reduced to an equation with constant coefficients by using the transformation $x = e^z$. In either case the transformed equation can be solved using the methods of Chapter 5.

Example 8.14 Solve by reducing to the *normal form* the equation

$$y'' - \frac{2}{x}y' + \left[1 + \frac{2}{x^2}\right]y = xe^x \qquad [1]$$

Solution 8.14 Here, we have $P = -2/x, Q = 1 + 2/x^2$ and $R = xe^x$. We seek the general solution of [1] in the form

$$y = uv \qquad [2]$$

$$\text{(where)} \quad u = e^{-\frac{1}{2}\int P\,dx} = e^{-\frac{1}{2}\int(-2dx)/x} = x. \qquad [3]$$

The function v has to be determined satisfying [8.27] viz., $v'' + Iv = R/u$ where, from [8.28], we have

$$I = Q - \frac{1}{4}P^2 - \frac{1}{2}\frac{dP}{dx} = \left[1 + \frac{2}{x^2}\right] - \frac{1}{4}\left[\frac{4}{x^2}\right] - \frac{1}{2}\left[\frac{2}{x^2}\right] \quad \text{or} \quad I = 1.$$

Therefore, [8.27] becomes, on using [3],

$$\frac{d^2v}{dx^2} + v = e^x \qquad [4]$$

The complete solution of [4], using the methods of Chapter 5, can be found to be

$$v = c_1 \cos x + c_2 \sin x + \frac{1}{2}e^x. \qquad [5]$$

Substituting $u = x$, from [3], and for v from [5] in [2], the general solution of [1] is

$$y = x\left[c_1 \cos x + c_2 \sin x + \frac{1}{2}e^x\right].$$

Example 8.15 Solve: $y'' + 4y' \operatorname{cosec} 2x + 2y\tan^2 x = e^x \cot x$.

Solution 8.15 Here, we have $P = 4$ cosec $2x, Q = 2\tan^2 x$ and $R = e^x \cot x$. We seek the general solution in the form

$$y = uv \qquad [1]$$

where $u = e^{-\frac{1}{2}\int P dx} = e^{-\frac{1}{2}\int (4 \text{ cosec} 2x) dx}$

or $u = \cot x. \qquad [2]$

The function $v(x)$ in [1] satisfies equation [8.27] $v'' + Iv = R/u$ where

$$I = Q - \frac{1}{4}P^2 - \frac{1}{2}\frac{dP}{dx}$$

$$= 2\tan^2 x - \frac{1}{4}[16 \text{ cosec}^2 2x] - \frac{1}{2}[-8 \text{ cosec } 2x. \cot 2x]$$

or $I = -2.$

Therefore, the normal form of the given equation is

$$v'' - 2v = e^x. \qquad [3]$$

The general solution of [3] is

$$v = c_1 e^{\sqrt{2}x} + c_2 e^{-\sqrt{2}x} - e^x. \qquad [4]$$

Using [2] and [4] in [1], the complete solution of the given equation is

$$y = \cot x [c_1 e^{\sqrt{2}x} + c_2 e^{-\sqrt{2}x} - e^x].$$

Example 8.16 Solve: $y'' + x^{-1/3} y' + \left[\frac{1}{4x^{2/3}} \quad \frac{1}{6x^{4/3}} - \frac{6}{x^2}\right] y = 0.$

Solution 8.16 We have here

$$P = x^{-1/3}, \quad Q = \left[\frac{1}{4x^{2/3}} - \frac{1}{6x^{4/3}} - \frac{6}{x^2}\right] \quad \text{and} \quad R = 0.$$

We seek the solution of the given equation in the form

$$y = uv \qquad [1]$$

where $u = e^{\frac{1}{-2}\int P dx} = e^{\frac{1}{-2}\int x^{-1/3} dx}$

or $u = \exp\left(-\frac{3}{4}x^{2/3}\right). \qquad [2]$

The function $v(x)$ in [1] satisfies $v'' + Iv = R/u$ where

$$I = Q - \frac{1}{4}P^2 - \frac{1}{2}\frac{dp}{dx} \quad \text{or} \quad I = -\frac{6}{x^2}.$$

Therefore, the normal form of the equation becomes $v'' - (6/x^2)v = 0$, whose general solution is

$$v = c_1 x^3 + \frac{c_2}{x^2}. \qquad [3]$$

The solution [3] is obtained by the Cauchy–Euler method (see Chapter 6). Now, using [2] and [3] in [1], the general solution of the given equation is

$$y = \left[c_1 x^3 + \frac{c_2}{x^2}\right] \exp\left(-\frac{3}{4}x^{2/3}\right).$$

Example 8.17 If u and su are any two solutions of the equation $v'' + Iv = 0$ prove that $s''/s' = -2u'/u$ and hence deduce that $s'''/s'' - 3/2[s''/s'] = 2I$.

Solution 8.17 Since u and su are two solutions of the equation $u'' + Iv = 0$ we have

$$u'' + Iu = 0 \quad \text{or} \quad I = -\frac{u''}{u} \qquad [1]$$

and $(s''u + 2s'u' + su'') + I(su) = 0$

$$\text{or} \quad I = -\frac{s''u + 2s'u' + su''}{su}. \qquad [2]$$

We get $\dfrac{s''u + 2s'u' + su''}{su} = \dfrac{u''}{u}$ (from [1] and [2]).

Simplifying this expression, we get $s''u + 2s'u' = 0$ or

$$\frac{s''}{s'} = -2\frac{u'}{u}. \qquad [3]$$

Differentiating [3], we get

$$s''' + 3s''\frac{u'}{u} + 2s'\frac{u''}{u} = 0. \qquad [4]$$

Using [1] and [3] in [4], we get the result.

Example 8.18 Remove the second term and hence solve the equation

$$y'' - 4xy' + (4x^2 - 1)y = -3\sin 2x \exp\,(x^2). \qquad [1]$$

Solution 8.18 Here, we have $P = -4x, Q = 4x^2 - 1$ and $R = -3\sin 2x \exp(x^2)$. For *removing the second term* i.e. to reduce [1] to the *normal form,* we choose

$$u = e^{-\frac{1}{2}\int P dx} = e^{-\frac{1}{2}\int (-4x) dx} = \exp(x^2). \quad [2]$$

Let the general solution of [1] be

$$y = uv \quad [3]$$

where v satisfies the equation

$$v'' + Iv = \frac{R}{u} \quad [4]$$

with $I = Q - \frac{1}{4}P^2 - \frac{1}{2}\frac{dP}{dx} = (4x^2 - 1) - \frac{1}{4}(16x^2) - \frac{1}{2}(-4)$ or $I = 1.$

Substituting the above relationship in [4], we have

$$v'' + v = -3\sin 2x. \quad [5]$$

The auxiliary equation of [5] is $m^2 + 1 = 0$ whose roots are $\pm i$ so that the complementary function of [5] is

$$v_h = c_1 \cos x + c_2 \sin x. \quad [6]$$

The particular integral of [5] is given by

$$v_p = \frac{1}{D^2+1}[-3\sin 2x] = \frac{-3}{-(2)^2+1}\sin 2x$$

or $v_p = \sin 2x.$ [7]

The general solution of [5] is therefore

$$v = c_1 \cos x + c_2 \sin x + \sin 2x. \quad [8]$$

from [6] and [7]. Hence the general solution of [1], using [2] and [8] in [3], is

$$y = \exp(x^2)[c_1 \cos x + c_2 \sin x + \sin 2x].$$

Exercise 8D

Solve the following equations by reducing to the *normal form* :

8.36 $y'' - 2xy' + (x^2+2)y = \exp\left(\dfrac{x^2+2x}{2}\right)$

8.37 $y'' - 2y'\tan x + 5y = 0$

8.38 $x^2y'' - 2(x^2+x)y' + (x^2+2x+2)y = 0$

8.39 $y'' - 2y'\tan x + 5y = e^x \sec x$

8.40 $y'' + 2xy' + (x^2+5)y = x \exp(-x^2/2)$

8.41 $(x^2+1)y'' + 4xy' + (9x^2+11)y = 0$

8.42 $x^2y'' - 4x^3y' + (4x^4 - 2x^2 + 5)y = 0, \ x > 0$

8.43 $y'' - 4xy' + 4x^2y = xe^x$

8.44 $x^2y'' - 2x(1+x)y' + 2(1+x)y = x^3, \ x > 0$ given that

$y = 1/2$ and $y' = 1$ when $x = 1$.

8.45 $x^2y'' - 2xy' + (x^2+2)y = x^3e^x, x > 0$

8.5 Exact differential equations

Let us consider a differential equation in the form [8.1] viz.

$$a_0y'' + a_1y' + a_2y = f(x)$$

where a_0, a_1, a_2 and $f(x)$ are all known functions of x.

Suppose there exists a *first order linear* equation

$$b_0(x)y' + b_1(x)y = g(x) \qquad [8.29]$$

such that [8.1] can be obtained from [8.29] directly by differentiation with respect to x. Then equation [8.1] is called an *exact equation* and [8.29] is called a *first integral* of [8.1]. We have the following necessary and sufficient condition for [8.1] to be exact.

Theorem 8.4 A necessary and sufficient condition that the equation [8.1] given as, $a_0y'' + a_1y' + a_2y = f(x)$ may be *exact* is

$$a_0'' - a_1' + a_2 = 0$$

where a_0, a_1, a_2 and $f(x)$ are known functions of x.

Proof: (i) Suppose equation [8.1] is *exact*. Then there exists a first integral of the form [8.29] viz. $b_0y' + b_1y = g(x)$. Differentiating [8.29] with respect to x, we get

$$b_0y'' + [b_0' + b_1]y' + b_1'y = g'(x). \qquad [8.30]$$

Comparing equations [8.1] and [8.30], we get $b_0 = a_0, b_0' + b_1 = a_1, b_1' = a_2$ and $g' = f$ which gives $a_2 = b_1' = a_1' - b_0'' = a_1' - a_0''$ or $a_0'' - a_1' + a_2 = 0$ and hence [8.30] is a *necessary* condition for [8.1] to be exact.

(ii) Conversely, suppose the condition [8.30] viz. $a_0'' - a_1' + a_2 = 0$ holds. Then, equation [8.1] can be written as

$$a_0y'' + a_1y' + (a_1' - a_0'')y = f(x) \quad \text{or} \quad a_0y'' - a_0''y + a_1y' + a_1'y = f(x)$$

$$\text{or} \quad \frac{d}{dx}[a_0y' + (a_1 - a_0')y] = f(x)$$

$$\text{or} \quad \frac{d}{dx}\left[a_0y' + (a_1 - a_0')y - \int f(x)dx - c\right] = 0$$

where c is an arbitrary constant. From this result it follows that equation [8.1] may be obtained by differentiating the equation

$$a_0y' + (a_1 - a_0')y = \int f(x)dx + c \qquad [8.31]$$

which is of the form [8.29]. This proves that [8.1] is *exact* and hence [8.30] is a *sufficient* condition.

Note: From Theorem 8.4 it follows that if the given equation [8.1] satisfies condition [8.30], then [8.31] is a *first integral* of [8.1]. Solving the first order differential equation [8.31], the general solution of [8.1] can thus be obtained.

Example 8.19 Solve: $x^2y'' + 3xy' + y = \dfrac{1}{(1-x)^2}, \quad 0 < x < 1.$

Solution 8.19 Comparing the equation with its standard form [8.1], we have $a_0 = x^2$, $a_1 = 3x, a_2 = 1$ and $f(x) = 1/(1-x)^2$. We have $a_0'' - a_1' + a_2 = (2) - (3) + 1 = 0$ and hence the given equation is *exact* . Therefore, a *first integral* of it is given by [8.31]:

$$a_0y' + (a_1 - a_0')y = \int f(x)dx$$

$$\text{i.e.,} \quad x^2y' + (3x - 2x)y = \int \frac{1}{(1-x)^2}dx$$

$$\text{or} \quad x^2y' + xy = \frac{1}{1-x} + c_1$$

or $y' + \frac{1}{x}y = \frac{1}{x^2(1-x)} + \frac{c_1}{x^2}$. [1]

Equation [1] is a first order linear equation. Its integrating factor is

$$e^{\int \frac{1}{x}dx} = e^{\log x} = x$$

so that the general solution of [1], and hence of the given equation, is

$$(y)(x) = \int \left[\frac{1}{x^2(1-x)} + \frac{c_1}{x^2}\right] x\,dx$$

or $xy = \int \frac{1}{x(1-x)}dx + c_1 \int \frac{1}{x}dx = \int \left[\frac{1}{x} + \frac{1}{1-x}\right]dx + c_1 \int \frac{1}{x}dx$

or $xy = \log x - \log(1-x) + c_1 \log x + c_2 = \log\left(\frac{x}{1-x}\right) + c_1 \log x + c_2$.

Example 8.20 Solve: $(2x^2+3x)y'' + (6x+3)y' + 2y = (x+1)e^x$.

Solution 8.20 Here, we have $a_0 = 2x^2+3x$, $a_1 = 6x+3$, $a_2 = 2$ and $f(x) = (x+1)e^x$. Let us consider $a_0'' - a_1' + a_2 = (4) - (6) + 2 = 0$ so that the given equation is *exact* . A first integral of it is given by

$$a_0 y' + (a_1 - a_0')y = \int f(x)dx$$

i.e., $(2x^2+3x)y' + \{(6x+3) - (4x+3)\}y = \int (x+1)e^x dx$

or $(2x^2+3x)y' + 2xy = (x+1)e^x - \int (1)e^x dx$ or $y' + \frac{2}{2x+3}y = \frac{xe^x + c_1}{x(2x+3)}$

or $y' + \frac{2}{2x+3}y = \frac{e^x}{2x+3} + \frac{c_1}{x(2x+3)}$. [1]

Equation [1] is a first order linear differential equation whose integrating factor is

$$e^{\int (2dx/2x+3)} = e^{\log(2x+3)} = 2x+3.$$

Therefore, the general solution of [2] is

$$y(2x+3) = \int \left(e^x + \frac{c_1}{x}\right)dx \quad \text{or} \quad (2x+3)y = e^x + c_1 \log x + c_2.$$

This is also the general solution of the given equation.

Example 8.21 Solve $(1-x^2)y''-3xy'-y=0$ given that $y=1, y'=0$ when $x=0$.

Solution 8.21 Here $a_0=1-x^2, a_1=-3x, a_2=-1$ and $f(x)=0$. Consider

$$a_0''-a_1'+a_2=(-2)-(-3)+(-1)=0$$

so that the given equation is exact. Its first integral is given by

$$a_0y'+(a_1-a_0')y=\int f(x)dx$$

i.e. $(1-x^2)y'+\{(-3x)-(-2x)\}y=c_1$ or $(1-x^2)y'-xy=c_1$

Since $y=1$ and $y'=0$ for $x=0$, we get $c_1=0$ so that the first integral of the given equation becomes

$$(1-x^2)y'-xy=0 \qquad [1]$$

Separating the variables in [1], we have $\dfrac{dy}{y}=\dfrac{x}{1-x^2}dx.$

On integration, we get

$$\log y=-\frac{1}{2}\log(1-x^2)+\log c_2 \quad \text{or} \quad y\sqrt{1-x^2}=c_2.$$

Putting $x=0$ and $y=1$, we get c_2 so that the required solution is

$$y\sqrt{1-x^2}=1.$$

Example 8.22 Solve $y''\cos x+2y'\sin x+3y\cos x=\tan^2 x$.

Solution 8.22 Here, we have $a_0=\cos x, a_1=2\sin x, a_2=3\cos x$ and $f(x)=\tan^2 x$. Consider $a_0''-a_1'+a_2=(-\cos x)-(2\cos x)+(3\cos x)=0$ so that the given equation is *exact*. A first integral of it is given by

$$a_0y'+(a_1-a_0')y=\int f(x)dx \quad \text{i.e.,} \quad (\cos x)y'+(2\sin x+\sin x)y=\int \tan^2 x dx$$

or $(\cos x)y'+(3\sin x)y=\tan x-x+c_1$

or $$y'+(3\tan x)y=\int\frac{\tan x-x+c_1}{\cos x}dx. \qquad [1]$$

Equation [1] is a first order linear equation whose integrating factor is

$$e^{\int 3\tan x dx} = e^{3\log\sec x} = \sec^3 x$$

so that the general solution of [1] is

$$y(\sec^3 x) = \int \left[\frac{\tan x - x + c_1}{\cos x}\right] \sec^3 x dx$$

or $\quad y\sec^3 x = \int [\tan x \sec^4 x - x\sec^4 x + c_1 \sec^4 x] dx$

or $\quad y\sec^3 x = I_1 - I_2 + c_1 I_3 + c_2$ [2]

where $\quad I_1 = \int \tan x \sec^4 x dx, I_2 = \int x\sec^4 x dx \quad$ and $\quad I_3 = \int \sec^4 x dx.$

We have (i) $\quad I_1 = \int \tan x \sec^4 x dx$

$$= \int \tan x \sec^2 x \sec^2 x dx$$

$$= \int \tan x (1 + \tan^2 x) \sec^2 x dx.$$

Putting $t = \tan x$, we have $dt = \sec^2 x dx$ so that

$$I_1 = \int t(1+t^2)dt = \frac{t^2}{2} + \frac{t^4}{4}$$

or $\quad I_1 = \frac{1}{2}\tan^2 x + \frac{1}{4}\tan^4 x.$ [3]

(ii) $\quad I_2 = \int x\sec^4 x dx = \int x\sec^2 x \cdot \sec^2 x dx.$

Substitute $t = \tan x$ so that $dt = \sec^2 x dx$ and $x = \tan^{-1} t$. Then, we have

$$I_2 = \int (\tan^{-1} t)(1+t^2)dt = (\tan^{-1} t)\left(t + \frac{t^3}{3}\right) - \int \left(t + \frac{t^3}{3}\right)\left(\frac{1}{1+t^2}\right) dt$$

$$= \frac{1}{3}t(3+t^2)\tan^{-1} t - \frac{1}{3}\int \frac{3t+t^3}{1+t^2} dt$$

$$= \frac{1}{3}t(3+t^2)\tan^{-1} t - \frac{1}{3}\int \frac{t(t^2+1)+2t}{1+t^2} dt$$

$$= \frac{1}{3}t(3+t^2)\tan^{-1} t - \frac{1}{3}\int \left[t + \frac{2t}{1+t^2}\right] dt$$

$$= \frac{1}{3}\left[t(3+t^2)\tan^{-1}t - \frac{t^2}{2} - \log(1+t^2)\right]$$

or $$I_2 = \frac{1}{3}\left[x\tan x(3+\tan^2 x) - \frac{\tan^2 x}{2} - 2\log\sec x\right]. \qquad [4]$$

(iii) $$I_3 = \int \sec^4 x\,dx = \int \sec^2 x \sec^2 x\,dx$$

Substituting $t = \tan x$, we have $dt = \sec^2 x\,dx$ so that $I_3 = \int (1+t^2)dt = t + \frac{t^3}{3}$.

or $$I_3 = \tan x + \frac{1}{3}\tan^3 x. \qquad [5]$$

Substituting [3], [4] and [5] in [2], we have the general solution of [1], and hence of the given equation as

$$(\sec^3 x)y = \left[\frac{1}{2}\tan^2 x + \frac{1}{4}\tan^4 x\right] + c_1\left[\tan x + \frac{1}{3}\tan^3 x\right] + c_2$$

$$\frac{1}{3}\left[x\tan x(3+\tan^2 x) - \frac{\tan^2 x}{2} - 2\log\sec x\right].$$

Exercise 8E

Show that the following equations are *exact* and hence solve them:

8.46 $x^2y'' + 3xy' + y = \dfrac{1}{(1-x)^2}$

8.47 $xy'' + (1-x)y' - y = e^x$

8.48 $y'' + \sin x - y'\cos x + 2y\sin x = 0$

8.49 $x^2(1+x)y'' + 2x(2+3x)y' + 2(1+3x)y = 0$

8.50 $(1+x^2)y'' + 4xy' + 2y = \sec^2 x$

8.51 $(1+x+x^2)y'' + (2+4x)y' + 2y = k$

8.52 $xy'' + 3y' = \cos x$

8.53 $x^2y'' + y' - 2y = 2x$ given that $y = 2$ when $x = -1$ and $y = 0$ when $x = 1$.

8.54 $(1+x^2)y'' + 3xy' + y = 1 + 3x^2$

8.55 $x(1+x^2)y'' + 4x^2y' + 2xy = 0$

CHAPTER NINE

Power Series Solutions Of Differential Equations

9.1 Introduction

An infinite series of the form

$$\sum_{n=0}^{\infty} a_n x^n = a_0 + a_1 x + a_2 x^2 + \ldots$$

is called a *power series in* x. Familiar examples of power series are the Maclaurin series

$$e^x = \sum_{n=0}^{\infty} \frac{x^n}{n!} = 1 + x + \frac{x^2}{2!} + \frac{x^3}{3!} + \ldots$$

$$\cos x = \sum_{n=0}^{\infty} \frac{(-1)^n x^{2n}}{(2n)!} = 1 - \frac{x^2}{2!} + \frac{x^4}{4!} - \ldots$$

$$\sin x = \sum_{n=0}^{\infty} \frac{(-1)^n x^{2n+1}}{(2n+1)!} = x - \frac{x^3}{3!} + \frac{x^5}{5!} - \ldots$$

Homogeneous linear differential equations with constant coefficients can be solved by algebraic methods, and their solutions are elementary functions known from calculus. But many differential equations arising from physical problems may have variable coefficients and may not permit a general solution in terms of known functions.

We need a method to solve these equations whose solutions may be non-elementary functions. Bessel's equation, Legendre's equation and the hypergeometric equations are some equations which play an important role in engineering mathematics, and for which power series expansions can be used to obtain information about the form of solutions. The basic notations and definitions involved in solving second order linear differential equations by power series are discussed in this chapter.

9.2 Solution of differential equations in series

The basic idea of solving differential equations using the power series method is very simple. A differential equation being given, we first represent all given functions in the equation by power series. In many practical problems the given functions will be polynomials in which case, this first step may be omitted. Next, we assume a solution in the form of a power series, for example

$$y = a_0 + a_1 x + a_2 x^2 + \ldots = \sum_{n=0}^{\infty} a_n x^n. \qquad [9.1]$$

Differentiating the above equation we get,

$$y' = a_1 + 2a_2 x + 3a_3 x^2 + \ldots = \sum_{n=1}^{\infty} n a_n x^{n-1} \qquad [9.2]$$

$$y'' = 2a_2 + 3.2a_3 x + 4.3a_4 x^2 + \ldots = \sum_{n=2}^{\infty} n(n-1) a_n x^{n-2} \qquad [9.3]$$

and so on. Inserting these series into the equation, collecting like powers of x, and equating the sum of coefficients of each occuring power of x to zero, we get relations from which we can determine the unknown coefficients of [9.1] successively. We illustrate the procedure for some simple equations, which can also be solved otherwise.

Example 9.1 Solve: $y' - y = 0$.

Solution 9.1 Let us assume a solution of the form $y = \sum_{n=0}^{\infty} a_n x^n$.

Then, by [9.2], we have $y' = \sum_{n=1}^{\infty} n a_n x^{n-1}$. Substituting these values of y and y' in the example, we get

$$\sum_{n=1}^{\infty} n a_n x^{n-1} - \sum_{n=0}^{\infty} a_n x^n = 0$$

or $(a_1 + 2a_2 x + 3a_3 x^2 + \ldots) - (a_0 + a_1 x + a_2 x^2 + \ldots) = 0.$

Collecting like powers of x, we get

$$(a_1 - a_0) + (2a_2 - a_1)x + (3a_3 - a_2)x^2 + \ldots = 0.$$

Equating the coefficient of each power of x to zero, we have

$$a_1 - a_0 = 0; \quad 2a_2 - a_1 = 0; \quad 3a_3 - a_2 = 0; \quad \text{etc.}$$

Solving these equations, we get

$$a_1 = a_0; \quad a_2 = \frac{a_1}{2} = \frac{a_0}{2!}; \quad a_3 = \frac{a_2}{3} = \frac{a_0}{3!}; \quad \text{etc.}$$

Thus, substituting these values in $y = \sum_{n=0}^{\infty} a_n x^n$, we get

$$y = a_0 + a_0 x + \frac{a_0}{2!}x^2 + \frac{a_0}{3!}x^3 + \dots$$

$$\text{or} \quad y = a_0\left(1 + x + \frac{x^2}{2!} + \frac{x^3}{3!} + \dots\right) = a_0 e^x$$

which is the familiar general solution.

Example 9.2 Solve: $\dfrac{d^2y}{dx^2} + xy = 0$ or $y'' + xy = 0$.

Solution 9.2 Let the solution of the example be assumed as $y = a_0 + a_1 x + a_2 x^2 + \dots$
Then $y' = a_1 + 2a_2 x + 3a_3 x^2 + \dots$ and $y'' = 2a_2 + 3.2a_3 x + 4.3a_4 x^2 + \dots$
Substituting for y, y' and y'' in the given example, we get

$$2.1a_2 + 3.2a_3 x + 4.3a_4 x^2 + \dots + n(n-1)a_n x^{n-2} + \dots$$

$$+x(a_0 + a_1 x + a_2 x^2 + \dots + a_n x^n + \dots) = 0$$

$$\text{or} \quad 2.a_2 + (3.2a_3 + a_0)x + (4.3a_4 + a_1)x^2 + (5.4a_5 + a_2)x^3 + \dots$$

$$+\left[(n+2)(n+1)a_{n+2} + a_{n-1}\right]x^n + \dots = 0.$$

Equating the coefficients of various powers of x to zero, we have

$$a_2 = 0 \quad \text{and} \quad 3.2a_3 + a_0 = 0 \quad \text{or} \quad a_3 = -a_0/3!$$

and $4.3a_4 + a_1 = 0$ or $a_4 = -2a_1/4!$ etc. Thus $(n+2)(n+1)a_{n+2} + a_{n-1} = 0$

$$\text{or} \quad a_{n+2} = \frac{-a_{n-1}}{[(n+2)(n+1)]}. \qquad [1]$$

Substituting $n = 4, 5, 6 \dots$ in [1], we get

$$a_6 = \frac{4a_0}{6!}; a_7 = \frac{5.2a_1}{7!}; a_8 = 0; a_9 = \frac{7.4a_0}{9!}$$

etc., which yields the solution

$$y = a_0\left(1 - \frac{x^3}{3!} + \frac{4x^6}{6!} - \frac{7.4x^9}{9!} + \dots\right) + a_1\left(x - \frac{2x^4}{4!} + \frac{5.2x7}{7!} + \dots\right).$$

Note: Equating to zero the coefficient of the smallest power of x gives a quadratic equation known as the *indicial equation.* The roots of the indicial equation give the complete solution depending on whether they are

(i) distinct and not differing by an integer
(ii) equal
(iii) distinct and differing by an integer

as given in the examples below.

Example 9.3 Solve the following equation in series

$$2x^2y'' - xy' + (1 - x^2)y = 0. \qquad [1]$$

Solution 9.3 Let the series solution of [1] be

$$y = \sum_{n=0}^{\infty} a_n x^{k+n}, a_0 \neq 0. \qquad [2]$$

Then $y' = \sum_{n=0}^{\infty} a_n(k+n)x^{k+n-1}$ and $y'' = \sum_{n=0}^{\infty} a_n(k+n)(k+n-1)x^{k+n-2}$.

Substituting for y and y' in [1], we get

$$2x^2\sum_{n=0}^{\infty} a_n(k+n)(k+n-1)x^{k+n-2} - x\sum_{n=0}^{\infty} a_n(k+n)x^{k+n-1}$$
$$+(1-x^2)\sum_{n=0}^{\infty} a_n x^{k+n} = 0$$

or $$2\sum_{n=0}^{\infty} a_n(k+n)(k+n-1)x^{k+n} - \sum_{n=0}^{\infty} a_n(k+n)x^{k+n}$$
$$+\sum_{n=0}^{\infty} a_n x^{k+n} - \sum_{n=0}^{\infty} a_n x^{k+n+2} = 0$$

$$\text{or} \quad \sum_{n=0}^{\infty} a_n\{2(k+n)(k+n-1)-(k+n)+1\}x^{k+n} - \sum_{n=0}^{\infty} a_n x^{k+n+2} = 0$$

$$\text{or} \quad -\sum_{n=0}^{\infty} a_n\{2(k+n)^2-3(k+n)+1\}x^{k+n} - \sum_{n=0}^{\infty} a_n x^{k+n+2} = 0$$

$$\text{or} \quad \sum_{n=0}^{\infty} a_n\{(2k+2n-1)(k+n-1)\}x^{k+n} - \sum_{n=0}^{\infty} a_n x^{k+n+2} = 0.$$

Now, equating to zero the coefficient of the smallest power of x, i.e., x^k, gives the indicial equation (with $n = 0$):

$$a_0(2k-1)(k-1) = 0 \quad \text{or} \quad (2k-1)(k-1) = 0 \quad \text{so that} \quad k = 1, 1/2.$$

Thus the roots of the indicial equation are unequal and not differing by an integer. We also obtain the recurrence relation by equating to zero the coefficient of x^{k+n} and obtain the following relation

$$a_n(2k+2n-1)(k+n-1) - a_{n-2} = 0$$

$$\text{or} \quad a_n = \frac{1}{(2k+2n-1)(k+n-1)} a_{n-2}. \qquad [3]$$

To obtain a_1, we equate to zero the coefficient of x^{k+1}, obtaining

$$a_1(2k+1)k = 0$$

which means $a_1 = 0$ when $k = 1$ or $1/2$. Thus $a_1 = a_3 = a_5 \ldots = 0$. Again taking $n = 2, 4, \ldots$ in [3], we get

$$a_2 = \frac{1}{(2k+3)(k+1)} a_0$$

$$\text{and} \quad a_4 = \frac{1}{(2k+5)(k+3)} a_2 = \frac{1}{(k+1)(k+3)(2k+3)(2k+5)} a_0$$

and so on. Substituting these values in [2], we have

$$y = a_0 x^k \left[1 + \frac{x^2}{(k+1)(2k+3)} + \frac{x^4}{(k+1)(k+3)(2k+3)(2k+5)} + \ldots\right].$$

If we substitute $k = 1$ and $a_0 = a$ in the above equation, we get

$$y = ax\left[1 + \frac{x^2}{2.5} + \frac{x^4}{2.4.5.9} + \ldots\right] = au, \text{ say}$$

and taking $k = \frac{1}{2}$ and $a_0 = b$ in the above equation, we get

$$y = bx^{1/2}\left[1 + \frac{x^2}{2.3} + \frac{x^4}{2.3.4.7} + \ldots\right] = bv, \text{say}.$$

Then the required solution is

$$y = au + bv$$

where a and b are arbitrary constants

Example 9.4: Solve in series the equation $xy'' + y' + xy = 0$.

Solution 9.4 Consider the solution to be in the form

$$y = a_0x^n + a_1x^{n+1} + a_2x^{n+2} + \ldots \qquad [1]$$

Then $y' = na_0x^{n-1} + (n+1)a_1x^n + (n+2)a_2x^{n+1} + \ldots$

and $y'' = n(n-1)a_0x^{n-2} + (n+1)na_1x^{n-1} + (n+2)(n+1)a_2x^n + \ldots$

Substituting in the given equation, we get

$$x[n(n-1)a_0x^{n-2} + (n+1)na_1x^{n-1} + (n+2)(n+1)a_2x^n + \ldots]$$
$$+[na_0x^{n-1}+(n+1)a_1x^n+(n+2)a_2x^{n+1}+\ldots]$$
$$+x[a_0x^n + a_1x^{n+1} + a_2x^{n+2} + \ldots] = 0.$$

The lowest power of x is x^{n-1}. Its coefficient equated to zero gives

$$a_0[n(n-1) + n] = 0 \quad \text{or} \quad n^2 = 0 \quad \text{since} \quad a_0 \neq 0. \quad \text{So} \quad n = 0, 0.$$

These are equal roots of the indicial equation. Now taking $n = 1, 2, 3, 4, \ldots$ we get

$$a_1[(n+1)n + n + 1] = 0 \quad \text{or} \quad a_1 = 0 \quad \text{and} \quad a_2(n+2)^2 + a_0 = 0$$

or $\quad a_2 = -\dfrac{a_0}{(n+2)^2} \quad$ and $\quad a_3(n+3)^2 + a_1 = 0 \quad$ or $\quad a_3 = 0$

and so on. Thus, we have

$$a_1 = a_3 = a_5 = a_7 = \ldots = 0$$

and substituting for a_2, a_4, etc. in [1] we get

$$y_1 = a_0 x^n \left[1 - \frac{x^2}{(n+2)^2} + \frac{x^4}{(n+2)^2(n+4)^2} - \frac{x^6}{(n+2)^2(n+4)^2(n+6)^2} + \dots\right] \quad [2]$$

This gives only one solution instead of two, so the second solution is given by $\partial y_1/\partial n$ when $n = 0$. That is

$$y_2 = \left(\frac{\partial y_1}{\partial n}\right)_{n=0} = y_1 \log x + x^n \left\{ \frac{x^2}{(n+2)^2} \cdot \frac{2}{n+2} - \frac{x^4}{(n+2)^2(n+4)^2}\left[\frac{2}{n+2} + \frac{2}{n+4}\right] + \dots \right\} \quad \text{when} \quad n = 0$$

$$= y_1 \log x + \left\{ \frac{1}{2^2}x^2 - \frac{1}{2^2 4^2}\left(1 + \frac{1}{2}\right)x^4 + \frac{1}{2^2.4^2.6^2}\left(1 + \frac{1}{2} + \frac{1}{3}\right)x^6 + \dots \right\}. \quad [3]$$

Thus the complete solution is given by [2] and [3].

9.3 Frobenius method

The differential equation

$$P(t)\frac{d^2y}{dt^2} + Q(t)\frac{dy}{dt} + R(t)y = 0 \quad [9.4]$$

is said to be *singular* at $t = t_0$ if $P(t_0) = 0$. In some special cases, we can find the solutions $y(t)$ of [9.4] in the form

$$y(t) = \sum_{n=0}^{\infty} a_n (t - t_0)^{n+r}$$

This is known as the method of Frobenius* and we can illustrate it with the following example.

Example 9.5 Find two linearly independent solutions of the differential equation

* Georg Frobenius (1849–1917), German mathematician

$$2t\frac{d^2y}{dt^2}+\frac{dy}{dt}+ty=0;\quad 0<t<\infty. \qquad [1]$$

Solution 9.5 Let $y(t)=\sum_{n=0}^{\infty}a_nt^{n+r}; a_0\neq 0.$

Also $y'(t)=\sum_{n=0}^{\infty}(n+r)a_nt^{n+r-1}$ and $y''(t)=\sum_{n=0}^{\infty}(n+r)(n+r-1)a_nt^{n+r-2}.$

So $t^r\left[2\sum_{n=0}^{\infty}(n+r)(n+r-1)a_nt^{n-1}+\sum_{n=0}^{\infty}(n+r)a_nt^{n-1}+\sum_{n=0}^{\infty}a_nt^{n+1}\right]$

$$=t^r\left[2\sum_{n=0}^{\infty}(n+r)(n+r-1)a_nt^{n-1}+\sum_{n=0}^{\infty}(n+r)a_nt^{n-1}+\sum_{n=2}^{\infty}a_{n-2}t^{n-1}\right]$$

$$=\left[2r(r-1)a_0+ra_0\right]t^{r-1}+\left[2(1+r)ra_1+(1+r)a_1\right]t^r$$

$$+\sum_{n=1}^{\infty}\left[2(n+r)(n+r-1)a_n+(n+r)a_n+a_{n-2}\right]t^{n+r-1}.$$

Equating the coefficients of each power of t to zero, we get

$$2r(r-1)a_0+ra_0=r(2r-1)a_0=0 \qquad [2]$$

$$2(r+1)ra_1+(r+1)a_1=(r+1)(2r+1)a_1=0 \qquad [3]$$

$$2(n+r)(n+r-1)a_n+(n+r)a_n=(n+r)\left[2(n+r)-1\right]a_n=-a_{n-2}. \qquad [4]$$

From [2], we have $r=0$ or $r=1/2$. From [2] and [3], we have $a_1=0$ and from [4] we have the recurrence formula

$$a_n=\frac{-a_{n-2}}{n(2n-1)}; n\geqslant 2.$$

Since $a_1=0$, we see that all *odd* coefficients are zero, and *even* coefficients are determined from

$$a_2=\frac{-a_0}{2.3};\quad a_4=\frac{-a_2}{4.7}=\frac{a_0}{2.4.3.7};\quad a_6=\frac{-a_4}{6.11}=\frac{-a_0}{2.4.6.3.7.11}.$$

Taking $a_0=1$, we see that

$$y(t)=1-\frac{t^2}{2.3}+\frac{t^4}{2.4.3.7}+\cdots=1+\sum_{n=1}^{\infty}\frac{(-1)^nt^{2n}}{2^nn!3.7\cdots(4n-1)}$$

which is one solution of [1] on $0 < t < \infty$. Again we can take $r = 1/2$; then

$$a_n = \frac{-a_{n-2}}{(n+1/2)\,[2(n+1/2)-1]} = \frac{-a_{n-2}}{n(2n+1)}.$$

Here also all odd coefficients are zero. Then the even coefficients are determined from

$$a_2 = \frac{-a_0}{2.5}; \quad a_4 = \frac{-a_2}{4.9} = \frac{a_0}{2.4.5.9}; \quad a_6 = \frac{-a_4}{6.13} = \frac{-a_0}{2.4.6.5.9.13}.$$

Taking $a_0 = 1$, we get

$$y(t) = t^{1/2}\left[1 - \frac{t^2}{2.5} + \frac{t^4}{2.4.5.9} + \cdots\right] \Rightarrow t^{1/2}\left[1 + \sum_{n=1}^{\infty} \frac{(-1)^n t^{2n}}{2^n n! 5.9 \cdots (4n+1)}\right]$$

which is a second solution of [1] in $0 < t < \infty$.

Example 9.6 Use the method of Frobenius to solve Bessel's equation of order zero:

$$t^2\frac{d^2y}{dt^2} + t\frac{dy}{dt} + t^2 y = 0, \quad t > 0. \qquad [1]$$

Solution 9.6 Here, $P(t) = t^2$, $Q(t) = t$ and $R(t) = t^2$, so that

$$t\frac{Q(t)}{P(t)} = \frac{t^2}{t^2} = 1 \quad \text{or} \quad t^2\frac{R(t)}{P(t)} = t^2\frac{t^2}{t^2} = t^2.$$

Let $y(t) = \sum_{n=0}^{\infty} a_n t^{n+r}$ be a solution of [1]. Then

$$y'(t) = \sum_{n=0}^{\infty}(n+r)a_n t^{n+r-1} \quad \text{and} \quad y''(t) = \sum_{n=0}^{\infty}(n+r)(n+r-1)a_n t^{n+r-2}.$$

Therefore, [1] can be written as

$$t^r \sum_{n=0}^{\infty}(n+r)(n+r-1)a_n t^n + \sum_{n=0}^{\infty}(n+r)a_n t^n + \sum_{n=0}^{\infty} a_n t^{n+2}$$

$$= t^r\left[\sum_{n=0}^{\infty}(n+r)(n+r-1)a_n t^n + \sum_{n=0}^{\infty}(n+r)a_n t^n + \sum_{n=2}^{\infty} a_{n-2} t^n\right]$$

Taking the sum of coefficients of like powers of t equal to zero, we get

$$[r(r-1)+r]a_0 = r^2 a_0 = 0 \quad [2]$$

$$\big[(r+1)r+(r+1)\big]a_1 = (r+1)^2 a_1 = 0 \quad [3]$$

and $\big[(r+n)(r+n-1)+r+n\big]a_n = (r+n)^2 a_n = -a_{n-2};\ n \geqslant 2.$ [4]

Now [2] is the *indicial equation* which has a double root $r = 0$. From [3], we find $a_1 = 0$ and [4] gives the recurrence formula as

$$a_n = \frac{-a_{n-2}}{n^2}.$$

All odd coefficients are zero and

$$a_2 = \frac{-a_0}{2^2}; \quad a_4 = \frac{-a_2}{4^2} = \frac{a_0}{2^2.4^2}; \quad a_6 = \frac{-a_4}{6^2} = \frac{-a_0}{2^2.4^2.6^2}.$$

Taking $a_0 = 1$, we see that

$$y(t) = 1 - \frac{t^2}{2^2} + \frac{t^4}{2^2.4^2} + \ldots = \sum_{n=0}^{\infty} \frac{(-1)^n t^{2n}}{2^2.4^2 \ldots (2n)^2}$$

is one solution of [1]. This solution is known as the *Bessel function* of order zero and is denoted by $J_0(t)$. A second solution

$$y_2(t) = J_0(t) \int \frac{dt}{t J_0^2(t)}$$

can also be found. This solution is called the *Neumann function* of order zero and denoted by $Y_0(t)$.

Exercise 9

9.1 Apply power series method to solve

(i) $2x(1-x)\dfrac{d^2y}{dx^2} + (5-7x)\dfrac{dy}{dx} - 3y = 0$

(ii) $(x+1)y' - (x+2)y = 0$

9.2 Show that the equation

$$2(\sin t)y'' + (1-t)y' - 2y = 0$$

has two solutions $y_1(t)$ and $y_2(t)$ of the form

$$y_1(t) = \sum_{n=0}^{\infty} a_n t^n \quad \text{and} \quad y_2(t) = t^{1/2} \sum_{n=0}^{\infty} b_n t^n.$$

9.3 The differential equation $ty'' + (1-t)y' + \lambda y = 0$, where λ is a constant is called the *Laguerre differential equation.*

(i) Show that the indicial equation of the Laguerre equation is $r^2 = 0$.

(ii) Find a solution $y(t)$ of the Laguerre equation of the form

$$y(t) = \sum_{n=0}^{\infty} a_n t^n.$$

(iii) Show that this solution reduces to a polynomial of degree n if $\lambda = n$.

9.4 Find two linearly independent series solution when $|x| < 1$, for *Chebyshev's equation*

$$(1-x^2)y'' - xy' + p^2 y = 0$$

where p is a constant.

9.5 Show that *Hermite's equation*

$$y'' - 2xy' + 2py = 0$$

where p is a constant, has a general solution $y(x) = a_0 y_1(x) + a_1 y_2(x)$ where

$$y_1(x) = 1 - \frac{2p}{2!}x^2 + 2^2\frac{p(p-2)}{4!}x^4 - 2^3\frac{p(p-2)(p-4)}{6!}x^6 + \cdots$$

$$y_2(x) = x - \frac{2(p-1)}{3!}x^3 + 2^2\frac{(p-1)(p-3)}{5!}x^5 - 2^3\frac{(p-1)(p-3)(p-5)}{7!}x^7 + \cdots$$

9.6 Show that the following equation has two linearly independent solutions of the form $y = x^r \sum_{n=0}^{\infty} a_n x^n$, and find these solutions:

$$x^2 y'' - 2xy' + (2 - x^3)y = 0.$$

CHAPTER TEN

Legendre's Equation And Polynomials

10.1 Introduction

Sometimes, we feel that abstract ideas dominate mathematics. But the study of specific functions and their properties provide a healthy counterpoise, showing the many important applications of these functions to mathematical physics.

In this section we first seek solutions to several important classes of second order homogeneous equations, which have many important properties.

The physical background and several typical applications are discussed at the end of this chapter

10.1.1 Legendre's differential equation

A Legendre equation*, which is a special type of second order differential equation, is of the form

$$(1-x^2)\frac{d^2y}{dx^2} - 2x\frac{dy}{dx} + n(n+1)y = 0 \qquad [10.1]$$

where n is a real number. But in most applications, only integral values of n are needed. We can easily see that when $x = \pm 1$, [10.1] reduces to a first order equation which is easily solvable. So, we consider $x \neq \pm 1$. Assuming the solution to be in the series form, we substitute $y = a_0 x^m + a_1 x^{m+1} + a_2 x^{m+2} + \cdots$ in [10.1], which now takes the form

$$a_0 m(m-1)x^{m-2} + a_1(m+1)mx^{m-1} + \cdots + [a_{r+2}(m+r+2)(m+r+1)$$

$$-\{(m+r)(m+r+1) - n(n+1)\}a_r]x^{m+r} + \cdots = 0.$$

Equating the coefficients of the lowest power of x, i.e. x^{m-2}, to zero, we have $a_0 m(m-1) = 0$ which yields

$$m = 0 \quad \text{or} \quad 1 \quad [a_0 \neq 0]. \qquad [10.2]$$

---

*Named after the French mathematician, Adrien Marie Legendre (1752–1833).

Again, equating to zero the coefficients of x^{m-1} and x^{m+r}, we get

$$a_1 m(m+1) = 0 \tag{10.3}$$

$$a_{r+2}(m+r+2)(m+r+1) - [(m+r)(m+r+1) - n(n+1)]a_r = 0 \tag{10.4}$$

or $$a_{r+2} = \frac{(m+r-n)(m+r+n+1)}{(m+r+2)(m+r+1)} a_r. \tag{10.5}$$

When $m = 0$, [10.3] is satisfied and therefore $a_1 \neq 0$. Then [10.5] becomes

$$a_{r+2} = \frac{(r-n)(r+n+1)}{(r+1)(r+2)} a_r \text{ (for } \quad r = 0, 1, 2, \ldots, \infty)$$

or $$a_r = \frac{(r-2-n)(r-1+n)}{r(r-1)} a_{r-2} \quad \text{(for} \quad r = 2, 3, \ldots, \infty). \tag{10.6}$$

This is a recurrence formula, and for $r = 2, 3, \cdots$ etc., we have

$$a_2 = \frac{-n(n+1)}{2!} a_0; \; a_3 = -\frac{(n-1)(n+2)}{3!} a_1;$$

$$a_4 = \frac{-(n-2)(n+3)}{4.3} a_2 = \frac{n(n-2)(n+1)(n+3)}{4!} a_0;$$

$$a_5 = \frac{(n-3)(n+4)}{5.4} a_3 = \frac{(n-1)(n-3)(n+2)(n+4)}{5!} a_1$$

and so on. Hence, for $m = 0$, there are two independent solutions of [10.1], viz.,

$$\begin{aligned} y_1 &= a_0 \left\{ 1 - \frac{n(n+1)}{2!} x^2 + \frac{n(n-2)(n+1)(n+3)}{4!} x^4 \ldots \right\} \\ y_2 &= a_1 \left\{ x - \frac{(n-1)(n+2)}{3!} x^3 + \frac{(n-1)(n-3)(n+2)(n+4)}{5!} x^5 \ldots \right\}. \end{aligned} \tag{10.7}$$

When $m = 1$, [10.3] shows that $a_1 = 0$. Then [10.4] shows that

$$a_3 = a_5 = a_7 = \cdots = 0$$

and $$a_2 = -\frac{(n-1)(n+2)}{3!} a_0; \; a_4 = \frac{(n-1)(n-3)(n+2)(n+4)}{5!} a_0 \quad \text{etc.}$$

So for $m = 1$, we get the solution [10.7] again, showing that

$$y = c_1 y_1 + c_2 y_2 \qquad [10.8]$$

is the general solution of [10.1] in $-1 < x < 1$. Since the two solutions y_1 and y_2 are such that y_1 contains only even powers of x while y_2 contains only odd powers of x, the ratio y_1/y_2 is not a constant, and hence y_1 and y_2 are two linearly independent solutions. Hence [10.8] is a general solution of [10.1] in $-1 < x < 1$.

10.2 Legendre polynomials $P_n(x)$

In many applications, the parameter n is a positive integer. If n is odd, then clearly the second series in [10.7], viz. y_2, contains only a finite number of terms. If n is even, then the first series in [10.7], viz. y_1, contains only a finite number of terms. In either of these cases the series which reduces to a finite sum is known as a *Legendre polynomial* or *zonal harmonic* of order n, denoted by $P_n(x)$.

Since Legendre polynomials are of great practical importance, we can consider them in greater detail. The usual standard form for these polynomials is obtained by assigning to a_0 and a_1 values which will make the coefficient of the highest power of x in each series equal to $(2n)!/2^n(n!)^2$. So, we take

$$a_0 = (-1)^{n/2}\frac{1.3.5\ldots(n-1)}{2.4.6\ldots n} = (-1)^{n/2}\frac{n!}{2^n[(n/2)!]^2}$$

and $$a_1 = (-1)^{(n-1)/2}\frac{1.3.5\ldots(n)}{2.4.6\ldots(n-1)} = (-1)^{(n-1)/2}\frac{(n+1)!}{2^n\left[\left(\frac{n-1}{2}\right)!\right]\left[\left(\frac{n+1}{2}\right)!\right]}.$$

Since $P_n(x)$ is a polynomial of degree n that contains only even or only odd powers of x, according as n is even or odd, it can be written in the form

$$P_n(x) = a_n x^n + a_{n-2}x^{n-2} + a_{n-4}x^{n-4} + \cdots \qquad [10.9]$$

which ends with a_0 if n is even and $a_1 x$ if n is odd. Since the coefficient of the highest power of x in $P_n(x)$ should be equal to $(2n)!/2^n(n!)^2$ and since, from the recurrence formula [10.6], we know that

$$a_r = \frac{(r-2-n)(r-1+n)}{r(r-1)}a_{r-2} \quad \text{or} \quad a_{r-2} = \frac{r(r-1)}{(r-2-n)(r-1+n)}a_r$$

where $r = n, n-2\ldots$, we get

$$a_{n-2} = \frac{-n(n-1)}{2(2n-1)}a_n$$

$$\text{and} \quad a_{n-4} = -\frac{(n-2)(n-3)}{4(2n+3)}a_{n-2} = \frac{n(n-1)(n-2)(n-3)}{2.4(2n-1)(2n-3)}a_n.$$

We have, from [10.9]

$$P_n(x) = \frac{(2n)!}{2^n(n!)^2}\left[x^n - \frac{n(n-1)}{2(2n-1)}x^{n-2} + \frac{n(n-1)(n-2)(n-3)}{2.4(2n-1)(2n-3)}x^{n-4} + \dots\right.$$

$$\left. + \frac{(-1)^r n(n-1)\cdots(n-2r+1)}{2^r r!(2n-1)(2n-3)\cdots(2n-2r+1)}x^{n-2r} + \cdots\right]. \qquad [10.10]$$

The coefficient of x^{n-2r} in [10.10] can also be written as

$$\frac{(-1)^r n!\,(2n-2r)!\,2^r n!}{2^r r!\,(n-2r)!\,(2n)!\,(n-r)!} = \frac{(-1)^r (n!)^2(2n-2r)!}{r!\,(2n)!(n-r)!\,(n-2r)!} \qquad [10.11]$$

since $\quad n(n-1)\cdots(n-2r+1) = n!/(n-2r)!$ and

$$(2n-1)(2n-3)\cdots(2n-2r+3)(2n-2r+1)$$

$$= \frac{(2n-2r+1)(2n-2r+2)(2n-2r+3)\cdots(2n-3)(2n-2)(2n-1)2n}{(2n-2r+2)\cdots(2n-2)(2n)}$$

$$= \frac{(2n)!}{(2n-2r)!}\,\frac{1}{2^r(n-r+1)\cdots(n-1)n} = \frac{(2n)!(n-r)!}{(2n-2r)!2^r n!}.$$

Thus [10.10] can be written, using [10.11], as

$$P_n(x) = \sum_{r=0}^{m} \frac{(-1)^r(2n-2r)!}{2^n r!(n-r)!(n-2r)!}x^{n-2r} \qquad [10.12]$$

where $m = n/2$ if n is even and $m = (n-1)/2$ if n is odd. This expression [10.12] is called *Legendre's polynomial of order* n and is denoted by $P_n(x)$. It is easy to check that the first five Legendre polynomials (see Fig. 10.1) are

$$P_0(x) = 1;\, P_1(x) = x;\, P_2(x) = (1/2)(3x^2-1);\, P_3(x) = (1/2)(5x^3-3x);$$

$$P_4(x) = (1/8)(35x^4 - 30x^2 + 3).$$

As these results illustrate, $P_n(1) = 1$, and $P_n(-1) = (-1)^n$ for all values of n.

10.3 Rodrigues' formula

Theorem 10.1: Legendre polynomials can also be expressed in the form

$$P_n(x) = \frac{1}{2^n n!}\frac{d^n}{dx^n}[(x^2-1)^n] \text{ for } n = 1, 2, 3, \dots$$

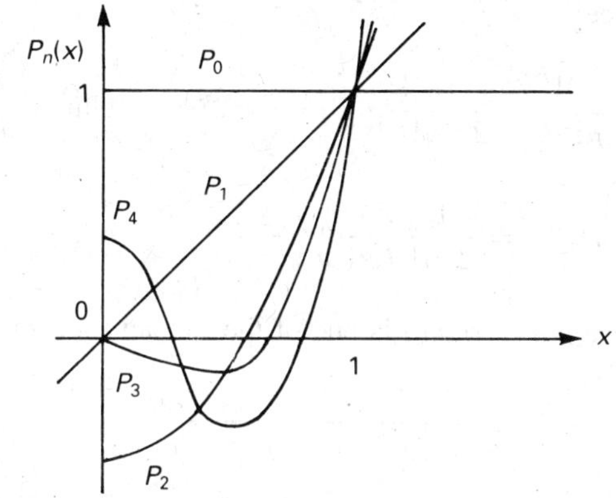

Fig. 10.1 *Legendre polynomials.*

Proof: This result can be proved by direct differentiation and induction, but it is perhaps more interesting to proceed in the following way. Let $v = (x^2 - 1)^n$. Then, we have $dv/dx = 2nx(x^2 - 1)^{n-1}$. Multiplying the last equation by $(x^2 - 1)$, we get

$$(x^2 - 1)\left(\frac{dv}{dx}\right) = 2nx(x^2 - 1)^n \quad \text{or} \quad (1 - x^2)\left(\frac{dv}{dx}\right) - 2nxv = 0.$$

If we differentiate this repeatedly with respect to x, we obtain by Leibnitz's theorem

$$(1 - x^2)v_{n+2} - 2xv_{n+1} + n(n+1)v_n = 0$$

after $(n + 1)$ differentiations. If we substitute $y = v_n$ in the above equation, we get

$$(1 - x^2)\frac{d^2y}{dx^2} - 2x\frac{dy}{dx} + n(n+1)y = 0$$

which is precisely Legendre's equation. But $y = v_n = d^n(x^2 - 1)^n/dx^n$ is obviously a polynomial of degree n. But there is only one polynomial solution of Legendre's equation, namely $P_n(x)$. Hence y must be a multiple of $P_n(x)$. So

$$y = \frac{d^n}{dx^n}(x^2 - 1)^n = cP_n(x) \quad \text{Thus} \quad \frac{d^n}{dx^n}(x^2 - 1)^n = cP_n(x).$$

Finally c has to be determined and this can be done by equating the coefficients of x^n on both sides of the above equation. The coefficient of x^n on the left hand side arises from the n-fold differentiation of the term of the highest degree x^{2n} and is

$$(2n)(2n - 1)\cdots[2n - (n - 1)] = (2n)!/n!.$$

On the other hand, the coefficient of the highest power of x in $P_n(x)$ is $(2n)!/2^n(n!)^2$. Hence

$$\frac{(2n)!}{n!} = c\frac{(2n)!}{2^n(n!)^2} \quad \text{or} \quad c = 2^n n!. \quad \text{Thus} \quad \frac{d^n}{dx^n}(x^2-1)^n = 2^n n!\, P_n(x)$$

$$\text{or} \quad P_n(x) = \frac{1}{2^n n!}\frac{d^n}{dx^n}(x^2-1)^n. \qquad [10.13]$$

Rodrigues' formula is one of the fundamental identities involving the Legendre polynomials.

10.4 Legendre polynomials — an interpretation

Let a charge e be placed at a point A and let V be the potential at another point B as shown in Fig. 10.2.

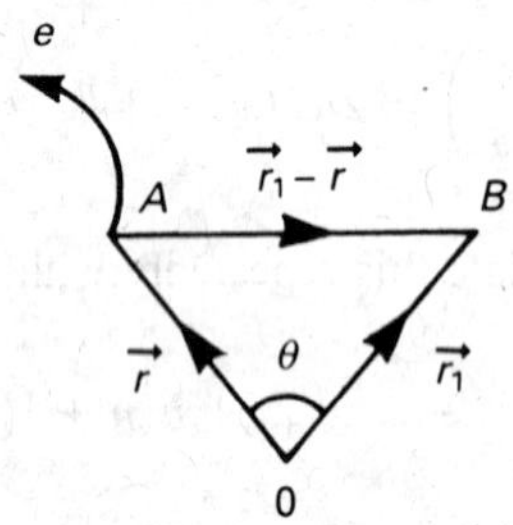

Fig.10.2 *An interpretation of Legendre polynomials.*

$$\text{Then } V = \frac{e}{AB} = \frac{e}{|\vec{r}_1 - \vec{r}|} = \frac{e}{[(\vec{r}_1 - \vec{r})\cdot(\vec{r}_1 - \vec{r})]^{1/2}} = \frac{e}{[r_1^2 - 2rr_1\cos\theta + r^2]^{1/2}}$$

$$= e(r_1^2 - 2rr_1\cos\theta + r^2)^{-1/2} \quad (\text{where} \quad r = |\vec{r}| \text{ and } r_1 = |\vec{r}_1|).$$

Then two cases arise:

Case (i): Let $r_1 > r$. Then

$$V = e(r_1^2)^{-1/2}\left[1 - \frac{2r}{r_1}\cos\theta + \left(\frac{r}{r_1}\right)^2\right]^{-1/2} = \frac{e}{r_1}[1 - 2zx + z^2]^{-1/2}$$

where $x = \cos\theta$ and $z = (r/r_1) < 1$. We can also see that $|x| \le 1$.

Case (ii): Let $r_1 < r$. Then

$$V = \frac{c}{r}\left[\left(\frac{r_1}{r}\right)^2 - 2\cos\theta\left(\frac{r_1}{r}\right) + 1\right]^{-1/2} = \frac{c}{r}\left[1 - 2zx + z^2\right]^{-1/2}$$

where $x = \cos\theta$ and $z = (r_1/r) < 1$. Again $|x| \leq 1$.

From case (i) and case (ii) we have the function $[1 - 2zx + z^2]^{-1/2}$ which can be expanded in a binomial series iff

$$|z^2 - 2zx| < 1 \quad \text{or} \quad |z(z - 2x)| < 1 \quad \text{or} \quad |z||z - 2x| < 1.$$

In such a case, we can expand

$$[1 - 2zx + z^2]^{-1/2} = 1 + \frac{1}{2}(2zx - z^2) + \frac{3}{8}(2zx - z^2)^2 + \frac{15}{48}(2zx - z^2)^3 + \cdots$$

$$= 1 + zx - \frac{z^2}{2} + \frac{3}{2}z^2x^2 + \cdots = 1 + zx + \left[-\frac{1}{2} + \frac{3}{2}x^2\right]z^2 + \cdots$$

$$= 1 + zx + \left[\frac{3}{2}x^2 - \frac{1}{2}\right]z^2 + \cdots = \sum_{n=0}^{\infty} P_n z^n.$$

We can now define P_n as the coefficient of z^n in the expansion of $[1 - 2zx + z^2]^{-1/2}$.

Example 10.1 Show that $\dfrac{1+z}{z\sqrt{1 - 2xz + z^2}} - \dfrac{1}{z} = \displaystyle\sum_{n=0}^{\infty}(P_n + P_{n+1})z^n.$

Solution 10.1 We know that

$$(1 - 2xz + z^2)^{-1/2} = \sum_{n=0}^{\infty} P_n z^n. \qquad [1]$$

Now,
$$\frac{1}{z\sqrt{1 - 2xz + z^2}} + \frac{1}{\sqrt{1 - 2xz + z^2}} - \frac{1}{z} = \frac{1}{z}\sum_{n=0}^{\infty} P_n z^n + \sum_{n=0}^{\infty} P_n z^n - \frac{1}{z}$$

$$= \frac{1}{z}P_0 z^0 + \frac{1}{z}\sum_{n=1}^{\infty} P_n z^n + \sum_{n=0}^{\infty} P_n z^n - \frac{1}{z}$$

$$= \frac{1}{z}\sum_{n=1}^{\infty} P_n z^n + \sum_{n=0}^{\infty} P_n z^n \quad (\text{since} \quad P_0 = 1)$$

$$= \frac{1}{z}\sum_{n=0}^{\infty} P_{n+1}z^{n+1} + \sum_{n=0}^{\infty} P_n z^n = \sum_{n=0}^{\infty} z^n(P_{n+1} + P_n).$$

Example 10.2 Show that $\dfrac{1-z^2}{(1-2xz+z^2)^{3/2}} = \sum_{n=0}^{\infty}(2n+1)P_n z^n.$

Solution 10.2 We know that

$$(1-2xz+z^2)^{-1/2} = \sum_{n=0}^{\infty} P_n z^n. \qquad [1]$$

Then, differentiating with respect to z, we get

$$\left(-\frac{1}{2}\right)(1-2xz+z^2)^{-3/2}(2z-2x) = \sum_{n=0}^{\infty} nP_n z^{n-1}$$

or $$(x-z)(1-2xz+z^2)^{-3/2} = \sum_{n=0}^{\infty} nP_n z^{n-1}. \qquad [2]$$

Multiplying both sides of [2] by $2z$, we have

$$2z(x-z)(1-2xz+z^2)^{-3/2} = 2\sum_{n=0}^{\infty} nP_n z^n. \qquad [3]$$

Adding [1] and [3], we get

$$\frac{1}{(1-2xz+z^2)^{1/2}} + \frac{2z(x-z)}{(1-2xz+z^2)^{3/2}} = \sum_{n=0}^{\infty} z^n P_n + \sum_{n=0}^{\infty} 2nz^n P_n$$

or $$\sum_{n=0}^{\infty}(2n+1)P_n z^n = \frac{1-2xz+z^2+2z(x-z)}{(1-2xz+z^2)^{3/2}} = \frac{1-z^2}{(1-2xz+z^2)^{3/2}}.$$

Note: $(1-2xz+z^2)^{-1/2}$ is called the *generating function* for the Legendre polynomials $P_n(x)$ since the coefficients in its expansion provide the expressions for $P_n(x)$.

10.5 Generating function for $P_n(x)$

The Legendre polynomials have a generating function from which various important properties can be derived. For the Legendre polynomials, the generating function is $f(x,z) = (1-2xz+z^2)^{-1/2}$ as the following theorem makes it clear.

Theorem 10.2: $$\frac{1}{\sqrt{1-2xz+z^2}} = P_0(x) + P_1(x)z + P_2(x)z^2 + \cdots + P_n(x)z^n + \cdots$$

Proof: Consider the left hand side of the above equation which can be expanded using the binomial theorem as follows:

$$\left[1 - 2xz + z^2\right]^{-1/2} = \left[1 - z(2x - z)\right]^{-1/2}$$

$$= 1 + \frac{1}{2}z(2x - z) + \frac{1.3}{2^2 2!}z^2(2x-z)^2 + \cdots$$

$$+ \frac{1.3\cdots(2n-3)}{2^{n-1}(n-1)!}z^{n-1}(2x-z)^{n-1} + \frac{1.3\cdots(2n-1)}{2^n n!}z^n(2x-z)^n + \cdots$$

Now, z^n occurs in the terms after and including the one containing $z^n(2x - z)^n$ and from these we find its total coefficient is

$$\frac{1.3\cdots(2n-1)}{2^n n!}(2x)^n - \frac{1.3\cdots(2n-3)}{2^{n-1}(n-1)!}\frac{n-1}{1!}(2x)^{n-2}$$

$$+ \frac{1.3\cdots(2n-5)}{2^{n-2}(n-2)!}\frac{(n-2)(n-3)}{2!}(2x)^{n-4} - \cdots$$

which can be written, after multiplying and dividing by factors needed to complete the factorials in the numerator, as

$$\frac{(2n)!x^n}{2^n n! n!} - \frac{(2n-2)!x^{n-2}}{2^{n-1}1!(n-1)!(n-2)!} + \frac{(2n-4)!\,x^{n-4}}{2^{n-2}2!(n-2)!(n-4)!} + \cdots$$

which is the expanded form of $P_n(x)$ as given by [10.10]. Thus, (as stated by the theorem)

$$[1 - 2xz + z^2]^{-1/2} = \sum_{n=0}^{\infty} P_n(x)z^n. \qquad [10.14]$$

Equation [10.14] is valid in the ranges $-1 \le x \le 1$ and $|z| < 1$ because of the region of convergence of the binomial expansion. It is of great importance in *potential theory*.

10.6 The Legendre coefficients

If we assume $x = \cos\theta = (e^{i\theta} + e^{-i\theta})/2$, where $i = \sqrt{-1}$, and substitute for x in

$$[1-2xz+z^2]^{-1/2}=\sum_{n=0}^{\infty}P_n(x)z^n$$

we get $[1-z(e^{i\theta}+e^{-i\theta})+z^2]^{-1/2}=\sum_{n=0}^{\infty}P_n(\cos\theta)z^n$. Now, we know that

$$[1-z(e^{i\theta}+e^{-i\theta})+z^2]^{-1/2}=(1-ze^{i\theta})^{-1/2}(1-ze^{-i\theta})^{-1/2}.$$

Using binomial theorem we can write

$$(1-ze^{i\theta})^{-1/2}=1+\frac{ze^{i\theta}}{2}+\frac{1.3}{2.4}z^2e^{2i\theta}+\cdots+\frac{1.3\cdots(2n-1)}{2.4\cdots 2n}z^ne^{in\theta}+\cdots\quad.$$

$$(1-ze^{-i\theta})^{-1/2}=1+\frac{ze^{-i\theta}}{2}+\frac{1.3}{2.4}z^2e^{-2i\theta}+\cdots+\frac{1.3\cdots(2n-1)}{2.4\cdots 2n}z^ne^{-in\theta}+\cdots$$

Multiplying the above two expressions and picking out the coefficient of z^n, we have

$$P_n(\cos\theta)=\frac{1.3\cdots(2n-1)}{2.4\cdots 2n}2\cos n\theta+\frac{1.1.3\cdots(2n-3)}{2.2.4\cdots(2n-2)}2\cos(n-2)\theta+\cdots$$

Example 10.3 If (R,ϕ,z) and (r,θ,ϕ) are cylindrical polar and spherical polar coordinates of the same point, show that

$$P_n(\cos\theta)=(-1)^n\,\frac{r^{n+1}}{n!}\,\frac{\partial^n}{\partial z^n}\left(\frac{1}{r}\right).$$

Solution 10.3 We have $r^2=(x^2+y^2+z^2)^{1/2}=(R^2+z^2)^{1/2}$. So

$$1/r=(R^2+z^2)^{-1/2}=f(R,z).$$

Then $f(R,z+h)=[R^2+(z+h)^2]^{-1/2}=[R^2+z^2+2hz+h^2]^{-1/2}$

$$=[r^2+2hr\cos\theta+h^2]^{-1/2}=\frac{1}{r}\left[1+\frac{2h\cos\theta}{r}+\frac{h^2}{r^2}\right]^{-1/2}$$

$$=\frac{1}{r}\sum_{n=0}^{\infty}(-1)^n\left(\frac{h}{r}\right)^n P_n(\cos\theta). \qquad [1]$$

Again, by Taylor's series, we get

$$f(R,z+h)=\sum_{n=0}^{\infty}\frac{h^n}{n!}\,\frac{\partial^n}{\partial z^n}f(R,z)=\sum_{n=0}^{\infty}\frac{h^n}{n!}\frac{\partial^n}{\partial z^n}\left(\frac{1}{r}\right). \qquad [2]$$

From [1] and [2] we have $\dfrac{(-1)^n}{r}\sum_{n=0}^{\infty}\dfrac{h^n}{r^n}P_n(\cos\theta)=\sum_{n=0}^{\infty}\dfrac{h^n}{n!}\dfrac{\partial^n}{\partial z^n}\left(\dfrac{1}{r}\right).$

Equating the coefficients of h^n on both sides of the above equation, we get

$$\frac{1}{n!}\frac{\partial^n}{\partial z^n}\left(\frac{1}{r}\right)=\frac{(-1)^n}{r^{n+1}}P_n(\cos\theta) \text{ or } P_n(\cos\theta)=\frac{(-1)^n}{n!}r^{n+1}\frac{\partial^n}{\partial z^n}\left(\frac{1}{r}\right).$$

Example 10.4 Prove: $1+\dfrac{1}{2}P_1(\cos\theta)+\dfrac{1}{3}P_2(\cos\theta)+\cdots=\log\left\{\dfrac{[1+\sin\theta/2]}{\sin\theta/2}\right\}.$

Solution 10.4 We know that

$$(1-2xz+z^2)^{-1/2}=\sum_{n=0}^{\infty}z^nP_n(x). \qquad [1]$$

Integrating w.r.t. z, from 0 to 1, we get

$$\sum_{n=0}^{\infty}\int_0^1 z^nP_n(x)dz=\int_0^1\frac{dz}{(1-2xz+z^2)^{1/2}}. \qquad [2]$$

Replacing x by $\cos\theta$ on both sides of [2] gives

$$\sum_{n=0}^{\infty}P_n(\cos\theta)\int_0^1 z^n dz=\int_0^1\frac{dz}{(1-2z\cos\theta+z^2)^{1/2}}$$

or $$\sum_{n=0}^{\infty}\frac{P_n(\cos\theta)}{n+1}=\left[\log\{(z-\cos\theta)+\sqrt{[(z-\cos\theta)^2+\sin^2\theta]}\}\right]_0^1$$

$$=\log\{(1-\cos\theta)+\sqrt{[(1-\cos\theta)^2+\sin^2\theta]}\}-\log(1-\cos\theta)$$

$$=\log\{(1-\cos\theta)+\sqrt{[2(1-\cos\theta)]}\}-\log(1-\cos\theta)$$

$$=\log\left[\frac{(1-\cos\theta)+\sqrt{2}\sqrt{(1-\cos\theta)}}{1-\cos\theta}\right]$$

$$=\log\frac{\sqrt{(1-\cos\theta)}+\sqrt{2}}{\sqrt{(1-\cos\theta)}}=\log\left[\frac{1+\sin(\theta/2)}{\sin(\theta/2)}\right].$$

$$\text{Thus}\quad 1+\frac{1}{2}P_1(\cos\theta)+\frac{1}{3}P_2(\cos\theta)+\cdots=\log\frac{1+\sin(\theta/2)}{\sin(\theta/2)}.$$

Note: Every coefficient of $P_n(\cos\theta)$ is positive and P_n is greatest when $\theta = 0$. But $P_n(\cos\theta) = P_n(1) = 1$. Thus $|P_n(\cos\theta)| \le 1$ for $n = 0, 1, 2, \cdots$ So

$$P_0(\cos\theta) = 1;\; P_1(\cos\theta) = \cos\theta;\; P_2(\cos\theta) = (1/4)(3\cos 2\theta + 1);$$

$$P_3(\cos\theta) = (1/8)(5\cos 3\theta + 3\cos\theta);\; P_4(\cos\theta) = (1/64)(35\cos 4\theta + 20\cos 2\theta + 9)$$

etc.

10.7 Some recurrence relations for $P_n(x)$

Given below are some basic recurrence relations from which several others follow.

1. $\boldsymbol{(n+1)P_{n+1}(x) = (2n+1)x\,P_n(x) - n\,P_{n-1}(x)}$

We know from Theorem 10.2 , that $\quad (1-2xz+z^2)^{-1/2} = \sum_{n=0}^{\infty} P_n(x)z^n.$

Differentiating both sides partially with respect to z,we get

$$\frac{x-z}{(1-2xz+z^2)^{3/2}} = \sum_{n=0}^{\infty} nP_n(x)z^{n-1}$$

$$\text{or}\quad (x-z)\sum_{n=0}^{\infty} P_n(x)z^n = (1-2xz+z^2)\sum_{n=0}^{\infty} nP_n(x)z^{n-1}.$$

This can be simplified to

$$\sum_{n=0}^{\infty}[xP_n(x)-P_{n-1}(x)]z^n = \sum_{n=0}^{\infty}[(n+1)P_{n+1}(x)-2nxP_n(x)+(n-1)P_{n-1}(x)]z^n.$$

Equating coefficients of z^n on both sides, we get

$$xP_n(x) - P_{n-1}(x) = (n+1)P_{n+1}(x) - 2nxP_n(x) + (n-1)P_{n-1}(x).$$

Simplifying, we have the required relation

$$(n+1)P_{n+1}(x) = (2n+1)xP_n(x) - nP_{n-1}(x).$$

2. $\boldsymbol{P_n(x) = P'_{n+1}(x) - 2xP'_n(x) + P'_{n-1}(x).}$

Differentiating

$$(1 - 2xz + z^2)^{-1/2} = \sum_{n=0}^{\infty} P_n(x) z^n \qquad [1]$$

w.r.t. x keeping z fixed, we get

$$z(1 - 2xz + z^2)^{-3/2} = \sum_{n=0}^{\infty} P'_n(x) z^n. \qquad [2]$$

Multiplying both sides of [2] by $(1 - 2xz + z^2)$, and using [1], we get

$$\sum_{n=0}^{\infty} P_n(x) z^n = (1 - 2xz + z^2) \sum_{n=0}^{\infty} P'_n(x) z^{n-1}$$

$$= \sum_{n=0}^{\infty} P'_n(x) z^{n-1} - 2x \sum_{n=0}^{\infty} P'_n(x) z^n + \sum_{n=0}^{\infty} P'_n(x) z^{n+1}.$$

Equating coefficients of z^n on both sides, we have

$$P_n(x) = P'_{n+1}(x) - 2xP'_n(x) + P'_{n-1}(x).$$

3. $\boldsymbol{nP_n(x) = x P'_n(x) - P'_{n-1}(x)}$

From relation **2** above we see that

$$z(1 - 2xz + z^2)^{-3/2} = \sum_{n=0}^{\infty} P'_n(x) z^n$$

or $\quad z(x - z)(1 - 2xz + z^2)^{-3/2} = (x - z) \sum_{n=0}^{\infty} P'_n(x) z^n.$

But $\quad \dfrac{x - z}{(1 - 2xz + z^2)^{3/2}} = \sum_{n=0}^{\infty} nP_n(x) z^{n-1}.$

So $\quad nP_n(x) = xP'_n(x) - P'_{n-1}(x).$

4. $\mathbf{(2n+1)(x^2-1)P'_n(x) = n(n+1)[P_{n+1}(x) - P_{n-1}(x)]}$

From relations 1 and 3 above , we have

$$(n+1)P_{n+1}(x) = (2n+1)xP_n(x) - nP_{n-1}(x); \quad [1]$$

$$nP_n(x) = xP'_n(x) - P'_{n-1}(x). \quad [2]$$

Differentiating [1] we get

$$(n+1)P'_{n+1}(x) = (2n+1)P_n(x) + (2n+1)xP'_n(x) - nP'_{n-1}(x).$$

Substituting for $xP'_n(x)$ from [2], we have

$$\begin{aligned}(n+1)P'_{n+1}(x) &= (2n+1)P_n(x) + (2n+1)[nP_n(x) + P'_{n-1}(x)] - nP'_{n-1}(x)\\ &= (2n+1)(n+1)P_n(x) + (n+1)P'_{n-1}(x)\end{aligned}$$

or $(2n+1)P_n(x) = P'_{n+1}(x) - P'_{n-1}(x).$

Subtracting [2] from the above relation, we obtain

$$(n+1)P_n(x) = P'_{n+1}(x) - xP'_n(x) \quad \text{or} \quad nP_{n-1}(x) = P'_n(x) - xP'_{n-1}(x)$$

by replacing $(n+1)$ by n. Multiplying [2] by x and subtracting it from the above equation, we get

$$n[P_{n-1}(x) - xP_n(x)] = (1-x^2)P'_n(x) \quad [3]$$

Relation [1] can also be written as

$$(n+1)[xP_n(x) - P_{n+1}(x)] = n[P_{n-1}(x) - xP_n(x)].$$

Using [3] this relation becomes

$$(1-x^2)P'_n(x) = (n+1)\big[xP_n(x) - P_{n+1}(x)\big]. \quad [4]$$

Also, from [3] we get

$$nxP_n(x) = nP_{n-1}(x) - (1-x^2)P'_n(x). \quad [5]$$

Multiplying [4] by n and substituting for $nxP_n(x)$ from [5], we get *Beltrami's relation*

$$n(1-x^2)P'_n(x) = (n+1)[nP_{n-1}(x) - (1-x^2)P'_n(x)] - n(n+1)P_{n+1}(x)$$

or $\quad (2n+1)(x^2-1)P_n'(x) = n(n+1)[P_{n+1}(x) - P_{n-1}(x)].$

10.8 Orthogonality of $P_n(x)$

If a sequence of functions $f_1(x), f_2(x) \cdots f_n(x) \cdots$ defined on an interval $a \le x \le b$ has the property that

$$\int_a^b f_m(x) f_n(x) dx = \begin{cases} 0 & \text{if} \quad m \neq n \\ \alpha_n & \text{if} \quad m = n \end{cases} \tag{10.15}$$

then f_i s are said to be orthogonal functions on this interval.

Theorem 10.3: The Legendre polynomials in algebraic form satisfy the orthogonality relations

$$\int_{-1}^{1} P_m(x) P_n(x) dx = \begin{cases} 0 & \text{if} \quad m \neq n \\ 2/(2n+1) & \text{if} \quad m = n \end{cases} \tag{10.16}$$

Proof: We know that the solutions of $(1-x^2)u'' - 2xu' + m(m+1)u = 0$ and $(1-x^2)v'' - 2xv' + n(n+1)v = 0$ are $P_m(x)$ and $P_n(x)$ respectively. Multiplying the first equation by v and the second by u and subtracting, we get

$$(1-x^2)(u''v - v''u) - 2x(u'v - v'u) + [m(m+1) - n(n+1)]uv = 0$$

or $\quad \dfrac{d}{dx}\{(1-x^2)(u'v - uv')\} + (m-n)(m+n+1)uv = 0.$

Integrating from -1 to 1, we have

$$(m-n)(m+n+1)\int_{-1}^{1} uv\,dx = \left[(1-x^2)(uv' - u'v)\right]_{-1}^{1} = 0.$$

Thus, $\displaystyle\int_{-1}^{1} P_m(x) P_n(x) dx = 0$ if $m \neq n$. When $m = n$, we have to establish that

$$\int_{-1}^{1} P_n(x) P_n(x) dx = \frac{2}{2n+1}.$$

Using Rodrigues' formula for $P_n(x)$, we have

$$\int_{-1}^{1}[P_n(x)]^2dx = \frac{(-1)^n(2n)!}{2^{2n}(n!)^2}\int_{-1}^{1}(x^2-1)^n dx$$

$$= \frac{(-1)^n 2(2n)!}{2^{2n}(n!)^2}\int_{0}^{1}(x^2-1)^n dx.$$

Changing the variable by taking $x = \sin\theta$ and using the formula

$$\int \cos^{2n+1}\theta d\theta = \frac{1}{2n+1}\cos^{2n}\theta\sin\theta + \frac{2n}{2n+1}\int \cos^{2n-1}\theta d\theta$$

we get $$\int_{-1}^{1}[P_n(x)]^2dx = \frac{(2n)!}{2^{2n}(n!)^2}\left\{2\int_0^{\pi/2}\cos^{2n+1}\theta d\theta\right\}$$

$$= \frac{(2n)!}{2^{2n}(n!)^2}\left\{2.\frac{2n}{2n+1}.\frac{2n-2}{2n-1}\cdots\frac{2}{3}\int_0^{\pi/2}\cos\theta d\theta\right\}$$

$$= \frac{(2n)!}{2^{2n}(n!)^2}2\left\{\frac{2^n n!}{1.3\cdots(2n-1)(2n+1)}\right\}$$

$$= \frac{(2n)!\ 2\ 2^{2n}(n!)^2}{2^{2n}\ (n!)^2(2n)!\ (2n+1)} = \frac{2}{2n+1}.$$

Thus, $$\int_{-1}^{1}P_n^2(x)dx = \frac{2}{2n+1}.$$

Note: In the trigonometric form, the Legendre polynomials satisfy the orthogonality relations

$$\int_0^{\pi}P_m(\cos\theta)P_n(\cos\theta)\sin\theta d\theta = \begin{cases}0 & \text{if } m\neq n\\ 2/(2n+1) & \text{if } m = n\end{cases}$$

Exercise 10

Prove the following relationships.

10.1 (i) $x^4 = \frac{8}{35}P_4(x)+\frac{4}{7}P_2(x)+\frac{1}{5}P_0(x)$ **(ii)** $x^5 = \frac{8}{63}P_5(x)+\frac{4}{9}P_3(x)+\frac{3}{7}P_1(x)$

(iii) $1+2x-3x^2 = 2[P_1(x)-P_2(x)]$

(iv) $3x + 5x^2 - 4x^3 = \frac{5}{3}P_0(x) + \frac{3}{5}P_1(x) + \frac{10}{3}P_2(x) - \frac{8}{5}P_3(x)$

10.2 Using Rodrigues' formula prove that

(i) $P_4(x) = \frac{1}{8}(35x^4 - 30x^2 + 3)$ (ii) $P_5(x) = \frac{1}{8}(63x^5 - 70x^3 + 15x)$

(iii) $P_n(-x) = (-1)^n \, P_n(x)$ (iv) $P_n(1) = 1$

10.3 $P_n(x) = \dfrac{(-1)^n}{n!} \dfrac{d^n}{dx^n}\left[(1-x)^n\left\{1 - \dfrac{1-x}{2}\right\}^n\right]$

10.4 (i) $P_n'(x) = xP_{n-1}'(x) + nP_{n-1}(x)$ (ii) $P_n'(x) - P_{n-2}'(x) = (2n-1)P_{n-1}(x)$

(iii) $P_{n+1}'(x) + P_n'(x) = P_0(x) + 3P_1(x) + 5P_2(x) + \cdots + (2n+1)P_n(x)$

(iv) $P_n'(x) = (2n-1)P_{n-1}(x) + (2n-5)P_{n-3}(x) + (2n-9)P_{n-5}(x) + \cdots + A$

where $A = 3P_1(x)$ or $P_0(x)$ according as n is even or odd.

10.5 (i) $\displaystyle\int_{-1}^{1} P_n(x)dx = \begin{cases} 0 & \text{for} \quad n \neq 0 \\ 2 & \text{for} \quad n = 0 \end{cases}$ (ii) $\displaystyle\int_{-1}^{1} x^m P_n(x)dx = 0$ for $m < n$

(iii) $\displaystyle\int_{0}^{1} P_{2n}(x)P_{2n+1}(x)dx = \int_{0}^{1} P_{2n}(x)P_{2n-1}(x)dx$

(iv) $\displaystyle\left(n + \frac{1}{2}\right)\int_{-1}^{1} P_n^2(x)dx = \left(n - \frac{1}{2}\right)\int_{-1}^{1} P_{n-1}^2(x)dx$

(v) $\displaystyle\int_{-1}^{1} xP_n(x)P_{n-1}(x)dx = \frac{2n}{4n^2-1}$ (vi) $\displaystyle\int_{-1}^{1} [P_n'(x)]^2 dx = n(n+1)$

(vii) $\displaystyle\int_{-1}^{1} xP_n(x)P_n'(x)dx = \frac{2n}{2n-1}$

(viii) $\displaystyle\int_{-1}^{1} x^2 P_n^2(x)dx = \frac{1}{8}\left[\frac{1}{2n-1} + \frac{6}{2n+1} + \frac{1}{2n+3}\right]$

(ix) $\displaystyle\int_{-1}^{1} (x^2-1)P_{n+1}(x)P_n'(x)dx = \frac{2n(n+1)}{(2n+1)(2n+3)}$

(x) $\displaystyle\int_{-1}^{1} (1 - 2x^2 + t^2)^{-1/2} P_n(x)dx = \frac{2t^n}{2n+1}$

10.6 $P_n(x) = \dfrac{1}{\pi}\displaystyle\int_0^{\pi} [x \pm \sqrt{x^2-1}\cos\theta]^{-(n+1)}\,d\theta.$

[**Note:** This formula is called *Laplace's integral formula* for $P_n(x)$.]

10.7 $P_n(\cos\theta) = \displaystyle\sum_{r=0}^{n} (-1)^r ({}^nC_r)^2 (\cos\theta)^{2n-2}(\sin\theta)^{2r}$

10.8 If m and n are both positive integers such that $(m+n)$ is *odd*, show that

$$\int_0^{\pi} P_n(\cos\theta)\cos(m\theta)\,d\theta = 0.$$

10.9 $\displaystyle\int_0^{\pi} P_{2n}(\cos\theta)\,d\theta = \left[\frac{\Gamma\left(n+\frac{1}{2}\right)}{\Gamma(n+1)}\right]^2$

10.10 $P_n\left(\dfrac{1}{2}\right) = \left(\dfrac{1}{2^n}\right)\displaystyle\sum_{r=0}^{n} (-1)^r ({}^nC_r)^2\, 3^{n-r}.$

10.11 If $I_n = \displaystyle\int_{-1}^{1} \frac{x^n\,dx}{(1-2xt+t^2)^{1/2}},$ show that

$$t(2n+1)I_n = n(1+t^2)I_{n-1} + A$$

where $A = 2/t$ or $-2/t^2$ according as n is *even* or *odd.*

CHAPTER ELEVEN

Bessel's Differential Equation

11.1 Introduction

The linear differential equation

$$x^2\frac{d^2y}{dx^2}+x\frac{dy}{dx}+(x^2-n^2)y=0 \qquad [11.1]$$

where n is a non-negative real number, is known as Bessel's differential equation of order n. Its solutions are known as *Bessel functions.* These functions have diverse applications in physics and engineering, in connection with the propagation of waves, elasticity, fluid motion and especially in problems of potential theory and diffusion involving cylindrical symmetry. These functions are named after the German mathematician and astronomer, *Friedrich Wilhelm Bessel* (1784–1846).

11.1.1 Series solution of Bessel's differential equation

It is easy to see that zero is a regular singular point for Bessel's equation, that is, $x=0$ yields the trivial solution $y=0$. So we consider the case $x\neq 0$. Thus, [11.1] has a solution of the form

$$y=\sum_{r=0}^{\infty}a_r x^{r+p}.$$

Hence, substituting for y' and y'' in [11.1], we can see that

$$x^2y''+xy'+(x^2-n^2)y=\sum_{r=0}^{\infty}[(r+p)(r+p-1)a_r+(r+p)a_r-n^2a_r]x^{r+p}$$

$$+\sum_{r=0}^{\infty}a_r\,x^{r+p+2}$$

$$=[p(p-1)+p-n^2]a_0x^p+[(1+p)p+(1+p)-n^2]a_1x^{(p+1)}$$

$$+\sum_{r=2}^{\infty}\left\{[(r+p)^2-n^2]a_r+a_{r-2}\right\}x^{r+p}.$$

Setting the coefficients of the powers of x to zero, we obtain

$$(p^2-n^2)a_0=0 \qquad [11.2]$$

and $\quad [(p+1)^2-n^2]a_1=0$

or $\quad [(p+1)-n][(p+1)+n]a_1=0 \qquad [11.3]$

and the recursion formula

$$a_r=\frac{-a_{r-2}}{(r+p+n)(r+p-n)} \quad (\text{for } r=2,3,4\ldots). \qquad [11.4]$$

Equation [11.2] gives the equation $p^2-n^2=0$, called the *indicial equation,* and hence $p=\pm n$ are its roots. Using the largest indicial root $p=n$, [11.3] implies that $a_1=0$, and hence $a_n=0$ for all odd positive integers n. Therefore, with $p=n$ and $r=2k$, the recursion formula [11.4] becomes

$$a_{2k}=\frac{-a_{2(k-1)}}{4k(k+n)} \quad \text{for} \quad k=1,2,3\ldots$$

From this formula it follows that

$$a_2=\frac{-a_0}{4(1+n)}; \quad a_4=\frac{-a_2}{4.2(2+n)}=\frac{a_0}{4^2.2!(1+n)(2+n)}$$

and, in general

$$a_{2k}=\frac{(-1)^k a_0}{4^k k!(1+n)(2+n)\cdots(k+n)} \quad \text{for} \quad k=1,2,3\ldots$$

Conventionally the arbitrary constant a_0 is chosen so that

$$a_0 2^n\int_0^{\infty}t^n e^{-t}dt=1 \qquad [11.5]$$

and we obtain the solution

$$J_n(x)=a_0x^n\left[1+\sum_{k=1}^{\infty}\frac{(-1)^k}{4^k k!(1+n)\cdots(k+n)}x^{2k}\right].$$

The solution $J_n(x)$ is called the *Bessel function of the first kind* of order n. If $n=p$, we can see from [11.5] that

$$a_0 = \frac{1}{2^p p!} \quad \text{and hence} \quad J_p(x) = \sum_{k=0}^{\infty} \frac{(-1)^k}{k!(k+p)!} \left(\frac{x}{2}\right)^{2k+p} \quad \text{for } p = 0, 1, 2, \ldots$$

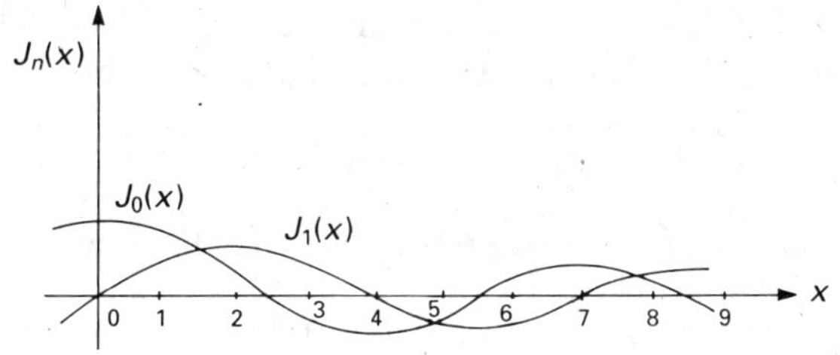

Fig. 11.1 *Bessel functions of first kind: $J_0(x)$ and $J_1(x)$.*

The most useful Bessel functions are of order 0 and 1, which are

$$J_0(x) = \sum_{k=0}^{\infty} \frac{(-1)^k}{(k!)^2} \left(\frac{x}{2}\right)^{2k} = 1 - \left(\frac{x}{2}\right)^2 + \frac{1}{2^2}\left(\frac{x}{2}\right)^4 - \frac{1}{6^2}\left(\frac{x}{2}\right)^6 + \cdots$$

$$= 1 - \frac{x^2}{2^2} + \frac{x^4}{2^2.4^2} - \frac{x^6}{2^2.4^2.6^2} + \cdots$$

and $$J_1(x) = \sum_{k=0}^{\infty} \frac{(-1)^k}{k!(k+1)!} \left(\frac{x}{2}\right)^{2k+1} = \frac{x}{2} - \frac{1}{1!2!}\left(\frac{x}{2}\right)^3 + \frac{1}{2!3!}\left(\frac{x}{2}\right)^5 - \cdots$$

Their graphs, shown in Fig.11.1, display several interesting properties such as the fact that each has a damped oscillatory behaviour producing an infinite number of positive zeroes, which occur alternately, in a manner suggesting the functions cos x and sin x. We also see that

$$J_{-n}(x) = \bar{a}_0 x^{-n} \left[1 + \sum_{k=1}^{\infty} \frac{(-1)^k}{4^k k!(1-n)\cdots(k-n)} x^{2k}\right]$$

exists for all $n > 0, n \neq 1, 2, \ldots$ and is a solution of Bessel's equation. But this definition involves the use of the gamma function, so if we consider $a_0 = 1/2^n\Gamma(n+1)$ by [11.5] we get

$$J_n(x) = \sum_{k=0}^{\infty} (-1)^k \left(\frac{x}{2}\right)^{n+2k} \frac{1}{k!\Gamma(n+k+1)} \qquad [11.6]$$

or $$J_n(x) = \sum_{k=0}^{\infty} (-1)^k \left(\frac{x}{2}\right)^{-n+2k} \frac{2}{k!\Gamma(-n+k+1)} \qquad [11.7]$$

Hence the complete solution of the Bessel's equation may be written in the form

$$y = AJ_n(x) + BJ_{-n}(x) \quad \text{(where } A \text{ and } B \text{ are arbitrary constants).}$$

Note: This is the complete solution when n is not integral or zero. When n is zero, we have $J_n(x) = J_{-n}(x)$, and the solution can be easily obtained. When n is integral, we proceed as follows. Let $y = u(x)J_n(x)$ be a solution of [11.1]. Now

$$y' = u'J_n + uJ_n' \quad \text{and} \quad y'' = u''J_n + 2u'J_n' + uJ_n''.$$

Substituting the above relationships in [11.1], we obtain

$$x^2(u''J_n + 2u'J_n' + uJ_n'') + x(u'J_n + uJ_n') + (x^2 - n^2)uJ_n = 0$$

or $\quad u\{x^2J_n'' + xJ_n' + (x^2 - n^2)J_n\} + x^2u''J_n + 2x^2u'J_n' + xu'J_n = 0.$

Since J_n is a solution of [11.1], we have $x^2J_n'' + xJ_n' + (x^2 - n^2)J_n = 0$. Thus

$$x^2u''J_n + 2x^2u'J_n' + xu'J_n = 0.$$

Dividing throughout by $x^2u'J_n$, the above equation becomes $\dfrac{u''}{u'} + 2\dfrac{J_n'}{J_n} + \dfrac{1}{x} = 0$

or $\quad \dfrac{d}{dx}(\log u') + 2\dfrac{d}{dx}(\log J_n) + \dfrac{d}{dx}(\log x) = 0$ or $\dfrac{d}{dx}\{\log(u'J_n^2x)\} = 0.$

On integration, we obtain $\log(u'J_n^2x) = \log C_2$, implying $u'J_n^2x = C_2$ or

$$u' = \frac{C_2}{xJ_n^2} \quad \text{or} \quad u = C_2\int\frac{dx}{xJ_n^2} + C_1.$$

Thus $\quad y = C_1J_n(x) + C_2J_n(x)\displaystyle\int\frac{dx}{x[J_n(x)]^2}.$

So, the complete solution of Bessel's equation [11.2] is

$$y = C_1J_n(x) + C_2Y_n(x) \quad \text{where} \quad Y_n(x) = J_n(x)\int\frac{dx/x}{[J_n(x)^2]}.$$

$Y_n(x)$ is called the *Bessel function of the second kind* of order n or *Neumann function.* When n is an integer, the function $J_n(x)$ is proportional to $J_{-n}(x)$. These two solutions are therefore not independent and the linear combination $C_1J_n(x) + C_2J_{-n}(x)$ is no longer a complete solution of Bessel's equation.

11.2 Identities for the Bessel functions J_n (x)

The Bessel functions are related by an amazing array of identities. Fundamental among these are discussed in this section.

1. $\boldsymbol{2n\, J_n(x) = x[J_{n+1}(x) + J_{n-1}(x)]}$

Proof: We have, by [11.6]

$$J_n(x) = \sum_{k=0}^{\infty} \frac{(-1)^k (x/2)^{n+2k}}{k!\Gamma(n+k+1)} = \frac{1}{2n}\left[\sum_{k=0}^{\infty}(-1)^k \frac{2n+2k-2k}{k!\Gamma(n+k+1)}\left(\frac{x}{2}\right)^{n+2k}\right]$$

$$= \frac{1}{2n}\left[\sum_{k=0}^{\infty}(-1)^k \frac{2(n+k)}{k!\Gamma(n+k+1)}\left(\frac{x}{2}\right)^{n+2k}\right.$$

$$\left. -\sum_{k=0}^{\infty} \frac{(-1)^k 2k}{k!\Gamma(n+k+1)}\left(\frac{n}{2}\right)^{n+2k}\right]$$

$$= \frac{1}{2n}\left[\sum_{k=0}^{\infty} \frac{(-1)^k x(x/2)^{n+2k-1}}{k!\Gamma(n+k)} + \sum_{k=1}^{\infty} \frac{(-1)^{k-1} x(x/2)^{n+2k-1}}{(k-1)!\Gamma(n+k+1)}\right]$$

(since $\Gamma(n+k+1) = (n+k)\Gamma(n+k)$). Thus, if we assume $p = k-1$, we get

$$2nJ_n(x) = x\left[\sum_{k=0}^{\infty} \frac{(-1)^k (x/2)^{n-1+2k}}{k!\Gamma(n-1+k+1)} + \sum_{P=0}^{\infty} \frac{(-1)^p (x/2)^{n+1+2p}}{p!\Gamma(n+1+p+1)}\right]$$

$$= x[J_{n-1}(x) + J_{n+1}(x)].$$

2. $\boldsymbol{\frac{d}{dx}[x^n J_n(x)] = x^n J_{n-1}(x)}$

Proof: We know

$$J_n(x) = \sum_{k=0}^{\infty} \frac{(-1)^k (x/2)^{2k+n}}{k!(k+n)!}$$

Multiplying both sides of the above equation by x^n and differentiating, we get

$$\frac{d}{dx}[x^n J_n(x)] = \frac{d}{dx}\sum_{k=0}^{\infty} \frac{(-1)^k x^{2n+2k}}{k!(k+n)!2^{2k+n}} = \sum_{k=0}^{\infty} \frac{(-1)^k x^{2n+2k-1}}{2^{2k+n-1}k!(k+n-1)!}$$

$$= x^n \sum_{k=0}^{\infty} \frac{(-1)^k (x/2)^{2k+n-1}}{k!(k+n-1)!} = x^n J_{n-1}(x).$$

Similarly, we can show that $\frac{d}{dx}[x^{-n}J_n(x)] = -x^{-n}J_{n+1}(x)$. The proof for this relationship is left as an exercise to students.

3. $\boldsymbol{J'_n(x) = \frac{1}{2}[J_{n-1}(x) - J_{n+1}(x)]}$

Proof: Starting again from $J_n(x) = \sum_{k=0}^{\infty} \frac{(-1)^k (x/2)^{n+2k}}{k!\Gamma(n+k+1)}$, we have

$$J'_n(x) = \sum_{k=0}^{\infty} \frac{(-1)^k (n+2k)(x/2)^{n+2k-1}}{2k!\Gamma(n+k+1)}$$

$$= \frac{1}{2}\sum_{k=0}^{\infty} \frac{(-1)^k (n+k)(x/2)^{n+2k-1}}{k!\Gamma(n+k+1)} + \sum_{k=0}^{\infty} \frac{(-1)^k k(x/2)^{n+2k-1}}{k!\Gamma(n+k+1)}$$

$$= \frac{1}{2}\sum_{k=0}^{\infty} \frac{(-1)^k (x/2)^{n+2k-1}}{k!\Gamma(n+k)} - \frac{1}{2}\sum_{k=1}^{\infty} \frac{(-1)^{k-1} (x/2)^{n+2k-1}}{(k-1)!\Gamma(n+k+1)}$$

$$= \frac{1}{2}\sum_{k=0}^{\infty} \frac{(-1)^k (x/2)^{n-1+2k}}{k!\Gamma(n-1+k+1)} - \frac{1}{2}\sum_{p=0}^{\infty} \frac{(-1)^p (x/2)^{n+1+2p}}{p!\Gamma(n+2+p)}$$

where $p = k - 1$. Thus $J'_n(x) = \frac{1}{2}[J_{n-1}(x) - J_{n+1}(x)]$.

11.3 Generating function for $J_n(x)$

The Bessel functions $J_n(x)$ of integral order are linked together by this relation:

$$e^{(x/2)(t-1/t)} = J_0(x) + \sum_{n=1}^{\infty} J_n(x)\left[t^n + (-1)^n t^{-n}\right].$$

Since $J_{-n}(x) = (-1)^n J_n(x)$, we have

$$e^{(x/2)(t-1/t)} = \sum_{n=-\infty}^{\infty} J_n(x) t^n.$$

This can be proved by multiplying the two series

$$e^{(x/2)t} = \sum_{p=0}^{\infty} \frac{1}{p!}\frac{x^p}{2^p}t^p \quad \text{and} \quad e^{-(x/2)t^{-1}} = \sum_{k=0}^{\infty} \frac{(-1)^k}{k!}\frac{x^k}{2^k}t^{-k}.$$

The result is a double series containing all possible products of a term from the first series and a term from the second. For each $n \geqslant 0$, we obtain a term of the double series containing t^n, where $p = n + k$ so that the coefficient of t^n is

$$\sum_{k=0}^{\infty} \frac{1}{(n+k)!}\frac{x^{n+k}}{2^{n+k}}\frac{(-1)^k}{k!}\frac{x^k}{2^k} = \sum_{k=0}^{\infty} \frac{(-1)^k (x/2)^{n+2k}}{k!(n+k)!} = J_n(x). \qquad [11.8]$$

Similarly a term containing t^{-n} arises when $k = n + p$, so the coefficient of t^{-n} is

$$\sum_{p=0}^{\infty} \frac{1}{p!}\left(\frac{x}{2}\right)^p \frac{(-1)^{n+p}}{(n+p)!}\frac{x^{n+p}}{2^{n+p}} = (-1)^n \sum_{p=0}^{\infty} \frac{(-1)^p (x/2)^{n+2p}}{p!(n+p)!}$$

$$= (-1)^n J_n(x). \qquad [11.9]$$

Thus, from [11.8] and [11.9]

$$e^{(x/2)(t-1/t)} = J_0(x) + \sum_{n=1}^{\infty} J_n(x)\left[t^n + (-1)^n t^{-n}\right].$$

The term $e^{(x/2)(t-1/t)}$ is known as the generating function.

11.3.1 Continued fraction for tan x

A topic of great historical interest is *Lambert's continued fraction* for tan x, which led to the first proof of the fact that π is not a rational number. We know

$$\left(\frac{d}{dx}\right)[x^{-n}J_n(x)] = -x^{-n}J_{n+1}(x).$$

Dividing by x^{-n}, this becomes

$$J_n'(x) - \frac{n}{x}J_n(x) = -J_{n+1}(x). \qquad [11.10]$$

Similarly from $\frac{d}{dx}[x^n J_n(x)] = x^n J_{n-1}(x)$, we can obtain

$$J_n'(x) + \frac{n}{x}J_n(x) = J_{n-1}(x) \qquad [11.11]$$

by differentiating and dividing by x^n. When [11.10] is subtracted from [11.11], we get

$$\frac{2n}{x}J_n(x) = J_{n-1}(x) + J_{n+1}(x) \quad \text{or} \quad J_{n-1}(x) = \frac{2n}{x}J_n(x) - J_{n+1}(x).$$

Dividing this by $J_n(x)$, we get $\dfrac{J_{n-1}(x)}{J_n(x)} = \dfrac{2n}{x} - \dfrac{1}{J_n(x)/J_{n+1}(x)}$.

When this formula is applied to the second denominator on the right with n replaced by $n+1$, and this process continued indefinitely, we get

$$\frac{J_{n-1}(x)}{J_n(x)} = \frac{2n}{x} - \cfrac{1}{\cfrac{2n+2}{n} - \cfrac{1}{\cfrac{2n+4}{n} - \dots}}$$

which is an infinite continued fraction. Let us now have a look at the above expression when $n = 1/2$. We have

$$\frac{J_{-1/2}(x)}{J_{1/2}(x)} = \cfrac{1}{\cfrac{1}{x} - \cfrac{1}{\cfrac{3}{x} - \cfrac{1}{\cfrac{5}{x} - \dots}}} \qquad [11.12]$$

Substituting for n in [11.6] and [11.7], we can easily show that

$$J_{1/2}(x) = \sqrt{\frac{2}{\pi x}}\sin x \quad \text{and} \quad J_{-1/2}(x) = \sqrt{\frac{2}{\pi x}}\cos x.$$

Since $\dfrac{J_{-1/2}(x)}{J_{1/2}(x)} = \cot x$, [11.12] becomes $\tan x = \cfrac{1}{\cfrac{1}{x} - \cfrac{1}{\cfrac{3}{x} - \cfrac{1}{\cfrac{5}{x} - \dots}}}$

11.4 Orthogonality of Bessel functions

We can show that the functions $\sqrt{x}J_n(\lambda x)$ are orthogonal on the interval $0 \le x \le 1$ in the theorem stated below.

Theorem 11.2: If λ and μ are zeroes of $J_n(x)$, i.e. $J_n(\lambda) = J_n(\mu) = 0$, then

$$\int_0^1 xJ_n(\lambda x)J_n(\mu x)dx = \begin{cases} 0 & \text{if } \lambda \ne \mu \\ \frac{1}{2}[J_n'(\lambda)]^2 & \text{if } \lambda = \mu \end{cases}$$

Proof: We know $y = J_n(x)$ is a solution of

$$y'' + \frac{1}{x}y' + \left(1 - \frac{n^2}{x^2}\right)y = 0.$$

If λ and μ are distinct positive constants, it follows that equations $u(x) = J_n(\lambda x)$ and $v(x) = J_n(\mu x)$ satisfy

$$u'' + (1/xu') + (\lambda^2 - (n^2/x^2))u = 0 \quad \text{and} \quad v'' + (1/xv') + (\mu^2 - (n^2/x^2))v = 0.$$

Multiplying these equations by u and v and subtracting, we obtain

$$\frac{d}{dx}(u'v - v'u) + \frac{1}{x}(u'v - v'u) = (\lambda^2 - \mu^2)uv.$$

Multiplying by x, this becomes $\frac{d}{dx}[x(u'v - v'u)] = (\lambda^2 - \mu^2)xuv.$

Integrating with limits 0 to 1, we get $x\left[u'v - v'u\right]_0^1 = (\lambda^2 - \mu^2)\int_0^1 xuvdx.$

If λ and μ are distinct positive zeroes of $J_n(x)$, the right hand side of the above expression vanishes. Since $u = J_n(\lambda x)$, we have

$$u' = \frac{d}{dx}[J_n(\lambda x)] = \frac{d}{d(\lambda x)}[J_n(\lambda x)].\frac{d(\lambda x)}{dx} = \lambda J_n'(\lambda x).$$

Similarly $v = J_n(\mu x)$ and $v' = \mu J_n'(\mu x)$. Thus

$$\int_0^1 xJ_n(\lambda x)J_n(\mu x)dx = \frac{\lambda J_n'(\lambda)J_n(\mu) - \mu J_n'(\mu)J_n(\lambda)}{\lambda^2 - \mu^2}$$

and $J_n(\lambda) = J_n(\mu) = 0$ or $\int_0^1 xJ_n(\lambda x)J_n(\mu x)dx = 0$ if $\lambda \neq \mu$.

If $\lambda = \mu$, we prove the result as follows. We know

$$u'' + \frac{1}{x}u' + \left(\lambda^2 - \frac{n^2}{x^2}\right)u = 0.$$

Multiplying by $2x^2u'$, it becomes $2x^2u'u'' + 2xu'^2 + 2\lambda^2x^{2u}u' - 2n^2uu' = 0$

or $\frac{d}{dx}(x^2u'^2) + \frac{d}{dx}(\lambda^2x^2u^2) - 2\lambda^2xu^2 - \frac{d}{dx}(n^2u^2) = 0$

or $2\lambda^2\int_0^1 xu^2dx = \left[x^2u'^2 + (\lambda^2x^2 - n^2)u^2\right]_0^1$ (integrating from 0 to 1)

When $x = 1$, since $u'(1) = \lambda J_n'(\lambda)$, we get

$$\int_0^1 xJ_n(\lambda x)^2 dx = \frac{1}{2}\left[J_n'(\lambda)\right]^2 + \frac{1}{2}\left[1 - \frac{n^2}{\lambda^2}\right][J_n(\lambda)]^2.$$

On using the relation $J_n'(x) - \frac{n}{x}J_n(x) = -J_{n+1}(x),$ we get

$$\int_0^1 x[J_n(\lambda x)]^2 dx = 1/2[J_n'(\lambda)]^2.$$

Note: The above result is also known as the *Lommel integral formula.*

11.5 Bessel's integral formula

We shall prove that $J_n(x) = \frac{1}{\pi}\int_0^\pi \cos(n\theta - x\sin\theta)d\theta, (n = 0, 1, 2, \ldots).$

We know that $e^{(x/2)(z-1/z)} = \sum_{n=-\infty}^{\infty} J_n(x)z^n.$

Substitute $z = e^{i\theta} = \cos\theta + i\sin\theta.$ Then $(1/z) = e^{-i\theta} = \cos\theta - i\sin\theta.$ Also

$$z + \frac{1}{z} = 2\cos\theta \quad \text{and} \quad z - \frac{1}{z} = 2i\sin\theta.$$

Also $z^n = \cos n\theta + i\sin n\theta$ and $z^{-n} = \cos n\theta - i\sin n\theta.$ So

$$z^n + \frac{1}{z^n} = 2\cos n\theta \quad \text{and} \quad z^n - \frac{1}{z^n} = 2i\sin n\theta.$$

$$\therefore \frac{1}{2}\left(z - \frac{1}{z}\right) = i\sin\theta \quad \text{and} \quad e^{(x/2)(z-1/z)} = e^{xi\sin\theta} = \cos(x\sin\theta) + i\sin(x\sin\theta).$$

When n is even, say $n = 2k$,

$$z^n + (-1)^n z^{-n} = z^{2k} + (-1)^{2k}z^{-2k} = e^{i2k\theta} + e^{-i2k\theta} = 2\cos 2k\theta.$$

When n is odd, say $n = 2k - 1$,

$$z^n + (-1)^n z^{-n} = z^{2k-1} + (-1)^{2k-1}z^{-2k+1}$$

$$= e^{i(2k-1)\theta} - e^{-i(2k-1)\theta} = 2i\sin(2k-1)\theta.$$

Also $e^{(x/2)[z-1/z]} = J_0(x) + \sum_{n=1}^{\infty} J_n(x)\left[z^n + (-1)^n z^{-n}\right].$

So $e^{xi\sin\theta} = \cos(x\sin\theta) + i.\sin(x\sin\theta)$

$$= J_0 + 2\sum_{k=1}^{\infty} J_{2k}(x)\cos 2k\theta + 2i\sum_{k=1}^{\infty} J_{2k-1}(x)\sin(2k-1)\theta.$$

Equating the real and imaginary parts in the last expression, we obtain the identities

$$\cos(x\sin\theta) = J_0(x) + 2\sum_{k=1}^{\infty} J_{2k}(x)\cos 2k\theta$$

and $\sin(x\sin\theta) = 2\sum_{k=1}^{\infty} J_{2k-1}(x)\sin(2k-1)\theta.$

We multiply both sides of the above equations by $\cos n\theta$ and $\sin n\theta$ respectively, and integrate each identity with respect to θ from 0 to π. We know

$$\int_0^{\pi} \cos m\theta\cos n\theta d\theta = \int_0^{\pi} \sin m\theta\sin n\theta d\theta = 0 \ (m \neq n)$$

and $\int_0^{\pi} \cos^2 n\theta d\theta = \int_0^{\pi} \sin^2 n\theta d\theta = \frac{\pi}{2}$

$$\therefore \quad \int_0^{\pi} \cos n\theta\cos(x\sin\theta)d\theta = \begin{cases} \pi J_n(x) & \text{if } n \text{ is even} \\ 0 & \text{if } n \text{ is odd} \end{cases}$$

and $\int_0^{\pi} \sin n\theta\sin(x\sin\theta)d\theta = \begin{cases} 0 & \text{if } n \text{ is even} \\ \pi J_n(x) & \text{if } n \text{ is odd} \end{cases}$

If we add these two expressions and divide by π, we have for all integral values of n

$$J_n(x) = \frac{1}{\pi}\int_0^{\pi} [\cos n\theta\cos(x\sin\theta) + \sin n\theta\sin(x\sin\theta)]d\theta$$

since for every value of n, one of the integrals vanishes while the remaining one contributes to $J_n(x)$. Thus

$$J_n(x) = 1/\pi\int_0^{\pi} \cos(n\theta - x\sin\theta)d\theta \quad \text{(whenever } n \text{ is an integer).}$$

Example 11.1 Verify that $J_0(x) = \frac{1}{2\pi}\int_0^{2\pi} e^{ix\sin\theta}d\theta.$

Solution 11.1 We can write

$$\frac{1}{2\pi}\int_0^{2\pi} e^{ix\sin\theta}d\theta = \frac{1}{2\pi}\int_0^{\pi} e^{ix\sin\theta}d\theta + \frac{1}{2\pi}\int_{\pi}^{2\pi} e^{ix\sin\theta}d\theta. \qquad [1]$$

In the second integral on the right hand side, taking $\theta = 2\pi - \phi$, and making the appropriate change of limits, we get

$$\frac{1}{2\pi}\int_{\pi}^{0} e^{ix\sin(2\pi-\phi)}(-d\phi) = \frac{1}{2\pi}\int_0^{\pi} e^{-ix\sin\phi}d\phi.$$

So the right hand side of [1] becomes

$$\frac{1}{2\pi}\int_0^{\pi}(e^{ix\sin\theta} + e^{-ix\sin\theta})d\theta = \frac{1}{\pi}\int_0^{\pi}\cos(x\sin\theta)d\theta = J_0(x).$$

Example 11.2 Show that (i) $J_{-1/2}(x) = \sqrt{2/\pi x}\ \cos x$

(ii) $J_{1/2}(x) = \sqrt{2/\pi x}\sin x$ (iii) $\left[J_{1/2}(x)\right]^2 + \left[J_{-1/2}(x)\right]^2 = 2/\pi x$

Solution 11.2 (i) We know

$$J_n(x) = \frac{x^n}{2^n\Gamma(n+1)}\left[1 - \frac{x^2}{2(2n+2)} + \frac{x^4}{2\cdot 4(2n+2)(2n+4)} + \cdots\right]$$

Substituting $n = -1/2$ in $J_n(x)$, we have

$$J_{-1/2}(x) = \frac{x^{-1/2}}{2^{-1/2}\Gamma(1/2)}\left[1 - \frac{x^2}{2(-1+2)} + \frac{x^4}{2.4(-1+2)(-1+4)} - \cdots\right]$$

$$= \sqrt{\frac{2}{\pi x}}\left[1 - \frac{x^2}{2!} + \frac{x^4}{4!} - \cdots\right] = \sqrt{\frac{2}{\pi x}}\cos x.$$

(ii) Substituting $n = 1/2$ in $J_n(x)$, we have

$$J_{1/2}(x) = \frac{x^{1/2}}{2^{1/2}\Gamma(3/2)}\left[1 - \frac{x^2}{2(1+2)} + \frac{x^4}{2.4(1+2)(1+4)} - \cdots\right]$$

$$= \sqrt{\frac{x}{2}}\frac{1}{(1/2)\sqrt{\pi}}\cdot\frac{1}{x}\left[x - \frac{x^3}{3!} + \frac{x^5}{5!} - \cdots\right] = \sqrt{\frac{2}{\pi x}}\sin x.$$

(iii) We square and add (i) and (ii) to obtain (iii).

Example 11.3 Prove that $\dfrac{d}{dx}\left[J_n^2 + J_{n+1}^2\right] = 2\left(\dfrac{n}{x}J_n^2 - \dfrac{n+1}{x}J_{n+1}^2\right).$

Solution 11.3 We know

$$\frac{d}{dx}\left(J_n^2 + J_{n+1}^2\right) = 2J_n J_n' + 2J_{n+1}\, J_{n+1}'. \qquad [1]$$

From $\quad J_n(x) = \displaystyle\sum_{k=0}^{\infty}(-1)^k\left(\frac{x}{2}\right)^{n+2k}\frac{1}{k!\Gamma(n+k+1)}$, differentiating w.r.t. x, we have

$$J_n'(x) = \sum_{k=0}^{\infty}(-1)^k\frac{(n+2k)}{k!\Gamma(n+k+1)}\left(\frac{x}{2}\right)^{n+2k-1}\left(\frac{1}{2}\right)$$

Therefore $\quad xJ_n'(x) = \displaystyle\sum_{k=0}^{\infty}(-1)^k\frac{(n+2k)}{k!\Gamma(n+k+1)}\left(\frac{x}{2}\right)^{n+2k}\left(\frac{1}{2}\right)$

$$= \sum_{k=0}^{\infty}(-1)^k\frac{n}{k!\Gamma(n+k+1)}\left(\frac{x}{2}\right)^{n+2k}\left(\frac{1}{2}\right)$$

$$+\sum_{k=0}^{\infty}(-1)^k\frac{2k}{k!\Gamma(n+k+1)}\frac{1}{2}\cdot\left(\frac{x}{2}\right)^{n+2k}$$

$$= nJ_n(x) - x\sum_{k=0}^{\infty}\frac{(-1)^{k-1}}{(k-1)!\Gamma[(n+1)+(k+1)-1]}\left(\frac{x}{2}\right)^{n-1+2k}$$

$$= nJ_n(x) - xJ_{n+1}(x).$$

Dividing throughout by x, we get

$$J_n'(x) = \frac{n}{x}J_n(x) - J_{n+1}(x). \qquad [2]$$

Also $\quad \left(\dfrac{d}{dx}\right)(x^{-n}J_n) = -x^{-n}J_{n+1}$ which gives

$$xJ_n'(x) = -nJ_n(x) + xJ_{n-1}(x) \quad \text{or} \quad J_n'(x) - \frac{-n}{x}J_n(x) + J_{n-1}(x). \qquad [3]$$

Replacing n by $n+1$, we get

$$J'_{n+1}(x) = -\frac{(n+1)}{x} J_{n+1}(x) + J_n(x). \qquad [4]$$

Replacing $J'_n(x)$ and $J'_{n+1}(x)$ in [1] from [2] and [4], we have

$$\frac{d}{dx}(J_n^2 + J_{n+1}^2) = 2J_n \left[\frac{n}{x} J_n - J_{n+1}\right] + 2J_{n+1}\left[-\frac{(n+1)}{x} J_{n+1} + J_n\right]$$

$$= \frac{2n}{x} J_n^2 - \frac{2(n+1)}{x} J_{n+1}^2.$$

Hence $\dfrac{d}{dx}(J_n^2 + J_{n+1}^2) = 2\left[\dfrac{n}{x} J_n^2 - \dfrac{(n+1)}{x} J_{n+1}^2\right].$

Example 11.4 Prove that $\sqrt{\dfrac{\pi x}{2}} J_{3/2}(x) = \dfrac{1}{x}\sin x - \cos x.$

Solution 11.4 We know that

$$J_n(x) = \frac{x^n}{2^n \Gamma(n+1)} \left[1 - \frac{x^2}{2(2n+2)} + \frac{x^4}{2.4(2n+2)(2n+4)} - \cdots\right].$$

Taking $n = 3/2$, we have

$$J_{3/2}(x) = \frac{x^{3/2}}{2^{3/2}\Gamma(5/2)} \left[1 - \frac{x^2}{2.5} + \frac{x^4}{2.4.5.7} - \frac{x^6}{2.4.5.6.7.9} + \cdots\right]$$

$$= \frac{x\sqrt{x}}{2\sqrt{2}.(3/2)(1/2)\sqrt{\pi}} \left[1 - \frac{x^2}{2.5} + \frac{x^4}{2.4.5.7} - \frac{x^6}{2.4.5.6.7.9} + \cdots\right] \left[\text{since } \Gamma\left(\frac{5}{2}\right) = \frac{3}{2}\frac{1}{2}\sqrt{\pi}\right]$$

$$\therefore \quad \sqrt{\frac{\pi x}{2}} J_{3/2}(x) = \frac{1}{3}\left[x^2 - \frac{x^4}{2.5} + \frac{x^6}{2.4.5.7} - \frac{x^8}{2.4.5.6.7.9} + \cdots\right]$$

$$= \frac{2x^2}{3!} - \frac{4x^4}{5!} + \frac{6x^6}{7!} - \frac{8x^8}{9!} + \cdots$$

$$= \left(\frac{1}{2!} - \frac{1}{3!}\right)x^2 - \left(\frac{1}{4!} - \frac{1}{5!}\right)x^4 + \left(\frac{1}{6!} - \frac{1}{7!}\right)x^6 - \left(\frac{1}{8!} - \frac{1}{9!}\right)x^6 + \cdots$$

$$= \left(\frac{x^2}{2!} - \frac{x^4}{4!} + \frac{x^6}{6!} - \cdots\right) + \left(-\frac{x^2}{3!} + \frac{x^4}{5!} - \frac{x^6}{7!} + \cdots\right)$$

$$= -\left(1 - \frac{x^2}{2!} + \frac{x^4}{4!} - \frac{x^6}{6!} + \cdots\right) + \frac{1}{x}\left(x - \frac{x^3}{3!} + \frac{x^5}{5!} - \frac{x^7}{7!} + \cdots\right)$$

$$= -\cos x + \frac{1}{x}\sin x.$$

Hence $\sqrt{\frac{\pi x}{2}} J_{3/2}(x) = \frac{1}{x}\sin x - \cos x.$

Example 11.5 Show that

$$\cos x = J_0 - 2J_2 + 2J_4 - \cdots$$

and $\sin x = 2J_1 - 2J_3 + 2J_5 - \cdots$

Solution 11.5 We know that

$$e^{x(z-(1/z))/2} = J_0 + \left(z - \frac{1}{z}\right)J_1 + \left(z^2 + \frac{1}{z^2}\right)J_2 + \left(z^3 - \frac{1}{z^3}\right)J_3 + \cdots$$

Taking $z = e^{i\theta}$, we have

$$e^{x(e^{i\theta}-e^{-i\theta}/2)} = J_0 + \left(e^{i\theta} - e^{-i\theta}\right)J_1 + \left(e^{2i\theta} + e^{-2i\theta}\right)J_2 + \cdots$$

or $e^{xi\sin\theta} = J_0 + 2i\sin\theta J_1 + (2\cos 2\theta)J_2 + (2i\sin 3\theta)J_3 + \cdots$

or $\cos(x\sin\theta) + i\sin(x\sin\theta) = J_0 + 2\cos 2\theta J_2 + (2\cos 4\theta J_4 + \cdots)$

$$+ i(2\sin\theta J_1 + 2\sin 3\theta J_3 + \cdots).$$

Equating the real and imaginary parts, we have

$$\cos(x\sin\theta) = J_0 + 2\cos 2\theta J_2 + 2\cos 4\theta J_4 + \cdots \quad [1]$$

$$\sin(x\sin\theta) = 2\sin\theta J_1 + 2\sin 3\theta J_3 + 2\sin 5\theta J_5 + \cdots \quad [2]$$

Taking $\theta = \pi/2$, we have

$$\cos x = J_0 - 2J_2 + 2J_4 - \cdots \quad \text{and} \quad \sin x = 2J_1 - 2J_3 + 2J_5 - \cdots$$

Example 11.6 Show that $J_0(x) = \frac{1}{\pi}\int_0^{\pi} \cos(x\sin\theta)\,d\theta = \frac{1}{\pi}\int_0^{\pi} \cos(x\cos\theta)\,d\theta.$

Solution 11.6 We have from [1] of Example 11.5,

$$\cos(x\sin\theta) = J_0 + 2J_2 \cos 2\theta + 2J_4 \cos 4\theta + \cdots \qquad [1]$$

Integrating [1] between the limits 0 and π, and using the result $\int_0^{\pi} \cos n\theta d\theta = 0$, when n is an even integer, we get the relation

$$\int_0^{\pi} \cos(x\sin\theta)d\theta = J_0(x)\int_0^{\pi} d\theta + 0 + 0 + \cdots = J_0(x)(\pi)$$

$$\text{or} \quad J_0(x) = \frac{1}{\pi}\int_0^{\pi} \cos(x\sin\theta)d\theta. \qquad [2]$$

Again, replacing θ by $\left(\frac{\pi}{2} - \theta\right)$ in [1], we get

$$\cos(x\cos\theta) = J_0 - 2J_2 \cos 2\theta + 2J_4 \cos 4\theta + \cdots \qquad [3]$$

Integrating [3] with respect to θ between the limits 0 and π, we get

$$\int_0^{\pi} \cos(x\cos\theta)d\theta = J_0(x)\int_0^{\pi} d\theta + 0 + 0 + \cdots = J_0(x)(\pi)$$

$$\text{or} \quad J_0(x) = \frac{1}{\pi}\int_0^{\pi} \cos(x\cos\theta)d\theta \cdots \qquad [4]$$

The required results are [2] and [4].

Example 11.7 Prove that $J_n(x) = 0$ has no repeated roots except at $x = 0$.

Solution 11.7 Let α be a repeated root of $J_n(x) = 0$. So $J_n(\alpha) = 0$ and $J_n'(\alpha) = 0$. From [2] and [3] of Example 11.3, we know

$$J_{n+1}(x) = \frac{n}{x}J_n(x) - J_n'(x) \quad \text{and} \quad J_{n-1}(x) = \frac{n}{x}J_n(x) + J_n'(x).$$

So, $J_n(x)$,$J_{n+1}(x)$ and $J_{n-1}(x)$ are all zero. This is absurd, since we cannot have two power series having the same sum function. Thus $J_n(x) = 0$ cannot have repeated roots except at $x = 0$.

Example 11.8 Prove that $\displaystyle\int_0^{\infty} e^{-ax} J_0(bx)dx = \frac{1}{\sqrt{a^2+b^2}}$, $a > 0$.

Solution 11.8 We know from [1] in Example 11.5 that

$$\cos(x\sin\theta) = J_0 + 2J_2 \cos 2\theta + \cdots \qquad [1]$$

taking $\theta = \phi$ in [1], we get $\cos(x\sin\phi) = J_0 + 2J_2 \cos 2\phi + 2J_4 \cos 4\phi - \cdots$. Integrating all the terms from 0 to π, we get

$$\int_0^\pi \cos(x\sin\phi)d\phi = J_0 \int_0^\pi d\phi + 2J_2 \int_0^\pi \cos 2\phi \; d\phi + \cdots = \pi J_0$$

or $$J_0 = \frac{1}{\pi}\int_0^\pi \cos(x\sin\phi)d\phi.$$

Therefore $$\int_0^\infty e^{-ax} J_0(bx)dx = \int_0^\infty e^{-ax}\left\{\frac{1}{x}\int_0^\pi \cos(bx\ \sin\phi)d\phi\right\}dx$$

$$= \frac{1}{\pi}\int_0^\pi \left[\int_0^\infty e^{-ax}\cos(bx\ \sin\phi)dx\right]d\phi$$

$$= \frac{1}{\pi}\int_0^\pi \left[\int_0^\infty e^{ax}.\frac{e^{i(bx\sin\phi)} + e^{-i(bx\ \sin\phi)}}{2}dx\right]d\phi$$

$$= \frac{1}{2\pi}\int_0^\pi \left[\int_0^\infty \left\{e^{-(a-ib\ \sin\phi)x} + e^{-(a+ib\ \sin\phi)x}\right\}dx\right]d\phi$$

$$= \frac{1}{2\pi}\int_0^\pi \left[\frac{e^{-(a-ib\ \sin\phi)x}}{-(a-ib\ \sin\phi)} - \frac{e^{-(a+ib\ \sin\phi)x}}{(a+ib\ \sin\phi)}\right]_0^\infty d\phi$$

$$= \frac{1}{2\pi}\int_0^\pi \left(\frac{1}{a-ib\ \sin\phi} + \frac{1}{(a+ib\sin\phi)}\right)d\phi = \frac{1}{2}\pi\int_0^\pi \frac{2ad\phi}{a^2+b^2\ \sin^2\phi}$$

$$= 2.\frac{a}{\pi}\int_0^{\pi/2} \frac{\text{cosec}^2\phi d\phi}{b^2 + a^2\text{cosec}^2\phi} = 2.\frac{a}{\pi}\int_0^{\pi/2}\frac{\text{cosec}^2\phi d\phi}{(a^2+b^2)+a^2\cot^2\phi}$$

$$= 2\frac{a}{\pi}\left(\frac{1}{a\sqrt{a^2+b^2}}\cot^{-1}\frac{(a\cot\phi)}{\sqrt{a^2+b^2}}\right)_0^{\pi/2}$$

$$= \frac{2}{\pi\sqrt{a^2+b^2}}(\cot^{-1}0 - \cot^{-1}\infty) = \frac{1}{\sqrt{a^2+b^2}}.$$

Exercise 11

Prove the following relationships.

11.1 (i) $J_{-3/2}(x) = -\sqrt{\dfrac{2}{\pi x}}[\sin x + \dfrac{1}{x}\cos x]$

(ii) $J_{5/2}(x) = \left[\dfrac{2}{\pi x}\right]^{1/2}\left[\dfrac{3-x^2}{x^2}\sin x - \dfrac{3}{x}\cos x\right]$

11.2 (i) $\displaystyle\int_0^x x^n J_{n-1}(x)dx = x^n J_n(x)$

(ii) $\displaystyle\int_0^x x^{n+1} J_n(x)dx = x^{n+1} J_{n+1}(x)$ (iii) $\displaystyle\int_0^\infty \frac{J_n(x)}{x}dx = \frac{1}{n}$

(iv) $\displaystyle\frac{2}{\pi}\int_0^{\pi/2} \cos(x\,\sin\theta)d\theta = J_0(x)$ (v) $\displaystyle\int_0^\infty J_0(x)dx = 1$

11.3 (i) $\displaystyle 4\int J_{n+1}(x)dx = \int J_{n-1}(x)dx - 2J_n(x)$

(ii) $\displaystyle 4\frac{d^2}{dx^2}[J_n(x)] = J_{n-2}(x) - 2J_n(x) + J_{n+2}(x)$

11.4 $\displaystyle\int_0^\infty e^{-x} J_0(x)dx = \frac{1}{\sqrt{2}}$

11.5 $\displaystyle J_n(x) = \frac{2(x/2)^{n-m}}{\Gamma(n-m)}\int_0^1 (1-t)^{n-m-1} J_n t^{m+1}(xt)dt$

11.6 Show by the use of recurrence formulae that (i) $J_0'' = (1/2)(J_2 - J_0)$

(ii) $J_2/J_1 = 1/x - J_0''/J_0'$ (iii) $J_0'''(x) + 3J_0'(x) + J_3(x) = 0$

11.7 Establish the *Jacobi series*

$\cos(x\cos\theta) = J_0 - 2J_2\cos 2\theta + 2J_4\cos 4\theta - \cdots$

$\sin(x\cos\theta) = 2[J_1\cos\theta - J_3\cos 3\theta + J_5\cos 5\theta - \cdots]$

Hence show that (i) $\sin x = 2[J_1 - J_3 + J_5 - \cdots]$

(ii) $\{J_0(x)\}^2 + 2\{J_1(x)\}^2 + 2\{J_2(x)\}^2 + 2\{J_3(x)\}^2 + \cdots = 1$

11.8 $\displaystyle\cos x = J_0(x) + 2\sum_{n=1}^{\infty}(-1)^n J_{2n}(x)$

11.9 $\displaystyle\lim_{x\to 0}\frac{J_n(x)}{x^n} = \frac{1}{2^n\Gamma(n+1)}$

11.10 $\displaystyle\int_0^x \{tJ_n(t)\}^2 dt = \frac{x^2}{2}\{J_n^2(x) - J_{n-1}(x)J_{n+1}(x)\}$

CHAPTER TWELVE

Total Differential Equations

12.1 Introduction

An ordinary differential equation of the form

$$Pdx + Qdy + Rdz = 0 \qquad [12.1]$$

or $$P + Q\frac{dy}{dx} + R\frac{dz}{dx} = 0 \qquad [12.2]$$

where P, Q and R are functions of x, y and z respectively is called a *total differential equation*. We shall now proceed to find the condition of *integrability* of [12.1].

12.2 Condition for integrability of *Pdx* + *Qdy* + *Rdz* = 0

Theorem 12.1: A necessary and sufficient condition that the equation [12.1] viz. $Pdx + Qdy + Rdz = 0$ may be integrable is

$$P\left(\frac{\partial Q}{\partial z} - \frac{\partial R}{\partial y}\right) + Q\left(\frac{\partial R}{\partial x} - \frac{\partial P}{\partial z}\right) + R\left(\frac{\partial P}{\partial y} - \frac{\partial Q}{\partial x}\right) = 0 \qquad [12.3]$$

Proof: *(i) Necessary condition*: Consider a function $u(x, y, z)$ whose total differential du is equal to $Pdx+Qdy+Rdz$ or its multiples (by factors that are functions of x, y, z). Then $u(x, y, z) = c$ is an integral of [12.1]. But we know that $du = (\partial u/\partial x)\ dx + (\partial u/\partial y)\ dy + (\partial u/\partial z)\ dz$. So, $\partial u/\partial x, \partial u/\partial y$ and $\partial u/\partial z$ must be proportional to P, Q and R, which means

$$\frac{\partial u/\partial x}{P} = \frac{\partial u/\partial y}{Q} = \frac{\partial u/\partial z}{R} = \lambda \quad \text{(where } \lambda \text{ is a function of } (x, y, z)).$$

Then, we have $\partial u/\partial x = \lambda P, \partial u/\partial y = \lambda Q$ and $\partial u/\partial z = \lambda R$.

We can easily see that, the first two relations above give

$$\frac{\partial}{\partial y}(\lambda P) = \frac{\partial}{\partial y}\left(\frac{\partial u}{\partial x}\right) = \frac{\partial^2 u}{\partial y \partial x} = \frac{\partial^2 u}{\partial x \partial y} = \frac{\partial}{\partial x}\left(\frac{\partial u}{\partial y}\right) = \frac{\partial}{\partial x}(\lambda Q)$$

or $$\frac{\lambda \partial P}{\partial y} + \frac{P \partial \lambda}{\partial y} = \frac{\lambda \partial Q}{\partial x} + \frac{Q \partial \lambda}{\partial x}.$$

Thus, we have

$$\lambda\left(\frac{\partial P}{\partial y} - \frac{\partial Q}{\partial x}\right) = Q\frac{\partial \lambda}{\partial x} - P\frac{\partial \lambda}{\partial y} \qquad [1]$$

$$\lambda\left(\frac{\partial R}{\partial x} - \frac{\partial P}{\partial z}\right) = P\frac{\partial \lambda}{\partial z} - R\frac{\partial \lambda}{\partial x} \qquad [2]$$

$$\lambda\left(\frac{\partial Q}{\partial z} - \frac{\partial R}{\partial y}\right) = R\frac{\partial \lambda}{\partial y} - Q\frac{\partial \lambda}{\partial z}. \qquad [3]$$

Multiplying [1], [2] and [3] by R, Q and P respectively and adding, we get

$$P\left(\frac{\partial Q}{\partial z} - \frac{\partial R}{\partial y}\right) + Q\left(\frac{\partial R}{\partial x} - \frac{\partial P}{\partial z}\right) + R\left(\frac{\partial P}{\partial y} - \frac{\partial Q}{\partial x}\right) = 0$$

which is the necessary condition for integrability.

(ii) Sufficient condition: Suppose P, Q and R satisfy the condition

$$P\left(\frac{\partial Q}{\partial z} - \frac{\partial R}{\partial y}\right) + Q\left(\frac{\partial R}{\partial x} - \frac{\partial P}{\partial z}\right) + R\left(\frac{\partial P}{\partial y} - \frac{\partial Q}{\partial x}\right) = 0.$$

The same condition is satisfied by $P_1 = \mu P, Q_1 = \mu Q$ and $R_1 = \mu R$; where μ is any function of x, y, z. We have

$$\frac{\partial Q_1}{\partial z} - \frac{\partial R_1}{\partial y} = \mu\frac{\partial Q}{\partial z} + Q\frac{\partial \mu}{\partial z} - \left(\mu\frac{\partial R}{\partial y} + R\frac{\partial \mu}{\partial y}\right)$$

or $$\frac{\partial Q_1}{\partial z} - \frac{\partial R_1}{\partial y} = \mu\left(\frac{\partial Q}{\partial z} - \frac{\partial R}{\partial y}\right) + Q\frac{\partial \mu}{\partial z} - R\frac{\partial \mu}{\partial y}.$$

Two other equations can be obtained in a similar manner. Then we can show that the same condition is satisfied by P_1, Q_1 and R_1 as by P, Q and R. Now $Pdx + Qdy$ can, without loss of generality, be regarded as an *exact differential*. Thus $\partial P/\partial y = \partial Q/\partial x$ and let $V = \int(Pdx + Qdy)$. Then $\partial V/\partial x = P$ and $\partial V/\partial y = Q$ and also

$$\frac{\partial P}{\partial z} = \frac{\partial^2 V}{\partial z \partial x} \quad \text{and} \quad \frac{\partial Q}{\partial z} = \frac{\partial^2 V}{\partial z \partial y}.$$

Substituting all these values in [12.3], we have

$$\frac{\partial V}{\partial x}\left(\frac{\partial^2 V}{\partial z \partial y} - \frac{\partial R}{\partial y}\right) + \frac{\partial V}{\partial y}\left(\frac{\partial R}{\partial x} - \frac{\partial^2 V}{\partial z \partial x}\right) = 0$$

or $$\frac{\partial V}{\partial x} \cdot \frac{\partial}{\partial y}\left(\frac{\partial V}{\partial z} - R\right) - \frac{\partial V}{\partial y} \cdot \frac{\partial}{\partial x}\left(\frac{\partial V}{\partial z} - R\right) = 0$$

or $$\begin{vmatrix} \dfrac{\partial V}{\partial x} & \dfrac{\partial}{\partial x}\left(\dfrac{\partial V}{\partial z} - R\right) \\ \dfrac{\partial V}{\partial y} & \dfrac{\partial}{\partial y}\left(\dfrac{\partial V}{\partial z} - R\right) \end{vmatrix} = 0.$$

This shows that a relation independent of x and y exists between V and $((\partial V/\partial z) - R)$. Consequently, $((\partial V/\partial z) - R)$ can be expressed as a function of z and V alone. That is, we can take $((\partial V/\partial z) - R) = f(z, V)$. Now

$$Pdx + Qdy + Rdz = \frac{\partial V}{\partial x}dx + \frac{\partial V}{\partial y}dy + \left(\frac{\partial V}{\partial z - f}\right)dz$$

$$= \left(\frac{\partial V}{\partial x}dx + \frac{\partial V}{\partial y}dy + \frac{\partial V}{\partial z}dz\right) - f dz = dV - f dz.$$

Thus [12.1] can be written as $dV - f dz = 0$. Therefore, its integral may be taken in the form

$$f(z, V) = 0.$$

Hence the condition is sufficient.

Note: (i) In the vector form, taking

$$\vec{A} = P\hat{i} + Q\hat{j} + R\hat{k} \quad \text{and} \quad \vec{r} = x\hat{i} + y\hat{j} + z\hat{k}$$

so that $d\vec{r} = dx\ \hat{i} + dy\ \hat{j} + dz\ \hat{k}$, we have

$$\vec{A} \cdot d\vec{r} = Pdx + Qdy + Rdz.$$

Therefore, the total differential equation $Pdx + Qdy + Rdz = 0$ is equivalent to the vector equation $\vec{A} \cdot d\vec{r} = 0$. Since we have

$$\text{curl } \vec{A} = \sum\left(\frac{\partial Q}{\partial z} - \frac{\partial R}{\partial y}\right)\hat{i}$$

it follows that a necessary and sufficient condition for integrability is $\vec{A} \cdot \text{curl } \vec{A} = 0$.
(ii) The condition can also be *remembered* as follows:

$$\begin{vmatrix} P & Q & R \\ P & Q & R \\ \partial/\partial x & \partial/\partial y & \partial/\partial z \end{vmatrix} = 0$$

(iii) The condition for the equation $Pdx + Qdy + Rdz = 0$ to be exact is

$$\frac{\partial P}{\partial y} = \frac{\partial Q}{\partial x}, \quad \frac{\partial Q}{\partial z} = \frac{\partial R}{\partial y} \quad \text{and} \quad \frac{\partial R}{\partial x} = \frac{\partial P}{\partial z}$$

or $P_y = Q_x$, $Q_z = R_y$ and $R_x = P_z$. Thus, every exact equation satisfies the condition of integrability [12.3].

12.2.1 Geometrical interpretation of *Pdx + Qdy + Rdz* = 0

Consider the surface $u(x, y, z) =$ constant. We know that $\partial u/\partial x, \partial u/\partial y$ and $\partial u/\partial z$ are proportional to the *direction cosines* of the normal at (x, y, z) on the surface. But

$$\frac{\partial u}{\partial x}dx + \frac{\partial u}{\partial y}dy + \frac{\partial u}{\partial z}dz = 0.$$

Comparing this with [12.1], viz. $Pdx + Qdy + Rdz = 0$, we see that $\partial u/\partial x, \partial u/\partial y, \partial u/\partial z$ are proportional to P, Q, R respectively. So [12.1] represents a single infinite system of surfaces so that the direction cosines of the normal at any point (x, y, z) on a surface belonging to this system are proportional to the value of P, Q, R at that point. Thus it follows that a point moving in such a manner that its coordinates and the direction cosines of its path always satisfies [12.1] can pass through any point in an infinity of directions and yet, while passing through any point must remain on the particular surface $u(x, y, z) = c$, which passes through the point.

12.3 Methods for solving *Pdx + Qdy + Rdz* = 0

We have the following methods:

Method 1: Solution by inspection
Method 2: One variable regarded as a constant

Method 3: Method of auxiliary equations
Method 4: Homogeneous equations

Any one of these methods may be adopted while solving total differential equations. Each of these methods is described in the following sections with some worked examples.

12.4 Solution by inspection

Sometimes by rearranging the terms of the given equation or by dividing by a suitable function of x, y and z, or by both, the equation may be altered to contain several parts which are *exact differentials*. For example, we have

(i) $$\frac{2x^2ydy - 2xy^2dx}{x^4} = d\left(\frac{y^2}{x^2}\right)$$ (ii) $$xydz + xzdy + yzdx = d(xyz)$$

(iii) $$\frac{xdy - ydx}{x^2 + y^2} = d\left(\tan^{-1}\frac{y}{x}\right)$$

In such cases it may not even be necessary to apply the condition of integrability.

Example 12.1 Solve : $(yz + xyz)dx + (zx + xyz)dy + (xy + xyz)dz = 0.$

Solution 12.1 We can easily show that the condition of integrability is satisfied. Dividing the equation by xyz, we have

$$\left(\frac{1}{x} + 1\right)dx + \left(\frac{1}{y} + 1\right)dy + \left(\frac{1}{z} + 1\right)dz = 0.$$

Integrating, we get the required solution as

$$\log x + x + \log y + y + \log z + z = c \quad \text{or} \quad \log(xyz) + (x + y + z) = c.$$

Example 12.2 Solve:

$$(y - z)(y + z - 2x)dx + (z - x)(z + x - 2y)dy + (x - y)(x + y - 2z)dz = 0.$$

Solution 12.2 The given equation is exact, and hence can be written as

$$(y^2dx + 2xydy) - (z^2dx + 2zxdz) + (z^2dy + 2zydz)-$$

$$(x^2dy + 2xydx) + (x^2dz + 2xzdx) - (y^2dz + 2yzdy) = 0$$

Integrating, we get $y^2x - z^2x + z^2y - x^2y + x^2z - y^2z = c$ or

$$x(y^2 - z^2) + y(z^2 - x^2) + z(x^2 - y^2) = c.$$

Example 12.3 Solve: $y^2dx - zdy + ydz = 0.$

Solution 12.3 Let us verify the condition for integrability for the equation. Here, we have

$$P = y^2, \quad \frac{\partial P}{\partial y} = 2y, \quad \frac{\partial P}{\partial z} = 0; \quad Q = -z, \quad \frac{\partial Q}{\partial x} = 0, \quad \frac{\partial Q}{\partial z} = -1;$$

and $R = y, \quad \dfrac{\partial R}{\partial x} = 0, \quad \dfrac{\partial R}{\partial y} = 1.$

Then $P\left(\dfrac{\partial Q}{\partial z} - \dfrac{\partial R}{\partial y}\right) + Q\left(\dfrac{\partial R}{\partial x} - \dfrac{\partial P}{\partial z}\right) + R\left(\dfrac{\partial P}{\partial y} - \dfrac{\partial Q}{\partial x}\right)$

$$= y^2(-1-1) - z(0-0) + y(2y-0) = 0.$$

Thus, the equation is integrable. Dividing throughout by y^2, the equation becomes

$$dx + \frac{(ydz - zdy)}{y^2} = 0.$$

The solution, on integration, is therefore

$$x + z/y = c.$$

Example 12.4 Solve:

$$(x^2y - y^3 - y^2z)dx + (xy^2 - x^2z - x^3)dy + (xy^2 + x^2y)dz = 0.$$

Solution 12.4 Dividing throughout by x^2y^2, the equation becomes

$$\left(\frac{1}{y} - \frac{y}{x^2} - \frac{z}{x^2}\right)dx + \left(\frac{1}{x} - \frac{z}{y^2} - \frac{x}{y^2}\right)dy + \left(\frac{1}{x} + \frac{1}{y}\right)dz = 0$$

or $$\frac{(ydx - xdy)}{y^2} + \frac{(xdy - ydx)}{x^2} + \frac{(xdz - zdx)}{x^2} + \frac{(ydz - zdy)}{y^2} = 0$$

$$\text{or}\quad d\left(\frac{x}{y}\right)+d\left(\frac{y}{x}\right)+d\left(\frac{z}{x}\right)+d\left(\frac{z}{y}\right)=0.$$

Integrating, we get $\frac{x}{y}+\frac{y}{x}+\frac{z}{x}+\frac{z}{y}=c$ or $x^2+y^2+z(x+y)=cxy$.

Example 12.5 Find $f(y)$ such that $\left(\frac{yz+z}{x}\right)dx - zdy + f(y)dz = 0$ is integrable, and hence solve the equation.

Solution 12.5 Multiplying throughout by x, the equation becomes $(yz+z)dx - xzdy + xf(y)dz = 0$. Comparing this equation with $Pdx + Qdy + Rdz = 0$, we have

$$P = yz + z, \quad Q = -xz, \quad R = xf(y). \qquad [1]$$

If the given equation is integrable, the condition of integrability must be satisfied, i.e.,

$$P\left(\frac{\partial Q}{\partial z}-\frac{\partial R}{\partial y}\right)+Q\left(\frac{\partial R}{\partial x}-\frac{\partial P}{\partial z}\right)+R\left(\frac{\partial P}{\partial y}-\frac{\partial Q}{\partial x}\right)=0.$$

Substituting from [1], we have

$$(yz+z)(-x-xf') - xz[f-(y+1)] + xf[z-(-z)] = 0$$

where f' is $\frac{df}{dy}$ so that $xz(1+y)f' = xzf$ or $\frac{df}{f}=\frac{dy}{(y+1)}$.

Integrating, we have

$$\log f = \log(y+1) + \log c \quad \text{or} \quad f = c(y+1)$$

where c is an arbitrary constant. The given equation now becomes

$$z(y+1)dx - xz\ dy + xc(y+1)dz = 0.$$

Dividing by $xz(1+y)$ and integrating, we get the required solution as

$$\log x - \log(y+1) + c\log z = \log k \quad \text{or} \quad xz^c = k(y+1).$$

12.5 One variable regarded as a constant

First, the condition of integrability is verified. If any two terms, like $Pdx + Qdy = 0$ can be solved easily, then we assume one of the variables (in this case z) as a constant so that $dz = 0$. Let

$$u(x,y) = f(z) \qquad [12.4]$$

be the solution of $Pdx + Qdy = 0$ where $f(z)$ is constant with respect to x and y. Then differentiate $u(x,y)$ with respect to x, y and z and compare with $Pdx+Qdy+Rdz = 0$. If the coefficients of df or dz involve functions of x and y, we can eliminate them using [12.4]. Thus, df/dz can be obtained, independent of x and y. By integrating df/dz, we can obtain the complete solution of the given equation.

The above method can be illustrated by means of the following examples:

Example 12.6 Solve: $2yzdx + zxdy - xy(1+z)dz = 0.$

Solution 12.6 To prove the condition of integrability, we have $P = 2yz$, $Q = zx$ and $R = -xy(1+z)$ so that

$$\frac{\partial P}{\partial y} = 2z, \frac{\partial Q}{\partial x} = z, \frac{\partial R}{\partial x} = -y(1+z), \frac{\partial P}{\partial z} = 2y, \frac{\partial Q}{\partial z} = x, \frac{\partial R}{\partial y} = -x(1+z).$$

Thus, $P\left(\dfrac{\partial Q}{\partial z} - \dfrac{\partial R}{\partial y}\right) + Q\left(\dfrac{\partial R}{\partial x} - \dfrac{\partial P}{\partial z}\right) + R\left(\dfrac{\partial P}{\partial y} - \dfrac{\partial Q}{\partial x}\right)$ becomes

$$2yz[x - \{-x(1+z)\}] + zx[-y(1+z) - 2y] - xy(1+z)(2z - z)$$

$$= 2yz(2x + xz) + xz(-3y - yz) - xy(1+z)z = 0.$$

Hence the equation is integrable. Taking z = constant, we have $dz = 0$ so that the given equation becomes $2yzdx + zxdy = 0$. Dividing by xyz, we get

$$2\frac{dx}{x} + \frac{dy}{y} = 0.$$

Integrating, we have $2\log x + \log y = \log f(z)$ or $f(z) = x^2y$. Differentiating this equation, we get

$$2xydx + x^2dy - f'(z)dz = 0 \quad \text{or} \quad 2yzdx + zxdy - (z/x)f'(z)dz = 0$$

Comparing this with the given equation, we have

$$\frac{z}{x}f'(z) = xy(1+z) \quad \text{or} \quad f'(z) = \frac{x^2y}{z}(1+z) \quad \text{or} \quad f'(z) = f(z)\frac{(1+z)}{z}$$

$$\text{or} \quad \frac{f'(z)}{f(z)} = \frac{1+z}{z} = \frac{1}{z} + 1.$$

Integrating, we get $\log f(z) = \log z + z + \log c$ or $f(z) = cze^z$. The solution is $x^2y = cze^z$.

Note: Example 12.6 can also be solved by the method of inspection as follows: Dividing throughout by xyz, we get

$$2\frac{dx}{x} + \frac{dy}{y} - \left(\frac{1}{z} + 1\right)dz = 0.$$

Integrating, we have the solution as

$$2\log x + \log y - (\log z + z) = \log c \quad \text{or} \quad x^2y = cze^z.$$

Example 12.7 Find $f(y)$ if

$$f(y)dx - zxdy - xy\log y dz = 0 \qquad [1]$$

is integrable. Find the corresponding integral.

Solution 12.7 Here, we have $P = f(y)$, $Q = -zx$ and $R = -xy\log y$. So, by the condition of integrability

$$f(y)[-x - (-x\log y - x)] - zx[-y\log y - 0] - xy\log y[f'(y) - (-z)] = 0$$

or $x\log y f(y) - xy\log y f'(y) = 0$ or $f(y) - yf'(y) = 0$ or $f'(y)/f(y) = 1/y$.

Integrating, we get $\log f(y) = \log y + \log c$ or $f(y) = cy$. Thus [1] becomes

$$cydx - zxdy - xy\log y dz = 0 \qquad [2]$$

Treating z as a constant, we get $dz = 0$. So, [2] becomes

$$cydx - zxdy = 0 \quad \text{or} \quad c\frac{dx}{x} - z\frac{dy}{y} = 0.$$

Integrating and treating z as a constant we get

$$c\log x - z\log y = F(z) \qquad [3]$$

where $F(z)$ is a function of z. Differentiating [3], we get

$$\frac{c}{x}dx - \frac{z}{y}dy - \log y\, dz = \left(\frac{\partial F}{\partial z}\right)dz$$

$$\text{or} \quad cydx - zxdy - \left[xy\log y + xy\left(\frac{\partial F}{\partial z}\right)\right]dz = 0. \qquad [4]$$

Comparing [2] and [4], we get

$$xy\log y + xy\frac{\partial F}{\partial z} = xy\log y \quad \text{or} \quad \frac{\partial F}{\partial z} = 0.$$

Integrating, we get $F = k$, a constant. Substituting for F in [3], we get the solution

$$c\log x - z\log y = k \text{ or } x^c = ay^z \quad (\text{taking } k = \log a).$$

Example 12.8 Verify that the following equation is integrable and find its primitive:

$$xydx + (x^2y - zx)dy + (x^2z - xy)dz = 0. \tag{1}$$

Solution 12.8 Here, we treat x as a constant so that $dx = 0$. Thus, [1] becomes

$$(x^2y - zx)dy + (x^2z - xy)dz = 0$$

or $$x^2(ydy + zdz) - x(zdy + ydz) = 0. \tag{2}$$

Integrating [2], treating x as a constant, we get

$$\frac{1}{2}x^2(y^2 + z^2) - xyz = f(x). \tag{3}$$

Differentiating [3] we get

$$\frac{1}{2}\cdot 2xdx(y^2 + z^2) + \frac{1}{2}x^2[2ydy + 2zdz] + xydz - yzdx - zxdy = \left(\frac{df}{dx}\right)dx$$

or $$\left[x(y^2 + z^2) - yz - \frac{df}{dx}\right]dx + (x^2y - xy)dy + (x^2z - xy)dz = 0.$$

Comparing the above equation with [1], we have

$$x(y^2 + z^2) - yz - \frac{df}{dx} = yz \quad \text{or} \quad x(y^2 + z^2) - 2yz = \frac{df}{dx}$$

or $$\frac{1}{2}x^2(y^2 + z^2) - xyz = \frac{1}{2}x\frac{df}{dx} \quad \text{or} \quad f = \frac{1}{2}x\frac{df}{dx} \quad (\text{from [3]})$$

or $$\frac{df}{f} = 2\frac{dx}{x} \quad \text{so that} \quad f = cx^2.$$

Substituting for f in [3], the required primitive is

$$\frac{1}{2}x^2(y^2 + z^2) - xyz = cx^2.$$

Example 12.9 Solve:

$$(e^x y + e^z)dx + (e^y z + e^x)dy + (e^y - e^x y - e^y z)dz = 0. \qquad [1]$$

Solution 12.9 We can easily see that the given equation is integrable. We now assume z as a constant and consider the following equation:

$$(e^x y + e^z)dx + (e^y z + e^x)dy = 0.$$

Integrating this equation, we get $e^x y + e^y z + e^z x = f(z)$. Differentiating, we get

$$(e^x y + e^z)dx + (e^y z + e^x)dy + (e^y + e^z x)dz = df. \qquad [2]$$

From [1], by re-arrangement, we get

$$(e^x y + e^z)dx + (e^y z + e^x)dy + (e^y + e^z x)dz = (e^x y + e^y z + e^z x)dx. \qquad [3]$$

Thus, from [2] and [3], we obtain

$$df = (e^x y + e^y z + e^z x)dz = f dz.$$

Hence $f = ce^z$. Thus, the required solution is

$$e^x y + e^y z + e^z x = ce^z.$$

12.6 Method of auxiliary equations

Let [12.1], viz. $Pdx + Qdy + Rdz = 0$, be the given equation. The condition for its integrability is as given in [12.3], viz.

$$P\left(\frac{\partial Q}{\partial z} - \frac{\partial R}{\partial y}\right) + Q\left(\frac{\partial R}{\partial x} - \frac{\partial P}{\partial z}\right) + R\left(\frac{\partial P}{\partial y} - \frac{\partial Q}{\partial x}\right) = 0.$$

Comparing [12.1] and [12.3], we get simultaneous equations known as *auxiliary equations:*

$$\frac{dx}{(\partial Q/\partial z) - (\partial R/\partial y)} = \frac{dy}{(\partial R/\partial x) - (\partial P/\partial z)} = \frac{dz}{(\partial P/\partial y) - (\partial Q/\partial x)}.$$

Let $u = a$ and $v = b$ be its two integrals. We find the values of $Adu + Bdv = 0$ and compare with [12.1] to find the values of A and B. Substituting these values of A and B in $Adu + Bdv = 0$ and solving, we get the required solution.

Note: This method fails if the given equation is not an exact equation. It is illustrated in the example given below.

Example 12.10 Solve: $(y^2 + yz)dx + (xz + z^2)dy + (y^2 - xy)dz = 0.$

Solution 12.10 Though the equation can be solved by other methods also, let us use the method of forming *auxiliary equations* to arrive at the solution. The given equation is integrable since it satisfies the condition of integrability,

$$(y^2 + yz)(x + 2z - 2y + x) + (xz + z^2)(-2y) + (y^2 - xy)(2y) = 0. \qquad [1]$$

Comparing [1] and the given equation, we get

$$\frac{dx}{2(x - y + z)} = \frac{dy}{-2y} = \frac{dz}{2y}.$$

From the last two expressions given above, we get $dy + dz = 0$ or $(y + z) = u$. Again, we have

$$\frac{dx + dz}{x + z} = \frac{dy}{-y}.$$

Integrating, we get $\log(x + z) + \log y = \log v$ or $y(x + z) = v$. Here, u and v are two integrals obtained by solving the auxiliary equations. Consider

$$Adu + Bdv = 0 \quad \text{or} \quad A(dy + dz) + B[dy(x + z) + y(dx + dz)] = 0$$

or $\quad Bydx + [A + B(x + z)]dy + (A + By)dz = 0.$

Comparing this with the given equation we have

$$By = y^2 + yz, \quad A + B(x + z) = xz + z^2, A + By = y^2 - xy.$$

Thus $\quad B = y + z = u$, and

$$A = -By + y^2 - xy = -(y + z)y + y^2 - xy = -y(z + x) = -v.$$

Therefore $\quad Adu + Bdv = 0$ becomes $-vdu + udv = 0$ or

$$\frac{du}{u} - \frac{dv}{v} = 0 \quad \text{or} \quad \log u - \log v = -\log c \quad \text{or} \quad \frac{v}{u} = c \quad \text{(on integrating).}$$

Thus the required solution is $c(y+z) = y(x+z)$.

12.7 Homogeneous equations

In case P, Q and R are *homogeneous* functions of x, y and z, then one variable may be separated from the other two by substituting (say) $x = zu$ and $y = zv$ so that $dx = zdu + udz$ and $dy = zdv + vdz$. So, two cases arise in which the given equation will be reduced to the form in which either the coefficient of dz is zero or not zero. In both cases the new equation can be easily integrated. We consider the following examples which make the above procedure clearer.

Example 12.11 Solve: $z^2dx + (z^2 - 2yz)dy + (2y^2 - yz - xz)dz = 0$.

Solution 12.11 The condition of integrability can be easily verified and is left as an exercise to students. Since the given equation is *homogeneous*, i.e., P, Q and R are homogeneous functions of x, y and z, we take $x = uz$ and $y = vz$. Then

$$\left.\begin{aligned} dx &= zdu + udz \\ dy &= zdv + vdz \end{aligned}\right\} \qquad [1]$$

Substituting [1] in the given equation, we get

$$z^2(zdu + udz) + z^2(1-2v)(zdv + vdz) + (2v^2 - v - u)z^2dz = 0$$

or $$z^3du + z^3(1-2v)dv + [u + v(1-2v) + (2v^2 - u - v)]z^2dz = 0$$

or $$z^3du + z^3(1-2v)dv + 0 \times z^2dz = 0.$$

Dividing by z^3, we get $du + (1-2v)dv = 0$. Integrating, we get

$$u + v - v^2 = c \quad \text{or} \quad \frac{x}{z} + \frac{y}{z} - \frac{y^2}{z^2} = c.$$

Therefore the required solution is

$$(x+y)z - y^2 = cz^2.$$

Example 12.12 Solve: $(y^2 + yz)dx + (xz + z^2)dy + (y^2 - xy)dz = 0$.

Solution 12.12 This is the same as the equation in Example 12.10, but we solve it using the method for homogeneous equations. Clearly, the condition of integrability is satisfied. We take $x = uz$ and $y = vz$ so that $dx = zdu + udz$ and $dy = zdv + vdz$. Substituting these in the given equation, we have

$$z^2(v^2 + v)(zdu + udz) + z^2(u + 1)(zdv + vdz) + z^2(v^2 - uv)dz = 0$$

$$\text{or}\quad z[(v^2 + v)du + (u + 1)dv] + [u(v^2 + v) + v(u + 1) + v^2 - uv]dz = 0$$

$$\text{or}\quad z[(v^2 + v)du + (u + 1)dv] + [uv^2 + uv + v + v^2]dz = 0$$

$$\text{or}\quad z[(v^2 + v)du + (u + 1)dv] + (v^2 + v)(u + 1)dz = 0$$

$$\text{or}\quad \frac{du}{u + 1} + \frac{dv}{v^2 + v} + \frac{dz}{z} = 0 \quad\text{or}\quad \frac{du}{u + 1} + \left(\frac{1}{v} - \frac{1}{v + 1}\right)dv + \frac{dz}{z} = 0.$$

Integrating, we get the required solution as

$$\log(u + 1) + \log v - \log(v + 1) + \log z = \log c \quad\text{or}\quad \frac{(u + 1)vz}{v + 1} = c$$

$$\text{or}\quad (uz + z)vz = (vz + z)c \quad\text{or}\quad (x + z)y = c(y + z).$$

12.8 The non-integrable single equation

If [12.1], viz. $Pdx + Qdy + Rdz = 0$, does not satisfy the condition of integrability [12.3], then there is no single relation between x, y and z to satisfy it. However we can find an infinite number of curves that lie on any given surface and satisfy [12.1].
To do this, we consider the surface

$$f(x, y, z) = c. \qquad [12.5]$$

Differentiating [12.5] and eliminating z and dz from the equation so obtained, we get an equation involving x and y only. The resulting solution and [12.5] together represent the desired curves. This is illustrated below.

Example 12.13 Find the curves represented by the solution of

$$ydx + (z - y)dy + xdz = 0 \qquad [1]$$

which lie in the plane

$$2x - y - z = 1. \qquad [2]$$

Solution 12.13 Equation [1] does *not* satisfy the condition of integrability. Differentiating [2], we have $2dx - dy - dz = 0$. Multiplying the above equation by x and adding to [1], we have

$$(y + 2x)dx + (z - y - x)dy = 0 \text{ or } (y + 2x)dx + (2x - y - 1 - y - x)dy = 0$$

or $\quad (y + 2x)dx + (x - 2y - 1)dy = 0$ or $(ydx + xdy) + 2xdx - 2ydy - dy = 0.$

Integrating, we get the required result

$$xy + x^2 - y^2 - y = c.$$

Example 12.14 Show that the curves of

$$xdx + ydy + c(\sqrt{1 - (x^2/a^2) + (y^2/b^2)})dz = 0 \qquad [1]$$

that lie on the *ellipsoid*

$$\frac{x^2}{a^2} + \frac{y^2}{b^2} + \frac{z^2}{c^2} = 1 \qquad [2]$$

also lie on the family of concentric spheres $x^2 + y^2 + z^2 = a^2$.

Solution 12.14 Equation [1] does not satisfy the condition of integrability. From [2], we have

$$1 - \frac{x^2}{a^2} - \frac{y^2}{b^2} = \frac{z^2}{c^2}.$$

Using this, [1] may be written as $xdx + ydy + zdz = 0$. Integrating, we get

$$x^2 + y^2 + z^2 = a^2 \quad \text{(where } a \text{ is an arbitrary constant. Hence the result.)}$$

Exercise 12

Examine the following equations for integrability and if integrable, solve them:

12.1 $\quad yzdx - 2xzdy + (xy - zy^3)dz = 0$

12.2 $\quad (y + z)dx + dy + dz = 0$

12.3 $zydx = zxdy + y^2dz$

12.4 $(2x^2 + 2xy + 2xz^2 + 1)dx + dy + 2zdz = 0$

12.5 $(y^2 + yz)dx + (xz + z^2)dy + (y^2 - xy)dz = 0$

12.6 $(2y - z)dx + 4dy - 2dz = 0$

12.7 $zdx + zdy + [2(x + y) + \sin z]dz = 0$

12.8 $(ydx + xdy)(a - z) + xydz = 0$

12.9 $(1 + yz)dx + x(z - x)dy - (1 + xy)dz = 0$

12.10 $(y^2 + yz + z^2)dx + (z^2 + zx + x^2)dy + (x^2 + xy + y^2)dz = 0$

12.11 $(y^2 + z^2 - x^2)dx - 2xydy - 2xzdz = 0$

12.12 $zdz + (x - a)dx = [b^2 - z^2 - (x - a)^2]^{1/2}dy$

12.13 $z^2dx + (z^2 - 2yz)dy + (2y^2 - yz - zx)dz = 0$

12.14 $(x^2y - y^3 - y^2z)dx + (xy^2 - x^2z - x^3)dy + (xy^2 + x^2y)dz = 0$

12.15 $yz(ax + y + z)dx + zx(x + ay + z)dy + xy(y + az + x)dz = 0$

CHAPTER THIRTEEN

Simultaneous Total Differential Equations

13.1 Introduction

Let us consider a system of total differential equations of the first order and first degree in the derivatives. Here, equations involving only three variables will be considered.

The general form of a system of two simultaneous total differential equations of the first order is

$$\left.\begin{aligned} P_1dx + Q_1dy + R_1dz = 0 \\ P_2dx + Q_2dy + R_2dz = 0 \end{aligned}\right\} \qquad [13.1]$$

where the coefficients are functions of x, y, z. Taking z as the independent variable and solving [13.1] for dx/dz and dy/dz , we get

$$\frac{dx}{dz} = \frac{Q_1R_2 - Q_2R_1}{P_1Q_2 - P_2Q_1} \quad \text{and} \quad \frac{dy}{dz} = \frac{R_1P_2 - P_1R_2}{P_1Q_2 - P_2Q_1}$$

or $$\frac{dx}{Q_1R_2 - Q_2R_1} = \frac{dy}{R_1P_2 - P_1R_2} = \frac{dz}{P_1Q_2 - P_2Q_1}$$

or $$\frac{dx}{P} = \frac{dy}{Q} = \frac{dz}{R}. \qquad [13.2]$$

13.2 Solution of simultaneous total differential equations

If one of the variables is absent from two members of [13.2], the procedure to solve the equations becomes quite simple. For example, suppose z is absent from P and Q. Then the solution of $dx/P = dy/Q$ gives a relation between x and y which is one equation of the complete solution. This equation may enable us to eliminate x or y from one of the other equations in [13.2]; and then another integral relation may be found. This will be the second equation of the solution. This method may be used in combination with the following method, making the equations easier to solve. By a well-known result in ratios and proportions, we have

$$\frac{dx}{P}=\frac{dy}{Q}=\frac{dz}{R}=\frac{l dx+m dy+n dz}{lP+mQ+nR}=\frac{l' dx+m' dy+n' dz}{l'P+m'Q+n'R}.$$

These relations are all satisfied by the same relations between x, y and z that satisfy [13.2]. It may be possible, by the proper choice of *multipliers* $l, m, n; l', m', n'$, etc., to obtain equations which are easily solved, and whose solutions are the solutions of [13.2]. In particular l, m, n may be found such that

$$lP+mQ+nR=0$$

so that $l dx+m dy+n dz=0$. And if

$$l dx+m dy+n dz$$

is an *exact differential* du, then we have $u=a$ as one equation of the complete solution. Similarly if

$$l' dx+m' dy+n' dz$$

is an *exact differential* dv, then we have $v=b$ as the second equation of the complete solution. The methods are illustrated by the examples given below.

Example 13.1 Solve the simultaneous equations: $\dfrac{dx}{x^2-y^2-z^2}=\dfrac{dy}{2xy}=\dfrac{dz}{2xz}$.

Solution 13.1 The equation formed by the last two fractions reduces to $dy/y=dz/z$ which has for its solution

$$y=cz. \qquad [1]$$

Using x, y, z as multipliers we get

$$\frac{dx}{x^2-y^2-z^2}=\frac{dy}{2xy}=\frac{dz}{2xz}=\frac{x dx+y dy+z dx}{x(x^2+y^2+z^2)}.$$

The equation formed by the last two fractions has for its solution

$$x^2+y^2+z^2=bz. \qquad [2]$$

The complete solution consists of [1] and [2].

Example 13.2 Solve: $\dfrac{dx}{mz - ny} = \dfrac{dy}{nx - lz} = \dfrac{dz}{ly - mx}$.

Solution 13.2 By using multipliers l, m and n one gets the equal fraction

$$\frac{ldx + mdy + ndz}{0} \quad \text{or} \quad ldx + mdy + ndz = 0$$

or $\quad lx + my + nz = C.$ [1]

Similarly, using multipliers x, y, z, we get $xdx + ydy + zdz = 0$ or

$$x^2 + y^2 + z^2 = k^2 \qquad [2]$$

The two integrals [1] and [2] form the complete solution.

Example 13.3 Solve: $\dfrac{dx}{z^2y} = \dfrac{dy}{z^2x} = \dfrac{dz}{yx}$.

Solution 13.3 Taking the first two fractions, we get $dx/y = dy/x$ or $xdx - ydy = 0$, which on integration gives

$$x^2 - y^2 = C_1. \qquad [1]$$

Again taking the second and third fractions, we get $dy/z^2 = dz/y^2$ or $y^2dy - z^2dz = 0$, which on integration gives

$$y^3 - z^3 = C_2. \qquad [2]$$

Equations [1] and [2] together give the required solution.

Example 13.4 Solve: $\dfrac{dx}{\cos(x+y)} = \dfrac{dy}{\sin(x+y)} = \dfrac{dz}{z}$.

Solution 13.4 Using $1, 1, 0$ and $1, -1, 0$ as multipliers we have

$$\frac{dz}{z} = \frac{dx + dy}{\cos(x+y) + \sin(x+y)} = \frac{dx - dy}{\cos(x+y) - \sin(x+y)}; \qquad [1]$$

$$x + y = t, \text{ (say)} \qquad [2]$$

so that $\quad dx + dy = dt \quad$ and $\quad dz/z = dt/(\cos t + \sin t)$. Now,

$$\cos t + \sin t = \sqrt{2}\left(\frac{1}{\sqrt{2}}\cos t + \frac{1}{\sqrt{2}}\sin t\right) = \sqrt{2}\sin\left(t + \frac{\pi}{4}\right)$$

$$\therefore \quad \frac{dz}{z} = \frac{1}{\sqrt{2}} \operatorname{cosec}\left(t + \frac{\pi}{4}\right) dt$$

or $\quad \log z = \frac{1}{\sqrt{2}}\left[\log \tan\left(\frac{1}{2}\left(t + \frac{\pi}{4}\right)\right) + \log C_1\right]$ (on integration)

or $\quad \sqrt{2} \log z = \log \tan \frac{1}{2}\left(t + \frac{\pi}{4}\right) + \log C_1$

or $\quad z^{\sqrt{2}} = C_1 \tan \frac{1}{2}\left(t + \frac{\pi}{4}\right)$

or $\quad z^{\sqrt{2}} \cot \frac{1}{2}\left(x + y + \frac{\pi}{4}\right) = C_1$ (from [2]). [3]

Again, from the last two fractions of [1], we get

$$[(dx + dy) \cos(x + y) - \sin(x + y)]/[\cos(x + y) + \sin(x + y)] = dx - dy$$

or $\quad [(dt)(\cos t - \sin t)]/[\cos t + \sin t] = dx - dy \quad$ [by taking $x + y = t$].

Integrating, we get

$$\log(\cos t + \sin t) - \log C_2 = x - y \text{ or } (\cos t + \sin t)/C_2 = e^{x-y}$$

or $\quad [\cos(x + y) \sin(x + y)]e^{y-x} = C_2.$ [4]

The complete solution is given by [3] and [4].

Example 13.5 Solve:

$$\frac{dx}{x(y^2 + z)} = \frac{dy}{-y(x^2 + z)} = \frac{dz}{z(x^2 - y^2)}. \quad [1]$$

Solution 13.5 Choosing $1/x$, $1/y$ and $1/z$ as multipliers, we get each fraction of [1] equal to

$$\frac{dx/x + dy/y + dz/z}{y^2 + z - (x^2 + z) + x^2 - y^2} = \frac{dx/x + dy/y + dz/z}{0}$$

so that we have $\quad dx/x + dy/y + dz/z = 0.\quad$ Integrating we get

$xyz = C_1.$ [2]

Again, choosing x, y and -1 as multipliers, each fraction of [1] is equal to

$$\frac{xdx + ydy - dz}{x^2(y^2+z) - y^2(x^2+z) - z(x^2-y^2)} = \frac{xdx + ydy - dz}{0}$$

or $2xdx + 2ydy - 2dz = 0$. So, on integration we have

$$x^2 + y^2 - 2z = C_2. \qquad [3]$$

Thus, [2] and [3] together give the complete solution.

Example 13.6 Solve:

$$\frac{dx}{1} = \frac{dy}{3} = \frac{dz}{5z + \tan(y - 3x)}. \qquad [1]$$

Solution 13.6 From the first two fractions we get $dy - 3dx = 0$, integrating which we get

$$y - 3x = C_1. \qquad [2]$$

Again from the first and third fractions of [1], we get $dx = dz/(5z + \tan C_1)$, and

$$x = \left(\frac{1}{5}\right) \log(5z + \tan C_1) + \left(\frac{1}{5}\right) C_2$$

or $$5x - \log\Big(5z + \tan(y - 3x)\Big) = C_2. \qquad [3]$$

Equations [2] and [3] together give the complete solution.

Exercise 13

Solve the following pairs of equations

13.1 $\dfrac{xdx}{y^3 z} = \dfrac{dy}{x^2 z} = \dfrac{dz}{y^3}$

13.2 $\dfrac{dx}{z} = \dfrac{dy}{-z} = \dfrac{dz}{z^2 + (x+y)^2}$

13.3 $$\frac{dx}{x^2} = \frac{dy}{y^2} = \frac{dz}{(x+y)z}$$

13.4 $$\frac{dx}{x^2} = \frac{dy}{y^2} = \frac{dz}{cxy}$$

13.5 $$\frac{dx}{x(x+y)} = \frac{dy}{-y(x+y)} = \frac{dz}{-(x-y)(2x+2y+z)}$$

13.6 $$\frac{dx}{z(x+y)} = \frac{dy}{z(x-y)} = \frac{dz}{(x^2+y^2)}$$

13.7 $$\frac{dx}{x^3+3xy^2} = \frac{dy}{y^3+3x^2y} = \frac{dz}{2(x^2+y^2)z}$$

13.8 $$dx = dy = \frac{dz}{(x+y)(1+2xy+3x^2y^2)}$$

13.9 $$\frac{dx}{x(2y^4-z^4)} = \frac{dy}{y(z^4-2x^4)} = \frac{dz}{z(x^4-y^4)}$$

13.10 $$\frac{pdx}{(q-r)yz} = \frac{qdy}{(r-p)xz} = \frac{rdz}{(p-q)xy}$$

13.11 $$\frac{dx}{y^2+yz+z^2} = \frac{dy}{z^2+zx+x^2} = \frac{dz}{x^2+xy+y^2}$$

13.12 $$\frac{dx}{x} = \frac{dy}{y} = \frac{dz}{z-a\sqrt{x^2+y^2+z^2}}$$

CHAPTER FOURTEEN

Laplace Transforms

14.1 Introduction

The theory of Laplace transforms is an essential part of the mathematical background required of engineers, physicists and mathematicians. It provides an easy and effective means for solving certain types of differential and integral equations. It is the foundation of the modern form of operational calculus, which was originated in an attempt to justify certain operational rules used by an English electrical engineer, Oliver Heaviside, in the latter part of the nineteenth century for solving equations in electromagnetic theory.

The Laplace transform reduces the problem of solving a differential equation to an algebraic problem. It is particularly useful in problems where the mechanical or electrical driving force has discontinuities, acts for a short time only or is periodic but not merely a sine or cosine function.

The process of solution consists of transforming the given problem to a simple algebraic one. This is solved by algebraic manipulation and the solution is again transformed back to obtain the required solution. It also takes care of initial conditions without the necessity of first constructing a complete solution and then specialising the arbitrary constants it contains.

At this juncture a few general remarks will be offered aimed at placing the ideas of this chapter in their proper context.

We know that differentiation and integration are important transformations of functions which can be denoted respectively by

$$D[f(x)] = f'(x) \quad \text{and} \quad I[f(x)] = \int_0^x f(y)dy.$$

Each of the transformations D and I operate on functions to produce other functions. When such transformations, T, are such that

$$T[C_1 f(x) + C_2 g(x)] = C_1 T[f(x)] + C_2 T[g(x)],$$

we say T is *linear* (here C_1 and C_2 are arbitrary constants).

A class of linear transformations of particular importance is that of the *integral transformations*. The Laplace transformation is a special case of an integral transformation and is defined by

$$\mathscr{L}\{f(t)\} = \int_0^\infty e^{-st} f(t)dt = F(s). \qquad [14.1]$$

14.2 Sufficient conditions for existence of Laplace transforms

We can now consider the circumstances under which a function has a Laplace transform. A grasp of the following simple statements will suffice for this purpose.

(i) *Piecewise continuity of a function*

A function is sectionally continuous (or *piecewise* continuous) in an interval $0 \le t \le b$ if $f(t)$ is continuous over every finite interval $0 \le t \le b$ except possibly at a finite number of points, where there are jump discontinuities and at which the function approaches different limits from left and right.

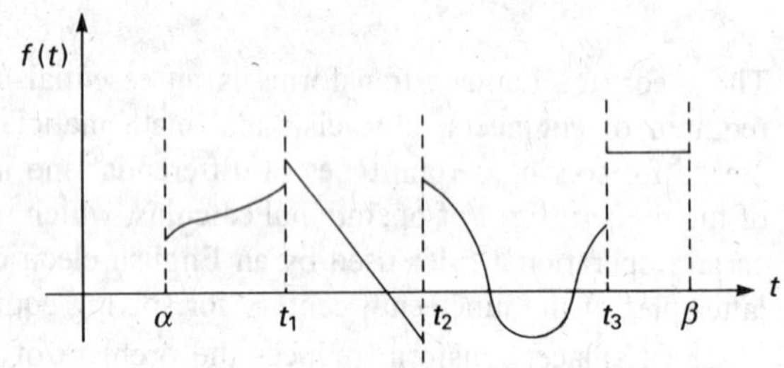

Fig. 14.1 *Sectionally continuous function.*

(ii) *Functions of exponential order*

A function $f(t)$ is said to be of exponential order σ, as $t \to \infty$, if there is an $M > 0$ such that

$$|f(t)| \leqslant M e^{\sigma t}, \quad \forall t \geqslant 0 \qquad [14.2]$$

or equivalently, $|e^{-\sigma t} f(t)| \leqslant M, \quad \forall\, t \geqslant 0$. Clearly e^{at} and $\sin bt$ are of exponential order, while e^{t^2} is not of exponential order since

$$|e^{-\sigma t} e^{t^2}| = e^{t^2 - \sigma t}$$

which can be made larger than any given constant by increasing t indefinitely.

Using the above remarks, we can prove the sufficient conditions for the existence of Laplace transforms.

Theorem 14.1: If $f(t)$ is piecewise continuous in every finite interval $[0, N]$ for $N > 0$, and of exponential order σ, that is by [14.2]

$$|f(t)| \leqslant M e^{\sigma t}, \forall\, t \geqslant 0 \quad \text{and} \quad M > 0$$

then the Laplace transform of $f(t)$ exists for all $s > \sigma$.

Proof: Consider

$$I = |F(s)| = \left|\int_0^\infty e^{-st} f(t)dt\right| \leqslant \int_0^\infty |f(t)|e^{-st}dt \leqslant \int_0^\infty Me^{\sigma t}e^{-st}dt$$

$$= \frac{M}{s-\sigma}.$$

Here, $s > \sigma$ since I is non-negative everywhere. This shows that the above conditions are sufficient for the existence of $\mathscr{L}\{f(t)\}$. However, they are *not necessary* as the following example shows.

Example 14.1 Show that the Laplace transform of $f(t) = t^{-1/2}$ does exist, even though $f(t)$ does not satisfy the sufficient conditions of Theorem 14.1.

Solution 14.1 Let $f(t) = t^{-1/2}$. At $t = 0$, $f(t)$ has an infinite discontinuity but its Laplace transform exists. We have

$$\mathscr{L}(t^{-1/2}) = \int_0^\infty e^{-st}t^{-1/2}dt.$$

or $$\mathscr{L}(t^{-1/2}) = \frac{1}{\sqrt{s}}\int_0^\infty e^{-p}p^{-1/2}dp = \frac{1}{\sqrt{s}}\Gamma\left(\frac{1}{2}\right) = \sqrt{\frac{\pi}{s}} \quad \text{(taking } st = p\text{)}$$

where Γ is the *gamma function* and $\Gamma(1/2) = \sqrt{\pi}$.

14.3 General properties of Laplace transforms

Linearity property Let $f(t)$ and $g(t)$ be two functions on $[0, \infty]$ such that the Laplace transforms $\mathscr{L}\{f(t)\}$ and $\mathscr{L}\{g(t)\}$ exist. If C_1 and C_2 are two constants, then

$$\mathscr{L}\{C_1 f(t) + C_2 g(t)\} = C_1\mathscr{L}\{f(t)\} + C_2\mathscr{L}\{g(t)\}. \qquad [14.3]$$

This property is valid since

$$\int_0^\infty \{C_1 f(t) + C_2 g(t)\}e^{-st}dt = C_1\int_0^\infty f(t)e^{-st}dt + C_2\int_0^\infty g(t)e^{-st}dt.$$

The extension of this result to linear combinations of more than two functions follows immediately.

Theorem 14.3: If $f(t)$ is piecewise continuous on $[0, N]$ for each $N > 0$, and $f(t)$ is of exponential order σ, that is, by [14.2] $|f(t)| \leqslant M e^{\sigma t}, \forall t \geqslant 0$ and $M > 0$, and if $\mathscr{L}\{f(t)\} = F(s)$ then we have

(a) $\mathscr{L}\{e^{at} f(t)\} = F(s-a)\ \forall s > \sigma + a, \forall a > 0$ [14.4]

(b) $\mathscr{L}\{g(t)\} = e^{-as} F(s)$ [14.5]

$$\text{where } g(t) = \begin{cases} f(t-a) & t > a \\ 0 & t < a \end{cases}, a > 0$$

(c) $\mathscr{L}\{f(at)\} = \dfrac{1}{a} F\left(\dfrac{s}{a}\right)$ (for any $a > 0$) [14.6]

Proof: **(a)** This property shows *shifting* on the s-axis and is called *first translation* (or *shifting*) *property* (Fig. 14.2).

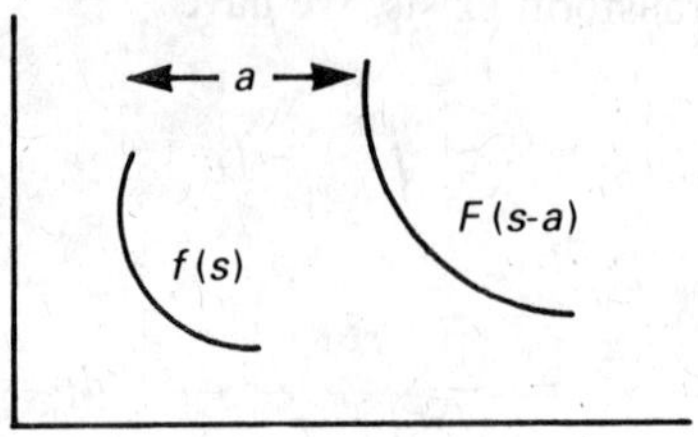

Fig. 14.2 *Curves for the functions $f(s)$ and $F(s-a)$.*

We can see that replacing s by $s-a$ in the transform corresponds to multiplying the original function by e^{at}. We know that

$$\mathscr{L}\{f(t)\} = \int_0^\infty e^{-st} f(t) dt = F(s)$$

so that $$\mathscr{L}\{e^{at} f(t)\} = \int_0^\infty e^{-st}\{e^{at} f(t)\} dt = \int_0^\infty e^{-(s-a)t} f(t) dt = F(s-a).$$

(b) This property shows shifting on the t-axis and is called *second translation (*or *shifting) property*. We know that

$$\mathscr{L}\{g(t)\} = \int_0^\infty e^{-st} g(t) dt = \int_0^a e^{-st} g(t) dt + \int_a^\infty e^{-st} g(t) dt$$

$$= \int_a^\infty e^{-st} f(t-a) dt = \int_0^\infty e^{-s(u+a)} f(u) du, \quad (\text{where } t = u + a)$$

$$= e^{-as} \int_0^\infty e^{-su} f(u) du = e^{-as} F(s).$$

(c) This property is termed as *change of scale property*. Here, we have

$$\mathscr{L}\{f(at)\} = \int_0^\infty e^{-st} f(at)dt = \int_0^\infty e^{-s(u/a)} f(u)\frac{du}{a} \quad \left[\text{taking } t = \frac{u}{a}\right]$$

$$= \frac{1}{a}\int_0^\infty e^{-su/a} f(u)du = \frac{1}{a}F\left(\frac{s}{a}\right).$$

14.4 Laplace transforms of some common functions

Laplace transforms of the functions in Table 14.1 are quite easy to compute:

The derivations for functions 3, 7 and 9 are provided below (as examples).

(3) $$\mathscr{L}(t^n) = \int_0^\infty e^{-st} t^n dt = -\left[\frac{t^n e^{-st}}{s}\right]_0^\infty + \frac{n}{s}\int_0^\infty e^{-st} t^{n-1} dt$$

$$= \frac{n}{s}\mathscr{L}(t^{n-1}) = \frac{n}{s}\left(\frac{n-1}{s}\right)\mathscr{L}(t^{n-2}) = \ldots = \frac{n!}{s^n}\mathscr{L}(1)$$

$$= \frac{n!}{s^{n+1}} \quad \text{when} \quad n \geqslant 0.$$

(7) $$\mathscr{L}(\sinh at) = \mathscr{L}\left(\frac{e^{at} - e^{-at}}{2}\right) = \int_0^\infty e^{-st}\left(\frac{e^{at} - e^{-at}}{2}\right)dt$$

$$= \frac{1}{2}\int_0^\infty e^{-st} e^{at} dt - \frac{1}{2}\int_0^\infty e^{-st} e^{-at} dt = \frac{1}{2}\mathscr{L}(e^{at}) - \frac{1}{2}\mathscr{L}(e^{-at})$$

$$= \frac{1}{2}\left[\frac{1}{s-a}\right] - \frac{1}{2}\left[\frac{1}{s+a}\right] = \frac{a}{s^2 - a^2} \quad \text{for} \quad s > |a|.$$

(9) By using the first shifting property and the result of (5), we get

$$\mathscr{L}\{e^{-bt}\sin at\} = F(s+b) = \frac{a}{(s+b)^2 + a^2}, \quad s > 0.$$

Table 14.1 *Laplace transforms of common functions.*

S.No	$f(t)$	$F(s) = \mathscr{L}\{f(t)\}$
1	1	$\frac{1}{s}$; $s > 0$
2	t	$\frac{1}{s^2}$; $s > 0$
3	t^n	$\frac{n!}{s^{n+1}}$; $s > 0$ $(n = 0, 1, 2, \ldots)$
4	e^{at}	$\frac{1}{s-a}$; $s > a$
5	$\sin at$	$\frac{a}{s^2+a^2}$; $s > 0$
6	$\cos at$	$\frac{s}{s^2+a^2}$; $s > 0$
7	$\sinh at$	$\frac{a}{s^2-a^2}$; $s >\mid a \mid$
8	$\cosh at$	$\frac{s}{s^2-a^2}$; $s >\mid a \mid$
9	$e^{-bt} \sin at$	$\frac{a}{\{(s+b)^2+a^2\}}$; $s > 0$
10	$e^{-bt} \cos at$	$\frac{s+b}{\{(s+b)^2+a^2\}}$; $s > 0$
11	$e^{-at} t^n$	$\frac{\Gamma(n+1)}{(s+a)^{n+1}}$

Exercise 14A

14.1 For each of the following functions (f) compute the Laplace transform $\mathscr{L}\{f(t)\}$ by direct evaluation of the integral. In each case specify the values for which $F(s)$ exists.

(i) $f(t) = 3e^{-2t}$ (ii) $f(t) = 3\cos 5t$

(iii) $f(t) = \begin{cases} \sin t & 0 \le t \le \pi \\ 0 & t > \pi \end{cases}$ (iv) $f(t) = \begin{cases} 0 & 0 < t < 2 \\ 4 & t > 2 \end{cases}$ (v) $f(t) = \sin 2t$

14.2 Compute $\mathscr{L}\{f(t)\}$ for each of the following:

(i) $f(t) = t^2e^{-4t}$ (ii) $f(t) = t\cos t$ (iii) $f(t) = e^{-t}(3\sinh\, 2t - 5\cosh\, 2t)$

(iv) $f(t) = e^{at+b}$ (v) $f(t) = (1 + te^{-t})^3$

14.3 Which of the following functions are of exponential order?

(i) t^n (ii) $\exp(t^2)$ (iii) $1/t$ (iv) $\tan t$ (v) t^2e^{3t}

14.5 Laplace transforms of derivatives and integrals

Consider the general Laplace transform formula

$$\mathscr{L}\,\{f(t)\} = \int_0^\infty e^{-st}f(t)dt = F(s).$$

Roughly speaking, differentiation of the function $f(t)$ here corresponds simply to *multiplication* of the transform $F(s)$ by s. This permits replacing operations of calculus by simple algebraic operations on transforms. And, since integration is the reverse of differentiation, integration of the function $f(t)$ corresponds to *division* of the transforms by s. These two properties are expressed in the form of the following theorems.

Theorem 14.4: If $f(t)$ is continuous $\forall\, t \geqslant 0$ and of exponential order, say σ, and has a derivative $f'(t)$ which is piecewise continuous on every finite interval $[0, N]$ for each $N > 0$, then the Laplace transform of the derivative $f'(t)$ exists for $s > \sigma$ and

$$\mathscr{L}\,\{f'(t)\} = s\,\mathscr{L}\,\{f(t)\} - f(0). \qquad [14.7]$$

Proof: The above result is a direct consequence of integration by parts since

$$\mathscr{L}\,\{f'(t)\} = \int_0^\infty e^{-st}f'(t)dt = \Big[e^{-st}f(t)\Big]_0^\infty + s\int_0^\infty e^{-st}f(t)dt.$$

Since $f(t)$ is of an exponential order, the integrated portion on the right is zero at the upper limit when $s > \sigma$ and is $-f(0)$ at the lower limit. Thus, we have

$$\mathscr{L}\,\{f'(t)\} = -f(0) + s\int_0^\infty e^{-st}f(t)dt \quad \text{or} \quad \mathscr{L}\,\{f'(t)\} = s\,\mathscr{L}\,\{f(t)\} - f(0).$$

Corollary 1: If in Theorem 14.2, $f(t)$ fails to be continuous at $t = 0$, but

$$\lim_{t \to 0} f(t) = f(0+)$$

exists (but is not equal to $f(0)$, which may or may not exist), then

$$\mathscr{L}\{f'(t)\} = s\,\mathscr{L}\{f(t)\} - f(0+).$$

The proof is left as an exercise to students.

Corollary 2: If in Theorem 14.2, $f(t)$ fails to be continuous at $t = a$, then

$$\mathscr{L}\{f'(t)\} = s\,\mathscr{L}\{f(t)\} - f(0) - e^{-as}\{f(a+) - f(a-)\}.$$

The proof is left as an exercise to students. By applying Theorem 14.4 to the second order derivative $f''(t)$, we obtain,

$$\begin{aligned}\mathscr{L}\{f''(t)\} &= s[s\,\mathscr{L}\{f(t)\} - f(0)] - f'(0)\\ &= s^2\,\mathscr{L}\{f(t)\} - sf(0) - f'(0). \end{aligned} \tag{14.8}$$

Similarly, by induction we can obtain the following *extension* of Theorem 14.4:

If $f(t)$ and its derivatives $f'(t), f''(t), \ldots, f^{n-1}(t)$ are continuous $\forall t \geqslant 0$ and are functions of exponential order for some $M > 0$ and σ, and the derivative $f^n(t)$ is piecewise continuous on every finite interval in the range $t \geqslant 0$, then the Laplace transform of $f^n(t)$ exists when $s > \sigma$ and is given by

$$\mathscr{L}\{f^n(t)\} = s^n\,\mathscr{L}\{f(t)\} - s^{n-1}f(0) - s^{n-2}f'(0) - \ldots - f^{n-1}(0). \tag{14.9}$$

Example 14.2 Show that $\mathscr{L}(\sin at) = a/(s^2 + a^2)$.

Solution 14.2 We can use [14.8] to prove this result. Let $f(t) = \sin at \Rightarrow f(0) = 0$. Then $f'(t) = a\cos at \Rightarrow f'(0) = a$ and $f''(t) = -a^2 \sin at$. Hence, we have

$$\mathscr{L}\{f''(t)\} = s^2\mathscr{L}\{f(t)\} - sf(0) - f'(0) \quad \text{(from [14.8])}$$

$$\therefore\ \mathscr{L}(-a^2 \sin at) = s^2\mathscr{L}(\sin at) - s(0) - a$$

or $\mathscr{L}(-a^2 \sin at) = s^2\,\mathscr{L}(\sin at) - a$ or $s^2\,\mathscr{L}(\sin at) + a^2\,\mathscr{L}(\sin at) = a$

or $\mathscr{L}(\sin at) = \dfrac{a}{s^2 + a^2}$.

Since we are concerned about differential equations, we are interested in results which express the Laplace transform of the derivative of a function in terms of the transform of the function itself. There seems to be no obvious reason as to why we should be concerned about the Laplace transform of the *integral* of a function.

For this, consider Kirchhoff's second law applied to an electrical circuit in series:

$$L\frac{di}{dt} + iR + \frac{Q}{C} = E$$

where L is the inductance, i is the current, R the resistance, Q the electric charge on a capacitor of capacitance C and E is the voltage (e.m.f.). There are two dependent variables i and Q, in the equation, of which we eliminate one. We have $i = dQ/dt$ or $Q = \int idt$. The first choice leads to a differential equation of the second order which we can solve. The second choice leads to the *integro–differential equation*

$$L\frac{di}{dt} + iR + \frac{1}{C}\int idt = E.$$

Once we obtain formulae for the Laplace transforms of integrals also, we can directly find i without first solving for Q. So, in order to help solve such integro-differential equations, we can use the formula for the Laplace transform of an integral given below.

Theorem 14.5: If f is piecewise continuous on $[0, N]$ for each $N > 0$ and is of exponential order σ, (that is by [14.2]) $|f(t)| \leqslant Me^{\sigma t}$ $, \forall t \geqslant 0, M > 0,$ then

$$\mathscr{L}\left\{\int_0^t f(u)du\right\} = \frac{1}{s}\mathscr{L}\{f(t)\}, \ \forall s > \sigma. \qquad [14.10]$$

Proof: Let $g(t) = \int_0^t f(u)du.$ Then we have $g'(t) = f(t)$ and $g(0) = 0.$

Also $\mathscr{L}\{g'(t)\} = s\,\mathscr{L}\{g(t)\} - g(0) = s\,\mathscr{L}\{g(t)\} = s\,\mathscr{L}\left\{\int_0^t f(u)du\right\}$

and so $\mathscr{L}\left\{\int_0^t f(u)du\right\} = \frac{1}{s}\mathscr{L}\{g'(t)\} = \frac{1}{s}\mathscr{L}\{f(t)\}(\text{from}[14.7]).$

Example 14.3 A function obeys the equation $f(t) + 2\int_0^t f(u)du = \cosh 2t.$ Find $F(s) = \mathscr{L}\{f(t)\}.$

Solution 14.3 We can see that $\mathscr{L}\{f(t) + 2\int_0^t f(u)du\} = \mathscr{L}\{\cosh 2t\}$ or by [14.3], we have

$$\mathscr{L}\{f(t)\} + 2\,\mathscr{L}\left\{\int_0^t f(u)du\right\} = \mathscr{L}\{\cosh 2t\}.$$

But $\mathscr{L}\{\cosh 2t\} = \dfrac{s}{s^2-4}$ and using [14.10], we get

$$\mathscr{L}\{f(t)\} + 2\frac{1}{s}\,\mathscr{L}\{f(t)\} = \frac{s}{s^2-4}.$$

Thus, we have $\mathscr{L}\{f(t)\} = \dfrac{s^2}{(s+2)^2(s-2)}$.

14.6 Further properties of Laplace transforms

1. If $\mathscr{L}\{f(t)\} = F(s)$, then

$$\mathscr{L}\{t^n f(t)\} = (-1)^n \frac{d^n}{ds^n} F(s) \quad (\text{for } n = 1, 2, 3, \ldots). \qquad [14.11]$$

We know that $\mathscr{L}\{f(t)\} = \int_0^\infty e^{-st} f(t)dt$. By Leibnitz's rule for differentiating under the integral sign, we have

$$\frac{dF}{ds} = \frac{d}{ds}\int_0^\infty e^{-st} f(t)dt = \int_0^\infty \frac{\partial}{\partial s} e^{-st} f(t)dt$$

$$= \int_0^\infty -te^{-st} f(t)dt = -\int_0^\infty e^{-st}\{tf(t)\}dt = -\mathscr{L}\{tf(t)\}.$$

Thus, we get $\mathscr{L}\{tf(t)\} = -\dfrac{d}{ds}\{F(s)\}$.

This proves the theorem for $n = 1$. By mathematical induction, if we assume that the theorem is true for $n = m$, then

$$\mathscr{L}\{t^m f(t)\} = (-1)^m \frac{d^m}{ds^m}\{F(s)\}.$$

Again, by Leibnitz's rule, we get $-\int_0^\infty e^{-st}\{t^{m+1} f(t)\}dt = (-1)^m F^{m+1}(s)$

or $\mathscr{L}\{t^{m+1} f(t)\} = (-1)^{m+1} \dfrac{d^{m+1}}{ds^{m+1}}\{F(s)\}$.

Since, whenever the theorem is true for $n = m$, it is true for $n = m + 1$ also, and the theorem holds for all positive integral values of n.

2. If $\mathscr{L}\{f(t)\} = F(s)$, then

$$\mathscr{L}\left\{\frac{f(t)}{t}\right\} = \int_s^{\infty} F(u)du. \qquad [14.12]$$

This shows that the integration of the transform of a function $F(s)$ corresponds to division of the function $f(t)$ by t.

From the definition of Laplace transform, it follows that

$$\int_s^{\infty} F(u)du = \int_s^{\infty}\left[\int_0^{\infty} e^{-ut} f(t)dt\right]du.$$

Here we may reverse the order of integration so that

$$\int_s^{\infty} F(u)du = \int_0^{\infty}\left[\int_s^{\infty} e^{-ut} f(t)du\right]dt = \int_0^{\infty} f(t)\left\{\frac{e^{-ut}}{-t}\right\}_s^{\infty} dt$$

$$= \int_0^{\infty} e^{-st}\frac{f(t)}{t}dt = \mathscr{L}\left\{\frac{f(t)}{t}\right\}.$$

Here we assume that $f(t)$ satisfies the conditions of the existence Theorem 14.1.

14.7 Laplace transforms of periodic functions

A function $f(t)$ is said to be *periodic* with period $p > 0$ if

$$f(t + p) = f(t) \text{ for all } t > 0.$$

Since periodic functions appear in many practical problems and in most cases are more complicated than single sine or cosine functions, we find it useful to establish the following result.

Theorem 14.6: The Laplace transform of a piecewise continuous *periodic* function $f(t)$ with period p is

$$\mathscr{L}\left\{f(t)\right\} = \frac{1}{1 - e^{-ps}}\int_0^{p} e^{-st} f(t)dt, \qquad s > 0. \qquad [14.13]$$

Proof: Since $f(t)$ is periodic and piecewise regular over every period, there exists a constant M such that

$$|f(t)| < M, \text{for all } t \geqslant 0.$$

Hence $f(t)$ is of exponential order $\sigma = 0$ by [14.2] and so $\mathscr{L}\{f(t)\}$ exists if $s > 0$. By definition,

$$\mathscr{L}\{f(t)\} = \int_0^\infty f(t)e^{-st}dt$$

$$= \int_0^p f(t)e^{-st}dt + \int_p^{2p} f(t)e^{-st}dt + \int_{2p}^{3p} f(t)e^{-st}dt + \ldots$$

Now in the second integral, let $t = T + p$; in the third integral, let $t = T + 2p$; and in general, let $t = T + np$ in the $(n+1)^{\text{th}}$ integral. In each case $dt = dT$ and the new limits become 0 and k. Hence, we get

$$\mathscr{L}\{f(t)\} = \int_0^p f(t)e^{-st}dt + \int_0^p f(T+p)e^{-s(T+p)}dT$$

$$+ \int_0^p f(T+2p)e^{-s(T+2p)}dT + \ldots$$

or $$\mathscr{L}\{f(t)\} = \int_0^p f(T)e^{-sT}dT + e^{-ps}\int_0^p f(T+p)e^{-sT}dT$$

$$+e^{-2ps}\int_0^p f(T+2p)e^{-sT}dT + \ldots$$

However, by hypothesis, since p is the period of $f(t)$,

$$f(T+p) = f(T+2p) = \ldots = f(T+np) = \ldots = f(T)$$

for all values of T. Thus, we have

$$\mathscr{L}\{f(t)\} = \int_0^p f(T)e^{-sT}dT + e^{-ps}\int_0^p f(T)e^{-sT}dT$$

$$+e^{-2ps}\int_0^p f(T)e^{-sT}dT + \ldots$$

or $$\mathscr{L}\{f(t)\} = \left(1 + e^{-ps} + e^{-2ps} + \ldots\right)\int_0^p f(T)e^{-sT}dT$$

$$= \frac{1}{1-e^{-ps}} \int_0^p f(T)e^{-sT}dT$$

since $1 + e^{-ps} + e^{-2ps} + \dots$ is a geometric series whose sum is $1/(1-e^{-ps})$.

Example 14.4 Find the transform of the rectangular wave shown in Fig 14.3.

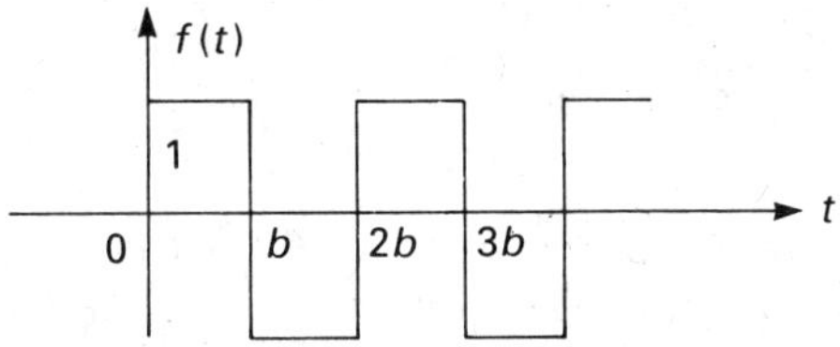

Fig. 14.3 *Rectangular wave.*

Solution 14.4 The period of the given function is $2b$. Hence

$$\mathscr{L}\{f(t)\} = \frac{1}{1-e^{-2bs}} \int_0^{2b} e^{-st} f(t)dt$$

$$= \frac{1}{1-e^{-2bs}} \left(\int_0^b 1e^{-st}dt + \int_b^{2b} (-1)e^{-st}dt \right)$$

or $$\mathscr{L}\{f(t)\} = \frac{1}{1-e^{-2bs}} \left(\left[\frac{e^{-st}}{-s}\right]_0^b - \left[\frac{e^{-st}}{-s}\right]_b^{2b} \right)$$

$$= \frac{1}{1-e^{-2bs}} \left(\frac{1-2e^{-bs}+e^{-2bs}}{s} \right) = \frac{(1-e^{-bs})^2}{s(1-e^{-bs})(1+e^{-bs})}$$

$$= \frac{(1-e^{-bs})}{s(1+e^{-bs})} = \frac{e^{bs/2}-e^{-bs/2}}{s(e^{bs/2}+e^{-bs/2})} = \frac{1}{s}\tanh\frac{bs}{2}.$$

Example 14.5 What is the Laplace transform of the *staircase function* (Fig.14.4 (a)) given by

$$f(t) = n+1, \quad np < t < (n+1)p, \quad n = 0, 1, 2, \dots$$

Solution 14.5 The simplest way of obtaining the Laplace transform of the staircase function is to consider it to be the difference between the two functions shown in Fig. 14.4(b). The transform of the linear function $(t+p)/p$ can easily be found by using properties [14.3] and (3) of Section 14.4. Thus, we have

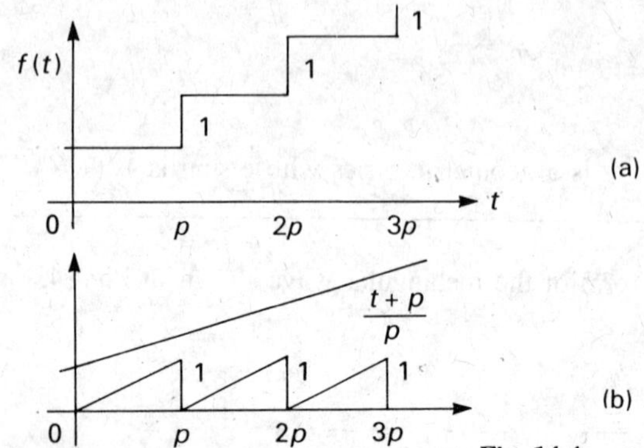

Fig. 14.4

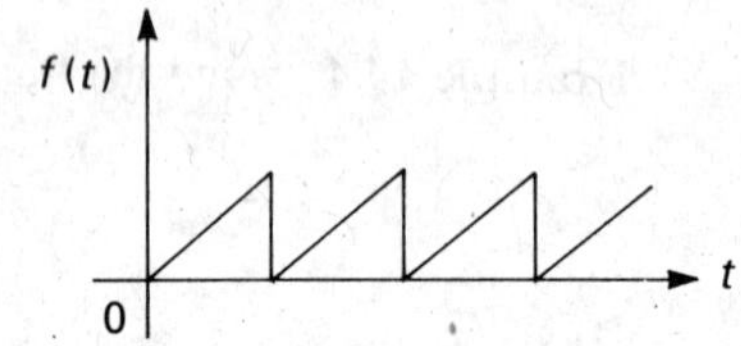

Fig. 14.5

Fig. 14.4(a) *Staircase function.* **(b)** *Solution of the staircase function.*
Fig. 14.5 *Saw-tooth function.*

$$\mathscr{L}\left(\frac{t+p}{p}\right) = \frac{1}{p}\left(\frac{1}{s^2} + \frac{p}{s}\right) \qquad [1]$$

The function $g(t) = t/p$ is called a *saw-tooth function* (Fig. 14.5) and its transform can easily be obtained as follows:

$$\mathscr{L}\{g(t)\} = \frac{1}{1-e^{-ps}} \int_0^p e^{-st} \frac{t}{p} dt$$

$$= \frac{1}{1-e^{-ps}} \frac{1}{p} \left[\frac{e^{-st}}{s^2}(-st-1)\right]_0^p = \frac{1-(1+ps)e^{-ps}}{ps^2(1-e^{-ps})}.$$

This can be simplified to

$$\mathscr{L}\{g(t)\} = \frac{1}{p}\left[\frac{1+ps}{s^2} - \frac{p}{s(1-e^{-ps})}\right] \qquad [2]$$

Thus from [1] and [2], we get

$$\mathscr{L}\{f(t)\} = \frac{1}{p}\left(\frac{1}{s^2} + \frac{p}{s}\right) - \frac{1}{p}\left[\frac{1+ps}{s^2} - \frac{p}{s(1-e^{-ps})}\right] = \frac{1}{s(1-e^{-ps})}.$$

14.8 Laplace transforms of some special functions

(i) Bessel function. We define a Bessel function of order n by

$$J_n(t) = \frac{t^n}{2^n\Gamma(n+1)}\left\{1 - \frac{t^2}{2(2n+2)} + \frac{t^4}{2.4(2n+2)(2n+4)} \cdots\right\}$$

from which we obtain $J_0(t) = 1 - \left(\frac{t}{2}\right)^2 + \frac{1}{(2!)^2}\left(\frac{t}{2}\right)^4 - \frac{1}{(3!)^2}\left(\frac{t}{2}\right)^6 + \dots$

and $J_1(t) = \frac{t}{2}\left[1 - \frac{1}{1!2!}\left(\frac{t}{2}\right)^2 + \frac{1}{2!3!}\left(\frac{t}{2}\right)^4 - \frac{1}{3!4!}\left(\frac{t}{2}\right)^6 + \dots\right]$

We can now obtain the transforms of $J_0(t)$ and $J_1(t)$. We have

$$\mathscr{L}\{J_0(t)\} = \mathscr{L}\left\{1 - \frac{t^2}{2^2} + \frac{t^4}{2^2 \cdot 4^2} - \frac{t^6}{2^2 \cdot 4^2 \cdot 6^2} + \dots\right\}$$

$$= \frac{1}{s} - \frac{1}{2^2}\frac{2!}{s^3} + \frac{1}{2^2 \cdot 4^2}\frac{4!}{s^5} - \frac{1}{2^2 \cdot 4^2 \cdot 6^2}\frac{6!}{s^7} + \dots$$

$$= \frac{1}{s}\left(1 + \frac{1}{s^2}\right)^{-1/2} = \frac{1}{(1+s^2)^{1/2}}.$$

Also, since $J_0'(t) = J_1(t)$, we get

$$\mathscr{L}\{J_1(t)\} = -\mathscr{L}\{J_0'(t)\} = -[s\mathscr{L}\{J_0(t)\} - 1] = 1 - \frac{s}{(s^2+1)^{1/2}}.$$

(ii) Error function. The error function or *probability integral* is defined as

$$\operatorname{erf}(\sqrt{t}) = \frac{2}{\sqrt{\pi}}\int_0^{\sqrt{t}} e^{-u^2}\,du = \frac{2}{\sqrt{\pi}}\sum_{n=0}^{\infty}\frac{(-1)^n}{n!}\frac{x^{n+(1/2)}}{2n+1}$$

The *complementary error function* is defined as $\operatorname{erfc}(\sqrt{t}) = \frac{2}{\sqrt{\pi}}\int_{\sqrt{t}}^{\infty} e^{-u^2}\,du.$

We show that $\mathscr{L}\{\operatorname{erf}(\sqrt{t})\} = \frac{1}{s\sqrt{s+1}}$ and hence deduce that

$$\mathscr{L}\{\operatorname{erfc}(\sqrt{t})\} = \frac{(s+1)^{1/2} - 1}{s(s+1)^{1/2}}.$$

Consider $\mathscr{L}\{\operatorname{erf}\sqrt{t}\} = \frac{2}{\sqrt{\pi}}\sum_{n=0}^{\infty}\frac{(-1)^n}{n!(2n+1)}\mathscr{L}\{t^{n+(1/2)}\}$

$$= \frac{2}{\sqrt{\pi}}\sum_{n=0}^{\infty}\frac{(-1)^n}{n!(2n+1)}\frac{\Gamma(n+1+1/2)}{s^{n+1+(1/2)}}.$$

Using function **11** from Table 14.1. Therefore, we have

$$\mathscr{L}\{\operatorname{erf}\sqrt{t}\} = \frac{2}{\sqrt{\pi}}\frac{1}{s^{3/2}}\sum_{n=0}^{\infty}\frac{(-1)^n}{n!(2n+1)}\frac{(n+\frac{1}{2})\Gamma(n+\frac{1}{2})}{s^n}$$

$$= \frac{1}{\sqrt{\pi}}\frac{\Gamma(1/2)}{s^{3/2}}\left\{1-\frac{1}{2}\frac{1}{s}+\frac{1}{2}\frac{3}{4}\frac{1}{s^2}-\frac{1}{2}\frac{3}{4}\frac{5}{6}\frac{1}{s^3}+\ldots\right\}$$

$$= \frac{1}{s^{3/2}}\left(1-\frac{1}{s}\right)^{-1/2} = \frac{1}{s\sqrt{s+1}} \text{ (since } \Gamma\left(\frac{1}{2}\right)=\sqrt{\pi}).$$

Thus, we have $\mathscr{L}\{\operatorname{erf}\sqrt{t}\} = 1/(s\sqrt{s+1})$. Also

$$\int_0^{\infty} e^{-u^2}\,du = \int_0^{\sqrt{t}} e^{-u^2}\,du + \int_{\sqrt{t}}^{\infty} e^{-u^2}\,du.$$

But $\int_0^{\infty} e^{-u^2}\,du = \frac{1}{2}\sqrt{\pi}$. Thus $\frac{2}{\sqrt{\pi}}\int_{\sqrt{t}}^{\infty} e^{-u^2}\,du = 1 - \frac{2}{\sqrt{\pi}}\int_0^{\sqrt{t}} e^{-u^2}\,du.$

That is, $\operatorname{erfc}\sqrt{t} = 1 - \operatorname{erf}\sqrt{t}$. Therefore

$$\mathscr{L}\{\operatorname{erfc}\sqrt{t}\} = \mathscr{L}\{1-\operatorname{erf}\sqrt{t}\} = \frac{1}{s} - \frac{1}{s\sqrt{s+1}} = \frac{\sqrt{s+1}-1}{s\sqrt{s+1}}.$$

Exercise 14B

14.4 If $\mathscr{L}\{f(t)\} = F(s)$, prove that

$$\mathscr{L}\{f'''(t)\} = s^3F(s) - s^2F(0) - sF'(0) - F''(0).$$

14.5 If $\mathscr{L}\{f''(t)\} = \tan^{-1}(1/s)$, $f(0) = 2$ and $f'(0) = -1$, find $\mathscr{L}\{f(t)\}$.

14.6 Find the Laplace transforms of the following functions

(i) $(\sin^2 t)/(t)$ (ii) $t^2(1-\cos t)$ (iii) $\sqrt{t}$ (iv) $tJ_0(at)$ (v) $t^3\sin t$

(vi) $\dfrac{(1-e^{-t})}{t}$ (vii) $\displaystyle\int_0^t \frac{1}{t}e^{-4t}\sin 3t\,dt$ (viii) $\displaystyle\int_0^t t\sinh t\,dt$

(ix) $t(\operatorname{erf}\sqrt{t})$ (x) $f(t) = \begin{cases} t & \text{for } 0 \le t \le 2 \\ 2 & \text{for } t > 2 \end{cases}$

14.7 A periodic function $f(t)$ of period $2a$ is defined by

$$f(t) = \begin{cases} t & \text{for} \quad 0 < t \le a \\ 2a - t & \text{for} \quad a < t < 2a \end{cases}$$

Show that $\mathscr{L}\{f(t)\} = (1/s^2)\tanh(as/2)$.

14.8 If $f(t) = t^2;\ 0 < t < 2$ and $f(t+2) = f(t)$, find $\mathscr{L}\{f(t)\}$.

14.9 If $f'(t) + f(t) + \int_0^t f(t)dt = 1,\ f(0) = 0$, find $\mathscr{L}\{f(t)\}$.

14.10 Find $\mathscr{L}\{f(t)\}$ if $f'''(t) + 2f''(t) - f'(t) - 2f(t) = 0$, and

$f(0) = f'(0) = 0;\ f''(0) = 1.$

14.11 Prove that $\mathscr{L}\{J_n(t)\} = \dfrac{[(\sqrt{s^2+1})^n - s]^n}{\sqrt{s^2+1}}$

14.12 If $F(s) = \mathscr{L}\{f(t)\}$, show that $\mathscr{L}\left\{\int_0^t dt_1 \int_0^{t_1} f(u)du\right\} = \dfrac{F(s)}{s^2}$ and hence generalise the result.

14.9 Laplace transform of the unit step function

The unit step function or the *Heaviside function* is defined as

$$H(t-a) = \begin{cases} 0 & \text{if} \quad t < a \\ 1 & \text{if} \quad t > a \end{cases}$$

where a is a non-negative constant. In particular, when $a = 0$, we have

$$H(t) = \begin{cases} 0 & \text{if} \quad t < 0 \\ 1 & \text{if} \quad t > 0 \end{cases}$$

The unit step function is the basic building block of certain functions whose knowledge greatly increases the usefulness of the Laplace transformation.

It is easy to show that the Laplace transform of $H(t-a)$ is given by

$$\mathscr{L}\{H(t-a)\} = \frac{e^{-as}}{s} \qquad [14.14]$$

This follows directly from the definition because

$$\mathscr{L}\{H(t-a)\} = \int_0^\infty e^{-st}H(t-a)dt$$

$$= \int_0^a e^{-st}0dt + \int_a^\infty e^{-st}1dt = \left.\frac{e^{-st}}{-s}\right]_a^\infty = \frac{e^{-as}}{s}.$$

In particular, when $a = 0$, we get $\mathscr{L}\{H(t)\} = 1/s$.

Heaviside shift theorem: If $\mathscr{L}\{f(t)\} = F(s)$, then

$$\mathscr{L}\{f(t-a)H(t-a)\} = e^{-as}F(s). \quad [14.15]$$

Proof: By definition, we have

$$\mathscr{L}\{f(t-a)H(t-a)\} = \int_0^\infty e^{-st}f(t-a)H(t-a)dt$$

$$= \int_0^a e^{-st}f(t-a)\,0dt + \int_a^\infty e^{-st}f(t-a)1dt$$

$$= \int_0^\infty e^{-s(u+a)}f(u)du = e^{-sa}\int_0^\infty e^{-su}f(u)du \quad (\text{where } u = t-a).$$

Example 14.6 Find the transform of the function $f(t) = \begin{cases} 1 & \text{if } 0 < t < \pi \\ 0 & \text{if } \pi < t < 2\pi \\ \sin t & \text{if } t > 2\pi \end{cases}$

Solution 14.6 We can represent $f(t)$ in terms of unit step functions as

$$f(t) = H(t) - H(t-\pi) + H(t-2\pi)\sin t$$

$$\mathscr{L}\{f(t)\} = \mathscr{L}\{H(t)\} - \mathscr{L}\{H(t-\pi)\} + \mathscr{L}\{H(t-2\pi)\sin t\}$$

$$= \frac{1}{s} - \frac{e^{-\pi s}}{s} + \frac{e^{-2\pi s}}{s^2+1}$$

14.10 Laplace transform of the unit impulse function

We shall now analyze the problem associated with an *instantaneous impulse* acting on a mass; for example, a hammer striking the mass initially. To discuss such cases a new concept is introduced as follows: For each $\in > 0$, we define $f_\in$ on $[0, \infty)$ by

$$f_\in(t) = \begin{cases} 1/\in & \text{for } 0 \le t < \in \\ 0 & \text{for } t \ge \in \end{cases}$$

that is, $f_\in(t) = (1/\in)[1 - H(t-\in)]$ (for $t \geqslant 0$). The function $f_\in(t)$ is called an $\in$ pulse, since

$$\int_0^\infty f_\in(t)dt = \int_0^\in \frac{1}{\in}dt = 1(\forall \in > 0).$$

It follows that if $f_\in(t)$ describes a force for $t \geqslant 0$, then $f_\in$ has impulse unity. Therefore $f_\in$ is called a *unit impulse* force. We have

$$\mathscr{L}\{f_\in(t)\} = \int_0^\in e^{-st}\frac{1}{\in}dt = \frac{1 - e^{-\in s}}{\in s}, \forall \in > 0.$$

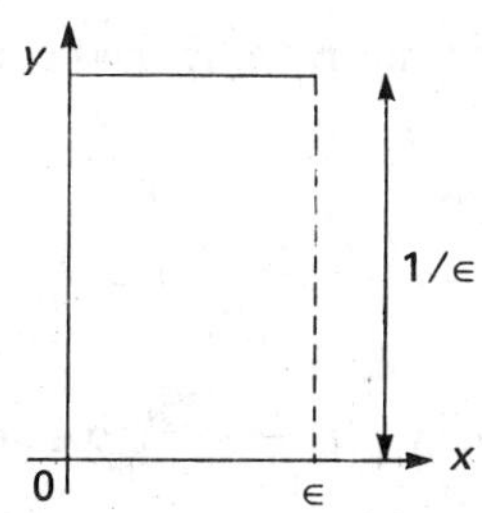

Fig. 14.6 *Unit impulse function.*

The graph of this function is shown in Fig. 14.6. Strictly speaking, $\lim_{\in \to 0} f_\in(t)$ does not exist as a function and hence $\mathscr{L}\{\lim_{\in \to 0} f_\in(t)\}$ is not defined; but if we throw caution to the winds, then

$$\delta(t) = \lim_{\in \to 0} f_\in(t)$$

is a *quasi-function* that is infinite for $t = 0$ and zero for $t > 0$ and can be shown to have the properties

$$\int_0^\infty \delta(t)dt = 1 \quad \text{and} \quad \mathscr{L}\ \{\delta(t)\} = 1.$$

This quasi-function $\delta(t)$ is called the *Dirac delta function* (or unit impulse function), named after the famous English physicist, P.A.M. Dirac (1902–84).

One important formal property of the δ-function is its ability to isolate or reproduce a particular value of a function $f(t)$ according to Theorem 14.8.

Theorem 14.8 If $f(t)$ is any function and $\delta(t-a)$ the unit impulse function, then

$$\int_0^{\infty} f(t)\delta(t-a)dt = f(a).$$

Proof: To prove this, we consider a function $g(t) = \int f(t)dt$, so that $g(t) = f(t)$.

Then we have $\int_0^{\infty} f(t)\delta(t-a)dt = \int_a^{a+\in} g'(t)\frac{1}{\in}dt$ obtained from

$$\delta(t-a) = \begin{cases} 1/\in & \text{for} \quad a \le t \le a+\in \\ 0 & \text{otherwise} \end{cases}$$

Thus $\int_0^{\infty} f(t)\delta(t-a)dt = \frac{1}{\in}[g(a+\in) - g(a)] = \frac{1}{\in}(a+\in -a)g'(a+\theta \in)$

by using the first mean value theorem where $0 < \theta < 1$. Taking the limit as $\in \to 0$, we get

$$\int_0^{\infty} f(t)\delta(t-a)dt = g'(a) = f(a).$$

In particular, when $f(t) = e^{-st}$, we get

$$\int_0^{\infty} e^{-st}\delta(t-a)dt = e^{-sa} \quad \text{or} \quad \mathscr{L}\{\delta(t-a)\} = e^{-sa}.$$

And, when $a = 0$, this becomes $\mathscr{L}\{\delta(t)\} = 1$.

Exercise 14C

14.13 Find the Laplace transform of $f(t)$ using the unit step function, where

$$f(t) = \begin{cases} 1 & \text{if} \quad 0 \le t < 1 \\ 2 & \text{if} \quad 1 \le t < 3 \\ 4 & \text{if} \quad 3 \le t < 4 \\ -2 & \text{if} \quad 4 \le t \end{cases}$$

14.14 Given that $f(t) = \begin{cases} 1 & \text{if} \quad 0 \le t < \pi/6, \\ \sin t & \text{if} \quad \pi/6 \le t \end{cases}$ find $L\{f(t)\}$.

[**Hint:** Write $f(t) = 1 + H(t-\pi/6)(\sin t - 1)$ and note $\sin t = (\sqrt{3}/2)\sin(t-\pi/6) + (1/2)\cos(t-\pi/6)$].

14.15 Determine the Laplace transform of the following functions

(i) $f(t) = 6 - 2H(t - \pi) - 4H(t - 2\pi)$

(ii) $f(t) = e^{2t}H(t - \log 5)$ (iii) $f(t) = (t - a)H(t - a)$

(iv) $f(t) = t^2H(t - 2)$ (v) $f(t) = H(1 - e^{-t})$

14.16 Evaluate $\int_0^{\infty} e^{-4t}\delta'(t-2)dt$, where $\delta(t-a) = \begin{cases} 1/\in & \text{for } a \le t \le a+\in \\ 0 & \text{otherwise} \end{cases}$

14.17 Compute the Laplace transform of $f(t) = tH(t-1) + t^2\delta(t-1)$.

14.11 Inverse Laplace transforms

Here, we shall consider the question of determining a function from its transform — that is, can the Laplace transform be inverted? Let us first consider the definition of the inverse Laplace transform. If $\mathscr{L}\{f(t)\} = F(s)$, then $f(t)$ is called the inverse transform of $F(s)$, and is denoted by $\mathscr{L}^{-1}\{F(s)\} = f(t)$.

The fact that the definition is well-defined depends on a type of uniqueness assertion for the Laplace transform known as *Lerch's theorem* which states:

If f and g are continuous functions on $[0,\infty)$ such that $\mathscr{L}\{f(t)\} = \mathscr{L}\{g(t)\}$ or $F(s) = G(s)(\forall s > s(0))$, then $f(t) = g(t)(\forall t \ge 0)$.

Some of the important properties of the inverse Laplace transforms are analogous with the corresponding properties of Laplace transforms:

1. *Linearity property.* If c_1 and c_2 are any constants while $F(s)$ and $G(s)$ are the Laplace transforms of $f'(t)$ and $g(t)$ respectively, then

$$\mathscr{L}^{-1}\{c_1F(s) + c_2G(s)\} = c_1f(t) + c_2g(t).$$

This result can be extended to more than two functions.

2. *First translation or shifting property.* If $\mathscr{L}^{-1}\{F(s)\} = f(t)$, then

$$\mathscr{L}^{-1}\{F(s-a)\} = e^{at}f(t).$$

3. *Second translation or shifting property.* If $\mathscr{L}^{-1}\{F(s)\} = f(t)$, then

$$\mathscr{L}^{-1}\{e^{-as}F(s)\} = \begin{cases} f(t-a) & \text{for} \quad t > a. \\ 0 \quad \text{for} & t < a. \end{cases}$$

4. *Change of scale property.* If $\mathscr{L}^{-1}\{F(s)\} = f(t)$, then

$$\mathscr{L}^{-1}\{F(as)\} = \frac{1}{a} f\left(\frac{t}{a}\right).$$

5. *Inverse Laplace transforms of derivatives.* If $\mathscr{L}^{-1}\{F(s)\} = f(t)$, then

$$\mathscr{L}^{-1}\{F^n(s)\} = \mathscr{L}^{-1}\left\{\frac{d^n}{ds^n}F(s)\right\} = (-1)^n t^n f(t).$$

6. *Inverse Laplace transform of integrals.* If $\mathscr{L}^{-1}\{F(s)\} = f(t)$, then

$$\mathscr{L}^{-1}\left\{\int_s^\infty F(u)du\right\} = \frac{f(t)}{t}.$$

7. *Multiplication by s^n.* If $\mathscr{L}^{-1}\{F(s)\} = f(t)$ and $f(0) = 0$, then

$$\mathscr{L}^{-1}\{sF(s)\} = f'(t).$$

8. *Division by s.* If $\mathscr{L}^{-1}\{F(s)\} = f(t)$, then

$$\mathscr{L}^{-1}\left\{\frac{F(s)}{s}\right\} = \int_0^t f(u)du.$$

This result can be extended to $\mathscr{L}^{-1}\left\{\frac{F(s)}{s^n}\right\} = \int_0^t \int_0^t \cdots \int_0^t f(u)du^n.$

The proofs of these properties are simple and left as an exercise. As specific illustrations, some examples are given below.

Example 14.7 Let $\mathscr{L}\{f(t)\} = 3/(s^2+4)$, for all $s > 0$. Find $f(t)$.

Solution 14.7 Rewriting $3/(s^2+4) = (3/2)\cdot[2/(s^2+2^2)]$ and using (5) of Table 14.1 we see that $f(t) = (3/2).\sin 2t$ or $\mathscr{L}^{-1}\{3/(s^2+4)\} = (3/2)\sin 2t$. One of the most useful techniques for determining inverse transforms is partial fraction expansion.

Example 14.8 If $\mathscr{L}\{F(s)\} = \dfrac{s+1}{s^3+s}$ for all $s > 0$, find $f(t)$.

Solution 14.8 By partial fraction expansion, we have

$$\frac{s+1}{s(s^2+1)} = \frac{A}{s} + \frac{Bs+C}{s^2+1} \quad \text{(where } A=1, \quad B=-1 \text{ and } C=1)$$

$$\therefore \quad \frac{s+1}{s(s^2+1)} = \frac{1}{s} + \frac{-s+1}{s^2+1}.$$

By the linearity property of $\mathscr{L}^{-1}$, we have

$$\mathscr{L}^{-1}\left\{\frac{s+1}{s^3+s}\right\} = \mathscr{L}^{-1}\left\{\frac{1}{s}\right\} - \mathscr{L}^{-1}\left\{\frac{s}{s^2+1}\right\} + \mathscr{L}^{-1}\left\{\frac{1}{s^2+1}\right\}$$

$$= 1 - \cos t + \sin t \quad \text{(from Table 14.1).}$$

Example 14.9 If $\mathscr{L}\{f(t)\} = \dfrac{2s}{4s^2+16}$, find $f(t)$.

Solution 14.9 By change of scale property

$$\mathscr{L}^{-1}\left\{\frac{2s}{4s^2+16}\right\} = \frac{1}{2}\cos 2t \quad \text{since} \quad \mathscr{L}^{-1}\left\{\frac{s}{s^2+16}\right\} = \cos 4t.$$

Example 14.10 If $\mathscr{L}\{f(t)\} = \dfrac{e^{-5s}}{(s-2)^4}$, find $f(t)$.

Solution 14.10 We know $\mathscr{L}^{-1}\left\{\dfrac{1}{(s-2)^4}\right\} = e^{2t}\,\mathscr{L}^{-1}\left\{\dfrac{1}{s^4}\right\} = e^{2t}\dfrac{t^3}{3!} = \dfrac{1}{6}t^3e^{2t}.$

Thus by second shifting property,

$$\mathscr{L}^{-1}\left\{\frac{e^{-5s}}{(s-2)^4}\right\} = \begin{cases} \frac{1}{6}(t-5)^3\, e^{2(t-5)} & \text{if } t > 5 \\ 0 & \text{if } t < 5 \end{cases}$$

14.12 Convolution theorem

Theorem 14.9 If $\mathscr{L}\{f(t)\} = F(s)$ and $\mathscr{L}\{g(t)\} = G(s)$, what is the inverse transform of $F(s)G(s)$? The inverse $h(t)$, called the convolution of f and g is defined as follows:

$$h(t) = f * g(t) = \int_0^t (t-u)g(u)du, \text{ for all } t \geqslant 0. \qquad [14.16]$$

The convolution operation $*$ has several properties that resemble those of multiplication. For instance, convolution is a commutative, associative and distributive operation:

$$f * g(t) \equiv g * f(t) \quad (\forall t \geqslant 0); \quad f * (g * h)(t) \equiv (f * g) * h(t) \quad (\forall t \geqslant 0);$$

$$f * (g + h)(t) = f * g(t) + f * h(t) \quad (\forall t \geqslant 0).$$

But one of the most important properties of convolution is the fact that the Laplace transform of the convolution of two functions is the product of the Laplace transform of the functions as stated in Theorem 14.10.

Theorem 14.10 Suppose $f(t)$ and $g(t)$ are the inverse transforms of $F(s)$ and $G(s)$ respectively, and satisfy the hypothesis of the existence theorem, then the inverse transform $h(t)$ of the product $F(s)G(s)$ is the convolution of $f(t)$ and $g(t)$ (written $f * g(t)$ and defined by

$$h(t) = f * g(t) = \int_0^t f(t-u)g(u)du$$

i.e, $$\mathscr{L}^{-1}\{F(s)G(s)\} = \int_0^t f(t-u)g(u)du \ldots \qquad [14.17]$$

Proof: It is sufficient if we show that

$$F(s)G(s) = \mathscr{L}\left\{\int_0^t f(t-u)g(u)du\right\}.$$

Now $$F(s)G(s) = \left[\int_0^\infty e^{-sv} f(v)dv\right]\left[\int_0^\infty e^{-su} g(u)du\right]$$

$$= \int_0^\infty \int_0^\infty e^{-s(u+v)} f(v)g(u)dvdu = \int_0^\infty \left[\int_0^\infty e^{-s(v+u)} f(v)dv\right] g(u)du$$

where the integration is extended over the first quadrant ($u \geqslant 0, v \geqslant 0$) in the uv-plane. We now introduce a new variable t in the inner integral of the last expression by taking $v = t - u$, so that $u + v = t$ and $dv = dt$(u being fixed during this integration). Thus

$$F(s)G(s) = \int_0^\infty \left[\int_u^\infty e^{-st} f(t-u)dt\right] g(u)du$$

$$= \int_0^\infty \int_u^\infty e^{-st} f(t-u)g(u)dtdu.$$

This integration is extended over the first half of the first quadrant ($t - u \geqslant 0$) in the tu-plane, and reversing the order as suggested in Fig 14.6, we get

$$F(s)G(s) = \int_0^\infty \left[\int_0^t e^{-st} f(t-u)g(u)du\right] dt$$

$$= \int_0^\infty e^{-st}\left[\int_0^t f(t-u)g(u)du\right] dt = \mathcal{L}\left\{\int_0^t f(t-u)g(u)du\right\}.$$

Example 14.11 Evaluate $\mathcal{L}^{-1}\left\{\frac{1}{s^2(s+1)^2}\right\}$, using the convolution theorem.

Solution 14.11 We have $\mathcal{L}^{-1}\left\{\frac{1}{s^2}\right\} = t,$ and $\mathcal{L}^{-1}\left\{\frac{1}{(s+1)^2}\right\} = te^{-t}.$

By the convolution theorem

$$\mathcal{L}^{-1}\left\{\frac{1}{s^2(s+1)^2}\right\} = \int_0^t (ue^{-u})(t-u)du = \int_0^t (ut-u^2)e^{-u}du$$

$$= (ut-u^2)\frac{e^{-u}}{-1} - (t-2u)e^{-u} + (-2)(-e^{-u})]_0^t = te^{-t} + 2e^{-t} + t - 2.$$

Exercise 14D

14.18 Determine the inverse Laplace transform of each of the following functions:

(i) $\frac{3}{s+4}$ (ii) $\frac{s}{s^2+2s+6}$ (iii) $\frac{2s+1}{s(s+1)}$ (iv) $\log\left(\frac{s+3}{s+2}\right)$

(v) $\frac{s-5}{s^2+6s+13}$ (vi) $\frac{e^{-2s}}{s(s+1)}$ (vii) $\frac{s^2}{(s^2+9)(s^2+16)}$

(viii) $\frac{2s^2+s-10}{(s-4)(s^2+2s+2)}$ (ix) $\log\left(1+\frac{1}{s^2}\right)$ (x) $\frac{s}{s^4+64}$

14.19 Find $\mathcal{L}^{-1}\left[\frac{1}{(s^2+a^2)^2}\right]$ by convolution.

14.20 Find (i) $\mathcal{L}^{-1}\left\{\log\left(\frac{s+2}{s+1}\right)\right\}$ (ii) $\mathcal{L}^{-1}\left\{\frac{1}{s}\log\left(\frac{s+2}{s+1}\right)\right\}$

using the properties of inverse Laplace transforms.

14.21 Find $\mathscr{L}^{-1}\left\{\dfrac{1}{(s+1)(s^2+1)}\right\}$ using the convolution theorem.

14.22 Using the convolution theorem, verify

$$\int_0^t \sin u \cos(t-u)du = \frac{1}{2}t \sin t.$$

14.23 Show that $\displaystyle\int_0^t\int_0^t\int_0^t f(t)dt^3 = \int_0^t \frac{(t-u)^2}{2!}f(u)du$ and hence derive

$$\int_0^t\int_0^t \cdots \int_0^t f(t)dt^n = \int_0^t \frac{(t-u)^{n-1}}{(n-1)!}f(u)du.$$

14.24 Show that (i) $\mathscr{L}^{-1}\left\{\dfrac{e^{-x\sqrt{s}}}{\sqrt{s}}\right\} = \dfrac{e^{-x^2/4t}}{\sqrt{\pi t}}$

(ii) $\displaystyle\int_0^t J_0(u)\sin(t-u)du = \frac{1}{2}tJ_1(t)$ (iii) $\mathscr{L}^{-1}\left\{\log \dfrac{s+\sqrt{s^2+1}}{2s}\right\} = \dfrac{1-J_0(t)}{t}$.

14.13 Applications to differential equations

Suppose we wish to find the particular solution of the differential equation

$$ay'' + by' + cy = f(t) \qquad [14.18]$$

which satisfies the initial conditions $y(0) = y_0$ and $y'(0) = y_0'$. We could find the general solution and then evaluate the arbitrary constants in accordance with the given initial conditions. But the use of Laplace transforms has several advantages. Taking the Laplace transform of each side of [1] and using its *linearity*, it follows

$$a\mathscr{L}(y'') + b\mathscr{L}(y') + c\mathscr{L}(y) = \mathscr{L}\{f(t)\}.$$

Using the formulas [14.7] and [14.8] of Theorem 14.4, we have

$$a[s^2\mathscr{L}(y) - sy_0 - y_0'] + b[s\mathscr{L}(y) - y_0] + c\mathscr{L}(y) = \mathscr{L}\{f(t)\}$$

and hence $\quad (as^2 + bs + c)\mathscr{L}(y) = \mathscr{L}\{f(t)\} + (as+b)y_0 + ay_0'.$

Solving for $\mathscr{L}(y)$ yields $\mathscr{L}(y) = \dfrac{L\{f(t)\} + (as+b)y_0 + ay_0'}{as^2 + bs + c}$.

The function $\mathscr{L}\ \{f(t)\}$ is a specific function of s, since $f(t)$ is known; and since a, b, c, y_0 and y_0' are known constants, $\mathscr{L}(y)$ is completely known as a function of s. If we can now find which function of y has the value

$$\frac{\mathscr{L}\{f(t)\} + (as+b)y_0 + ay_0'}{as^2 + bs + c}$$

then this function will be the solution of our problem. Of course, we assume that the functions $f(t), y, y'$ and y'' must have Laplace transforms.

Example 14.12 Find the solution of $y'' + y' - 2y = \sin t$ that satisfies the initial conditions $y(0) = 0$ and $y'(0) = 0$.

Solution 14.12 Applying the Laplace transform to both sides of the given equation and by linearity, we have

$$\mathscr{L}(y'') + \mathscr{L}\ (y') - 2\mathscr{L}(y) = \mathscr{L}(\sin t)$$

$$\mathscr{L}(\sin t) = \frac{1}{s^2+1}; s > 0 \quad \text{(by Table 14.1)}$$

$$\therefore\ [s^2\ \mathscr{L}(y) - s(0) - (0)] + [s\ \mathscr{L}(y) - (0)] - 2\mathscr{L}(y) = \frac{1}{s^2+1}$$

so that $\mathscr{L}(y) = \dfrac{1}{(s^2+s-2)(s^2+1)} = \dfrac{1}{(s-1)(s+2)(s^2+1)};\quad \forall\ s > 1.$

We have the partial fraction expansion

$$\frac{1}{(s-1)(s+2)(s^2+1)} = \frac{A}{s-1} + \frac{B}{s+2} + \frac{Cs+D}{s^2+1}$$

where $A = 1/6, B = -1/15, C = -1/10$ and $D = -3/10$. Therefore, we get

$$y(t) = \mathscr{L}^{-1}\left\{\frac{1}{(s-1)(s+2)(s^2+1)}\right\}$$

$$= \frac{1}{6}\mathscr{L}^{-1}\left\{\frac{1}{s-1}\right\} - \frac{1}{15}\mathscr{L}^{-1}\left\{\frac{1}{s+2}\right\} - \frac{1}{10}\mathscr{L}^{-1}\left\{\frac{s}{s^2+1}\right\} - \frac{3}{10}\mathscr{L}^{-1}\left\{\frac{1}{s^2+1}\right\}$$

$$= \frac{1}{6}e^t - \frac{1}{15}e^{-2t} - \frac{1}{10}\cos t - \frac{3}{10}\sin t$$

as the solution to the given initial value problem.

Example 14.13 Solve the initial value problem

$$\frac{d^2y}{dt^2} - 3\frac{dy}{dt} + 2y = e^{3t};\, y(0) = 1, y'(0) = 0.$$

Solution 14.13 Taking Laplace transforms of both sides of the differential equation gives

$$s^2\,\mathscr{L}\,\{y(t)\} - s - 3[s\,\mathscr{L}\,\{y(t)\} - 1] + 2\,\mathscr{L}\,\{y(t)\} = \frac{1}{s-3}$$

and this implies that $\mathscr{L}\,\{y(t)\} = \dfrac{1}{(s-3)(s^2-3s+2)} + \dfrac{s-3}{s^2-3s+2}$

$$= \frac{1}{(s-1)(s-2)(s-3)} + \frac{s-3}{(s-1)(s-2)}. \qquad [1]$$

To find $y(t)$, we expand each term on the right hand side of [1] in partial fractions. Thus,

$$\frac{1}{(s-1)(s-2)(s-3)} = \frac{A}{s-1} + \frac{B}{s-2} + \frac{C}{s-3}$$

or $\quad 1 = A(s-2)(s-3) + B(s-1)(s-3) + C(s-1)(s-2).$

Substituting $s = 1$ gives $A = 1/2$. Substituting $s = 2$ gives $B = -1$. Substituting $s = 3$ gives $C = 1/2$. Hence

$$\frac{1}{(s-1)(s-2)(s-3)} = \frac{1}{2(s-1)} - \frac{1}{s-2} + \frac{1}{2(s-3)}.$$

Similarly, we can write

$$\frac{s-3}{(s-1)(s-2)} = \frac{D}{s-1} + \frac{E}{s-2} \quad \text{or} \quad s - 3 = D(s-2) + E(s-1).$$

Substituting $s = 2$ gives $E = -1$. Substituting $s = 1$ gives $D = 2$. Hence

$$\mathscr{L}\{y(t)\} = \frac{1}{2(s-1)} - \frac{1}{s-2} + \frac{1}{2(s-3)} + \frac{2}{s-1} - \frac{1}{s-2}$$

$$= \frac{5}{2(s-1)} - \frac{2}{s-2} + \frac{1}{2(s-3)}$$

or $\quad \mathscr{L}\,\{y(t)\} = \mathscr{L}\left\{\frac{5}{2}e^{t} - 2e^{2t} + \frac{1}{2}e^{3t}\right\}$ or $y(t) = \frac{5}{2}e^{t} - 2e^{2t} + \frac{1}{2}e^{3t}.$

Example 14.14 Solve: $y'' + y = t\cos 2t;\ y = 0, y'' = 0$ when $t = 0.$

Solution 14.14 Taking the Laplace transform of both sides of the above differential equation, we get $\mathscr{L}\,\{y''\} + \mathscr{L}\,\{y\} = \mathscr{L}\,\{t\cos 2t\}$ or

$$s^2\,\mathscr{L}\,\{y(t)\} - sy(0) + \mathscr{L}\,\{y(t)\} = \mathscr{L}\{t\cos 2t\}. \qquad [1]$$

We know that $\quad \mathscr{L}\,\{\cos 2t\} = \dfrac{s}{s^2+4}$

$$\mathscr{L}\{t\cos 2t\} = \frac{1}{s^2+4} - \frac{8}{(s^2+4)^2}. \qquad [2]$$

Substituting in [1], we get

$$(s^2+1)\,\mathscr{L}\,\{y(t)\} = \frac{1}{s^2+4} - \frac{8}{(s^2+4)^2}$$

or $$\mathscr{L}\,\{y(t)\} = \frac{1}{(s^2+1)(s^2+4)} - \frac{8}{(s^2+1)(s^2+4)^2}$$

or $$\mathscr{L}\,\{y(t)\} = -\frac{5}{9(s^2+1)} + \frac{5}{9(s^2+4)} + \frac{8}{3(s^2+4)^2}$$

or $$y(t) = -\frac{5}{9}\sin t + \frac{5}{18}\sin 2t + \frac{8}{3}\,\mathscr{L}^{-1}\left\{\frac{1}{(s^2+4)^2}\right\}. \qquad [3]$$

From [2], $\quad \mathscr{L}\,\{t\cos 2t\} - \dfrac{1}{s^2+4} - \dfrac{8}{(s^2+4)^2} - \mathscr{L}\left\{\dfrac{\sin 2t}{2}\right\} - \dfrac{8}{(s^2+4)^2}:$

Now $\quad \dfrac{8}{(s^2+4)^2} = \mathscr{L}\left\{\dfrac{\sin 2t}{2} - t\cos 2t\right\}.$

Thus $\quad \mathscr{L}^{-1}\left\{\dfrac{1}{(s^2+4)^2}\right\} = \dfrac{1}{8}\left\{\dfrac{\sin 2t}{2} - t\cos 2t\right\}.$

$$y(t) = -\frac{5}{9}\sin t + \frac{5}{18}\sin 2t + \frac{1}{3}\left(\frac{1}{2}\sin 2t - t\cos 2t\right) \quad \text{(from[3])}$$

or $$y(t) = -\frac{5}{9}\sin t + \frac{4}{9}\sin 2t - \frac{t}{3}\cos 2t.$$

14.14 Applications in mechanics

Let us consider some elementary applications of Laplace transforms to the study of forced vibrations in linear springs.

Example 14.15 A mass m is attached to a linear spring having spring constant k and there is an applied force $f(t)$ for each $t \geqslant 0$. Assuming that no damping is present and that the mass is initially at equilibrium, find the position of the mass relative to equilibrium.

Solution 14.15 We know that the equation of motion is given by

$$my'' + ky = f(t)\ ;\ y(0) = y'(0) = 0. \qquad [1]$$

Taking the Laplace transform on both sides of [1] leads to

$$m\ \mathscr{L}\ (y'') + K\ \mathscr{L}(y) = \mathscr{L}\ \{f(t)\} \quad \text{or} \quad ms^2\ \mathscr{L}(y) + k\ \mathscr{L}(y) = \mathscr{L}\ \{f(t)\}$$

$$\text{or} \quad \mathscr{L}(y) = \frac{\mathscr{L}\{f(t)\}}{ms^2 + k} = \frac{1}{m}\frac{\mathscr{L}\{f(t)\}}{s^2 + (k/m)}$$

$$\text{and} \quad \mathscr{L}(y) = \frac{\mathscr{L}\{f(t)\}}{m(s^2 + \omega^2)}, \quad \text{where} \quad \omega = \sqrt{\frac{k}{m}}.$$

$\mathscr{L}(y)$ can be written as the product of two functions

$$\mathscr{L}\ \{f(t)\} \quad \text{and} \quad \frac{1}{m(s^2 + \omega^2)}.$$

So, from the convolution theorem the solution has the form

$$y(t) = \frac{1}{m\omega}\int_0^t \sin(\omega(t-u))f(u)du, \quad \forall t \geqslant 0.$$

Example 14.16 A particle p of mass 2 grams moves on the x-axis and is attracted towards the origin O with a force equal to $8x$. If it is initially at rest at $x = 10$, find its position at any subsequent time assuming
(a) no other forces act,
(b) a damping force numerically equal to 8 times the instantaneous velocity acts.

Solution 14.16 **(a)** When no other forces act by Newton's law, we know (Mass). (Acceleration) = Net force. The net force is given by $(-8x)$, whether $x > 0$ or $x < 0$. So, we have

$$2 \cdot \frac{d^2x}{dt^2} = -8x \quad \text{or} \quad \frac{d^2x}{dt^2} = -4x$$

with initial conditions $x(0) = 10$ and $x'(0) = 0$

$$\mathscr{L}(x'') + 4\mathscr{L}(x) = 0$$

or $\quad s^2\mathscr{L}(x) - 10s + 4\mathscr{L}(x) = 0 \quad \text{or} \quad \mathscr{L}(x) = \dfrac{10s}{s^2+4}$

or $\quad x = \mathscr{L}^{-1}\left(\dfrac{10s}{s^2+4}\right)$

or $\quad x = 10\cos 2t.$ [1]

The graph of the motion is shown in Fig.14.7. The amplitude is 10. The period is π and the frequency is $1/\pi$.

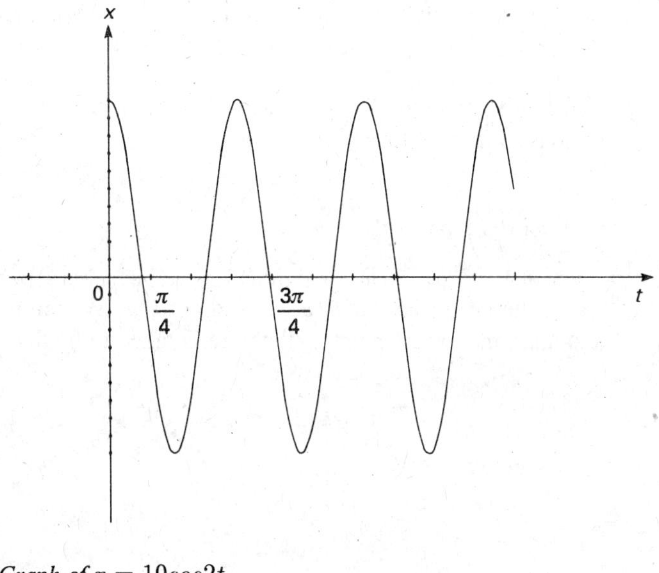

Fig. 14.7 *Graph of* $x = 10cos2t$.

(b) When $x > 0$ and $dx/dt > 0$. If P moves to the right, then the damping force is to the left and is given by $(-8dx/dt)$. Similarly when $x < 0$ and $dx/dt < 0$, the damping force is to the right and must be given by $(-8dx/dt)$. Then, Mass. Acceleration = Net force. We have

$$2\frac{d^2x}{dt^2} = -8x - 8\frac{dx}{dt}$$

or $\quad \dfrac{d^2x}{dt^2} + 4\dfrac{dx}{dt} + 4x = 0$ [2]

with initial conditions $x(0) = 10$ and $x'(0) = 0$. Taking the Laplace transform of [2] and using the initial conditions, we have

$$s^2\,\mathscr{L}(x) - 10s + 4[s\,\mathscr{L}(x) - 10] + 4\mathscr{L}x = 0 \quad \text{or} \quad \mathscr{L}(x) = \frac{10s+40}{s^2+4s+4}$$

$$\text{or} \quad x = \mathscr{L}^{-1}\left\{\frac{10s+40}{s^2+4s+4}\right\} = \mathscr{L}^{-1}\left\{\frac{10(s+2)+20}{(s+2)^2}\right\}$$

$$= 10\,\mathscr{L}^{-1}\left\{\frac{1}{s+2}\right\} + 20\,\mathscr{L}^{-1}\left\{\frac{1}{(s+2)^2}\right\} = 10e^{-2t} + 20te^{-2t} = 10e^{-2t}(1+2t).$$

The motion is non-oscillatory. The particle approaches O but never reaches it.

Example 14.17 Consider a wire bent into a smooth curve and let a bead of mass m start from rest and slide without friction down the wire to the origin under the action of its own weight. Find
(a) the time of descent to the lowest point of the wire, and
(b) the shape the wire must have if the time taken to reach the lowest point is a constant.

Solution 14.17 **(a)** To formulate this problem mathematically, we start with the principle of conservation of energy

$$\frac{1}{2}m\left(\frac{ds}{dt}\right)^2 = mg(v-y) \qquad [1]$$

where the bead starts at $P(u, v)$, and $Q(x, y)$ is some intermediate point in the motion, assuming the lowest point of the wire is taken to be the origin O (see Fig. 14.8).

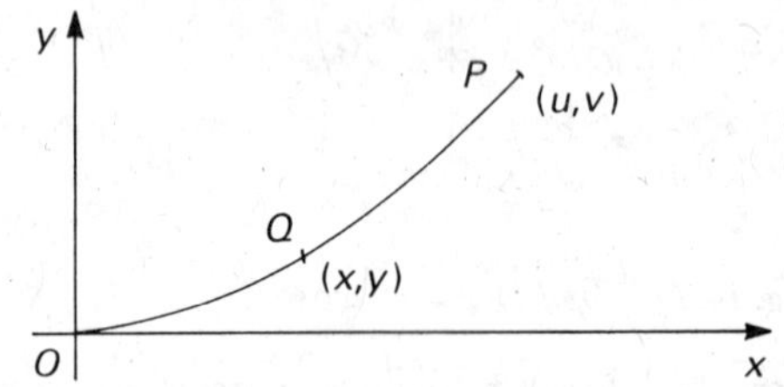

Fig. 14.8 *Path of descent of a bead.*

Now [1] can be written as $(ds/dt)^2 = 2g(v-y)$, or since s decreases as t increases

$$\frac{ds}{dt} = -\sqrt{2g(v-y)} \quad \text{or} \quad \frac{ds}{-\sqrt{2g(v-y)}} = dt.$$

The total time T taken for the bead to go from P to O is given by

$$T = \int_{y=v}^{y=0} dt = \int_v^0 \frac{-ds}{\sqrt{2g(v-y)}} = \int_0^v \frac{ds}{\sqrt{2g(v-y)}}$$

$$= \frac{1}{\sqrt{2g}} \int_0^v \frac{s'(y)dy}{\sqrt{v-y}}. \qquad [2]$$

(b) Here we have to find the curve when T is given i.e, T is a constant. This is called *Abel's mechanical problem*. Equation [2] is an integral equation of the convolution type and can be written as

$$T = \frac{1}{\sqrt{2g}}(f(y) * y^{-1/2}) \quad \text{so that} \quad \mathscr{L}(T) = \frac{1}{\sqrt{2g}} \mathscr{L}\{f(y)\} \mathscr{L}\{y^{-1/2}\}.$$

We can recall $\mathscr{L}\{y^{-1/2}\} = \sqrt{\pi/s}$ so that

$$\mathscr{L}\{f(y)\} = \sqrt{2g}\frac{\mathscr{L}\{T\}}{\sqrt{\frac{\pi}{s}}} = \sqrt{\frac{2g}{\pi}}\, s^{1/2} \mathscr{L}\{T\}.$$

We can find $f(y)$ by taking the inverse transform. When T is a constant, say T_0, the time of descent is independent of the starting point. The curve defined by this property is called the *tautochrone*; and

$$\mathscr{L}\{f(y)\} = \sqrt{\frac{2g}{\pi}}\, s^{1/2} \mathscr{L}\{T_0\} = \sqrt{\frac{2g}{\pi}}\, s^{1/2} \frac{T_0}{s} = \sqrt{\frac{2g}{\pi}} \frac{T_0}{s^{1/2}}. \qquad [3]$$

Since $ds/dy = \sqrt{1 + (dx/dy)^2}$, if we substitute $b = 2gT_0^2/\pi^2$ in [3], we get

$$\mathscr{L}\{f(y)\} = b^{1/2}\sqrt{\frac{\pi}{s}}. \qquad [4]$$

Now $\mathscr{L}^{-1}\sqrt{\pi/s} = y^{-1/2}$ so that

$$f(y) = \frac{b^{1/2}}{y^{1/2}} = \sqrt{\frac{b}{y}}. \qquad [5]$$

Now in [2], $s'(y)dy = f(y) = \sqrt{1 + (dx/dy)^2}$. Substituting this in [5] and squaring, we get

$$1 + \left(\frac{dx}{dy}\right)^2 = \frac{b}{y}$$

which is the differential equation of the curve. Simplifying and integrating we get

$$x = \int \sqrt{\frac{b-y}{y}}\, dy + C.$$

On substituting $y = b\sin^2\phi$, this becomes

$$x = 2b \int \cos^2 \phi d\phi + C = \frac{b}{2}[2\phi + \sin 2\phi] + C.$$

Therefore, we have

$$\left.\begin{aligned} x &= (b/2)\ [2\phi + \sin 2\phi] + C \\ y &= (b/2)\ (1 - \cos 2\phi) \end{aligned}\right\} \qquad [6]$$

Since the curve passes through the origin, $c = 0$ when $x = 0$ and $y = 0$. If we substitute $a = b/2$ and $\theta = 2\phi$, then [6] becomes

$$x = a(\theta + \sin\theta) \quad \text{and} \quad y = a(1 - \cos\theta).$$

These are the parametric equations of the *cycloid.* So, the *tautochrone* is a *cycloid.*

Exercise 14E

Solve the following equations using Laplace transforms.

14.25 $y'' + 4y = 4t;$ given $y(0) = 1,\ y'(0) = 5.$

14.26 $y''(t) - 2y'(t) + y(t) = e^t;$ given $y = 2, y' = -1$ at $t = 0.$

14.27 $y'' + k^2 y = \cos kt;$ given $y(0) = 0, y'(0) = 1/2.$

14.28 $y'' - y = 1;$ given $y(0) = -1, y'(0) = 1.$

14.29 $y'' + 2y' + 5y = 3e^{-t}\sin t;$ given $y(0) = 0,\ y'(0) = 3.$

14.30 $y'' - 6y' + 10y = 1;$ given $y(0) = 0,\ y'(0) = 0.$

14.31 $y'' + 2y' + 5y = 8\sin t + 4\cos t;$ given $y(0) = 1,\ y(\pi) = e^{-\pi}.$

14.32 $y'' - y = f(t) = \begin{cases} 0 & \text{if } 0 \le t < 2 \\ -1 & \text{if } 2 \le t \end{cases}$; given $y(0) = y'(0) = 0.$

14.33 $y'' + 2y' + 9y = \delta(t);$ given $y(0) = y'(0) = 0.$

14.34 $y'' - 4y' + 3y = f(t);$ given $y(0) = 1,\ y'(0) = 0.$

14.35 $y'' + 4y = \delta(t - 2);$ given $y(0) = 0,\ y'(0) = 1.$

14.36 $y'' - ty' + y = 1$; given $y(0) = 1,\ y'(0) = 2.$

14.37 $y''(t) + 9y(t) = 18t;$ given $y(0) = 0,\ y(\pi/2) = 0.$

CHAPTER FIFTEEN

Fourier Series

15.1 Introduction

Periodic functions f are defined as those functions which are such that

$$f(x+2p)=f(x)$$

for all x in the domain of f. Here, the period of f is said to be $2p$. It is easy to verify that if $2p$ is a period of f and n is a positive integer, then $2np$ is also a period of f. We call $2p$ the *fundamental period* or *primitive period* of $f(x)$, if it is the smallest period of $f(x)$. Some periodic functions have no fundamental period.

Familiar examples of periodic functions are the *sine* and *cosine* functions. Of course,

$$f(x)=\text{constant}=C$$

is also a periodic function in the sense of the definition. But, since there is no smallest positive number $2p$, such that

$$f(x+2p)=f(x)=\text{constant}$$

f has no fundamental period.

If ω is a non-zero constant, and $g(x)$ is defined on $(-\infty,\infty)$ by $g(x)=\cos\omega x$, then

$$g\left(x+\frac{2\pi}{\omega}\right)=\cos\left[\omega\left(x+\frac{2\pi}{\omega}\right)\right]=\cos\omega x=g(x).$$

Thus, $g(x)$ is a periodic function of period $2\pi/\omega$.

In fact, if m is any integer and p is positive, it can easily be shown that all functions of the type $\cos m\pi x/p$ and $\sin m\pi x/p$ are periodic functions of the time period $2p$. This fact raises the question of whether a general periodic function can be expressed as a series of *sine* and *cosine* terms. The series which will arise in this connection will be of the form

$$\frac{1}{2}a_0 + a_1 \cos\frac{\pi x}{p} + b_1 \sin\frac{\pi x}{p} + a_2 \cos\frac{2\pi x}{p} + b_2 \sin\frac{2\pi x}{p} + \ldots$$

where $a_0, a_1, a_2, \ldots, b_1, b_2, \ldots$ are real constants. Such a series is known as the *Fourier series*.

15.2 Euler's formulae

In order to obtain formulae for the coefficients a_0, a_n and b_n in the expansion of a function of period $2p$

$$f(x) = \frac{1}{2}a_0 + a_1 \cos\frac{\pi x}{p} + a_2 \cos\frac{2\pi x}{p} + \ldots + a_n \cos\frac{n\pi x}{p} + \ldots$$

$$+b_1 \sin\frac{\pi x}{p} + b_2 \sin\frac{2\pi x}{p} + \ldots + b_n \sin\frac{n\pi x}{p} + \ldots \qquad [15.1]$$

we need the following *definite integrals*, which are valid for all values of k, provided m and n satisfy the given restrictions:

1. $\displaystyle\int_k^{k+2p} \cos\frac{n\pi x}{p}dx = 0, n \neq 0$

2. $\displaystyle\int_k^{k+2p} \sin\frac{n\pi x}{p}dx = 0, n \neq 0$

3. $\displaystyle\int_k^{k+2p} \cos\frac{m\pi x}{p}\cos\frac{n\pi x}{p}dx = 0, m \neq n$

4. $\displaystyle\int_k^{k+2p} \cos^2\frac{n\pi x}{p}dx = p, n \neq 0$

5. $\displaystyle\int_k^{k+2p} \cos\frac{m\pi x}{p}\sin\frac{n\pi x}{p}dx = 0, m \neq n$

6. $\displaystyle\int_k^{k+2p} \sin\frac{m\pi x}{p}\sin\frac{n\pi x}{p}dx = 0, m \neq n$

7. $\displaystyle\int_k^{k+2p} \sin^2\frac{n\pi x}{p}dx = p, n \neq 0$ [15.2]

We can now proceed to determine a_n and b_n as follows:

To find a_0, assume that [15.1] can be integrated from $x = k$ to $x = k + 2p$ term by term so that we have

$$\int_k^{k+2p} f(x)dx = \frac{a_0}{2}\int_k^{k+2p} dx + a_1\int_k^{k+2p} \cos\frac{\pi x}{p}dx + \dots$$

$$\dots + a_n\int_k^{k+2p} \cos\frac{n\pi x}{p}dx + \dots + b_1\int_k^{k+2p} \sin\frac{\pi x}{p}dx + \dots$$

$$+ b_n\int_k^{k+2p} \sin\frac{n\pi x}{p}dx + \dots$$

Now, we know that $\displaystyle\frac{a_0}{2}\int_k^{k+2p} dx = \frac{a_0}{2}x\Big]_k^{k+2p} = pa_0.$

All other integrals on the right hand side vanish by 1 and 2 of [15.2]. Thus

$$\int_k^{k+2p} f(x)dx = pa_0$$

or $\displaystyle a_0 = \frac{1}{p}\int_k^{k+2p} f(x)dx.$ [15.3]

To find a_n, multiply each side of [15.1] by $\cos n\pi x/p$, and then integrate from k to $k + 2p$, assuming the validity of term by term integration. Then

$$\int_k^{k+2p} f(x)\cos\frac{n\pi x}{p}dx = \frac{1}{2}a_0\int_k^{k+2p} \cos\frac{n\pi x}{p}dx + \dots$$

$$+a_1\int_k^{k+2p} \cos\frac{\pi x}{p}\cos\frac{n\pi x}{p}dx + \dots + a_n\int_k^{k+2p} \cos^2\frac{n\pi x}{p}dx + \dots$$

$$+ b_1\int_k^{k+2p} \sin\frac{\pi x}{p}\cos\frac{n\pi x}{p}dx + \dots + b_n\int_k^{k+2p} \sin\frac{n\pi x}{p}\cos\frac{n\pi x}{p}dx + \dots$$

Again, all integrals on the right hand side above vanish, using [15.2], except

$$a_n\int_k^{k+2p} \cos^2\frac{n\pi x}{p}dx = pa_n. \quad \text{Thus} \quad \int_k^{k+2p} f(x)\cos\frac{n\pi x}{p}dx = pa_n$$

or $\displaystyle a_n = \frac{1}{p}\int_k^{k+2p} f(x)\cos\frac{n\pi x}{p}dx.$ [15.4]

To find b_n, we continue essentially the same procedure, but multiply [15.1] by $\sin\frac{n\pi x}{p}$ and then integrate term by term so that we get

$$\int_k^{k+2p} f(x)\sin\frac{n\pi x}{p}dx = \frac{1}{2}a_0\int_k^{k+2p} \sin\frac{n\pi x}{p}dx +$$

$$+a_1\int_k^{k+2p} \cos\frac{\pi x}{p}\sin\frac{n\pi x}{p}dx + \ldots$$

$$+ a_n\int_k^{k+2p} \cos\frac{n\pi x}{p}\sin\frac{n\pi x}{p}dx + \ldots + b_1\int_k^{k+2p} \sin\frac{\pi x}{p}\sin\frac{n\pi x}{p}dx + \ldots$$

$$+ b_n\int_k^{k+2p} \sin^2\frac{n\pi x}{p}dx + \ldots$$

As before, all integrals vanish on the right except one, and we get

$$\int_k^{k+2p} f(x)\sin(n\pi x/p)dx = pb_n$$

or $$b_n = \frac{1}{p}\int_k^{k+2p} f(x)\sin\frac{n\pi x}{p}dx. \qquad [15.5]$$

Formulae [15.3] [15.4] and [15.5] are called *Euler (Fourier) formulae* and the series [15.1], whose coefficients have these values, is called the *Fourier series* of $f(x)$.

Note: Actually, the formula [15.3] for a_0 need not be listed separately, since it can be obtained from the general expression for a_n by taking $n = 0$. However, the value of a_0 in a particular problem cannot be got by taking $n = 0$ in the integrated formula for a_n.

15.3 Dirichlet's conditions for a Fourier series expansion

At this stage, one point has to be carefully noted. It is not necessary that every periodic function $f(x)$ has a Fourier expansion. The analysis merely shows that if a function $f(x)$ has an expansion of the form [15.1], for which term by term integration is valid, then the coefficients in that series are given by the Euler formulae. We shall state the conditions which make it clear that a *function need not be continuous* in order to possess a valid Fourier expansion.

15.3.1 Statement of Dirichlet's conditions

Any function $f(x)$ can be expanded as a Fourier series

$$f(x) = \frac{1}{2}a_0 + \sum_{n=1}^{\infty} a_n \cos\frac{n\pi x}{p} + \sum_{n=1}^{\infty} b_n \sin\frac{n\pi x}{p} \tag{15.6}$$

if (i) $f(x)$ is periodic with period $2p$, i.e., $f(x+2p) = f(x)$;

(ii) $f(x)$ is a bounded function with at most a finite number of maxima and minima;

(iii) $f(x)$ has a finite number of points of discontinuity.

These conditions are called *Dirichlet's conditions* for a Fourier series expansion of a function.

Example 15.1 Find the Fourier expansion of the periodic function whose definition in one period is

$$f(x) = \begin{cases} 0 & \text{if } \; -\pi \le x \le 0 \\ \sin x & \text{if } \; 0 \le x \le \pi. \end{cases}$$

Solution 15.1 Using Euler's formulae and taking $k = -\pi$ and $p = \pi$, we get

$$a_0 = \frac{1}{p}\int_k^{k+2p} f(x)dx \quad \text{or} \quad a_0 = \frac{1}{\pi}\int_{-\pi}^{\pi} f(x)dx$$

$$= \frac{1}{\pi}\int_{-\pi}^{0} (0)dx + \frac{1}{\pi}\int_0^{\pi} \sin x\,dx$$

or $$a_0 = \frac{1}{\pi}\left[-\cos x\right]_0^{\pi} = \frac{2}{\pi} \tag{1}$$

$$a_n = \frac{1}{\pi}\int_{-\pi}^{\pi} f(x)\cos nx\,dx = \frac{1}{\pi}\int_{-\pi}^{0} (0)\cos nx\,dx + \frac{1}{\pi}\int_0^{\pi} \sin x \cos nx\,dx$$

$$= \frac{1}{\pi}\left[-\frac{1}{2}\left\{\frac{\cos(1-n)x}{1-n} + \frac{\cos(1+n)x}{1+n}\right\}\right]_0^{\pi}$$

$$= \frac{-1}{2\pi}\left[\frac{\cos(\pi - n\pi)}{1-n} + \frac{\cos(\pi + n\pi)}{1+n} - \left(\frac{1}{1-n} + \frac{1}{1+n}\right)\right]$$

$$= \frac{-1}{2\pi}\left[\frac{-\cos n\pi}{1-n} + \frac{-\cos n\pi}{1+n} - \frac{2}{1-n^2}\right]$$

or $$a_n = \frac{1 + \cos n\pi}{\pi(1-n^2)} \;(\text{if } n \neq 1). \tag{2}$$

And for $n = 1$, we have

$$a_1 = \frac{1}{\pi}\int_0^{\pi} \sin x \cos x dx = \left.\frac{\sin^2 x}{2\pi}\right]_0^{\pi} = 0. \qquad [3]$$

Also, $b_n = \frac{1}{\pi}\int_{-\pi}^{\pi} f(x) \sin nx dx$

$$= \frac{1}{\pi}\int_{-\pi}^{0} (0) \sin nx dx + \frac{1}{\pi}\int_0^{\pi} \sin x \sin nx dx$$

$$= \frac{1}{\pi}\left[\frac{1}{2}\left\{\frac{\sin(1-n)x}{1-n} - \frac{\sin(1+n)x}{1+n}\right\}\right]_0^{\pi}$$

or $b_n = 0$ (if $n \neq 1$) [4]

and $b_1 = \frac{1}{\pi}\int_0^{\pi} \sin^2 x dx = \frac{1}{\pi}\left[\frac{x}{2} - \frac{\sin 2x}{4}\right]_0^{\pi}$

or $b_1 = \frac{1}{2}$. [5]

Thus, from [1], [2], [3], [4] and [5], we get (by substituting $n = 1, 2, 3, \ldots$ in [15.6])

$$f(x) = \frac{1}{\pi} + \frac{\sin t}{2} - \frac{2}{\pi}\left(\frac{\cos 2t}{3} + \frac{\cos 4t}{15} + \frac{\cos 6t}{35} + \ldots\right).$$

Example 15.2 Find the Fourier expansion of the periodic function whose definition in one period is $f(x) = 4 - x^2, \quad -2 \leq x \leq 2$.

Solution 15.2 In this case $p = 2$. Hence taking $k = -2$ and using Euler's formulae, we get

(i) $a_0 = \frac{1}{2}\int_{-2}^{2} (4 - x^2) dx = \frac{1}{2}\left[4x - \frac{x^3}{3}\right]_{-2}^{2} = \frac{16}{3}$.

(ii) $a_n = \frac{1}{2}\int_{-2}^{2} (4 - x^2) \cos \frac{n\pi x}{2} dx$

$$= \frac{1}{2}\left[\frac{8}{n\pi}\sin\left(\frac{n\pi x}{2}\right)\right]_{-2}^{2} - \frac{1}{2}\left[\frac{8x}{n^2\pi^2}\cos\left(\frac{n\pi x}{2}\right) + \right.$$

$$\left. + \left(\frac{2x^2}{n\pi} - \frac{16}{n^3\pi^3}\right)\sin\left(\frac{n\pi x}{2}\right)\right]_{-2}^{2}$$

$$= \frac{-16}{n^2\pi^2}\cos n\pi \quad \text{or} \quad a_n = \frac{16}{n^2\pi^2}(-1)^{n+1}; \qquad n \neq 0.$$

(iii) $$b_n = \frac{1}{2}\int_{-2}^{2}(4-x^2)\sin\frac{n\pi x}{2}dx = \frac{1}{2}\left[\frac{-8}{n\pi}\cos\frac{n\pi x}{2}\right]_{-2}^{2}$$

$$-\frac{1}{2}\left[\frac{8x}{n^2\pi^2}\sin\frac{n\pi x}{2} - \left(\frac{2x^2}{n\pi} - \frac{16}{n^3\pi^3}\right)\cos\frac{n\pi x}{2}\right]_{-2}^{2} \quad \text{or } b_n = 0.$$

Substituting (i) – (iii) in the series [15.1], we get

$$f(x) = \frac{8}{3} + \frac{16}{\pi^2}\left(\cos\frac{\pi x}{2} - \frac{1}{2^2}\cos\frac{2\pi x}{2} + \frac{1}{3^2}\cos\frac{3\pi x}{2} - \frac{1}{4^2}\cos\frac{4\pi x}{2} + \ldots\right) \qquad [1]$$

Note: Interesting numerical series can often be obtained from the Fourier series by evaluating them for particular values of the independent variable. For instance if, in the series [1] we have just obtained, we substitute $x = 0$, we find

$$f(x) = f(0) = 4 = \frac{8}{3} + \frac{16}{\pi^2}\left(1 - \frac{1}{2^2} + \frac{1}{3^2} - \frac{1}{4^2} + \ldots\right)$$

or, solving for the series, we get $1 - \frac{1}{2^2} + \frac{1}{3^2} - \frac{1}{4^2} + \ldots = \frac{\pi^2}{12}$.

Exercise 15A

15.1 A periodic function f is defined for every real number x which is not an integer by the equation

$$f(x) = \begin{cases} 1 + kx & \text{for} \quad 0 < x < 1 \\ 1 & \text{for} \quad 1 < x < 2 \end{cases}$$

and $f(x+2) = f(x)$.
(i) Prove that if $k \neq 0$, the period of f is 2. (ii) If $k = 0$, what is the period of f?

15.2 Obtain the Fourier series for $f(x) = e^{-x}$ in the interval $0 < x < 2\pi$.

15.3 Determine the Fourier expansions of the following periodic functions over the given interval:

(i) $f(x) = \begin{cases} 1 \text{ for } & 0 < x < 1 \\ -1 \text{ for } & 1 < x < 2 \end{cases}$ (ii) $f(x) = x$ for $-1 < x < 1$.

(iii) $f(x) = x \sin x,\ 0 < x < 2\pi.$ **(iv)** $f(x) = \begin{cases} 0 & \text{for } -2 \le x < -1 \\ \sin(\pi x)/2 & \text{for } -1 < x < 1 \\ 0 & \text{for } 1 < x \le 2 \end{cases}$

(v) $f(x) = \begin{cases} 0 & \text{for } -2 \le x \le -1 \\ 1 + x & \text{for } -1 \le x \le 0 \\ 1 - x & \text{for } 0 \le x \le 1 \\ 0 & \text{for } 1 \le x \le 2 \end{cases}$

15.4 Prove that $x^2 = \dfrac{\pi^2}{3} + 4\displaystyle\sum_{n=1}^{\infty}(-1)^n \frac{\cos nx}{n^2}$ if $-\pi < x < \pi$, and hence show

(i) $\sum \dfrac{1}{n^2} = \dfrac{\pi^2}{6}$ (ii) $\sum \dfrac{(-1)^{n-1}}{n^2} = \dfrac{\pi^2}{12}$

15.5 Expand $f(x) = \left(\dfrac{\pi - x}{2}\right)^2$ if $0 < x < 2\pi$ in a Fourier series.

15.6 Find the Fourier series of the function

$$f(x) = \begin{cases} x & \text{if } -\pi/2 < x < \pi/2 \\ \pi - x & \text{if } \pi/2 < x < 3\pi/2 \end{cases}$$

15.7 Find the Fourier expansion of $f(x) = 1 + x$ in $-1 < x < 1$ and use it to find the sum of the series $\displaystyle\sum_{n=0}^{\infty} \frac{(-1)^n}{2n+1}$.

15.8 Find the Fourier series expansion to represent $f(x) = 2x - x^2, 0 < x < 3$.

15.4 Even and odd functions

When $f(x)$ possesses certain *symmetry* properties, the coefficients in its expansion become especially simple. Let us see in detail just what effect the symmetry of $f(x)$ has on the coefficients in the Fourier series of $f(x)$. Let us examine the expansions of even and odd functions.

A function $f(x)$ is said to be *even* if $f(-x) = f(x)$, e.g. $\cos x, x^2$.

A function $f(x)$ is said to be *odd* if $f(-x) = -f(x)$, e.g. $\sin x, x^3$.

Geometrically, an odd function is *symmetric* about the origin, while an even function is symmetric about the y-axis. The functions portrayed graphically in Figures 15.1 and 15.2 below are odd and even respectively, but that of Figure 15.3 is neither odd nor even.

We know that a periodic function $f(x)$ can be represented by the Fourier series

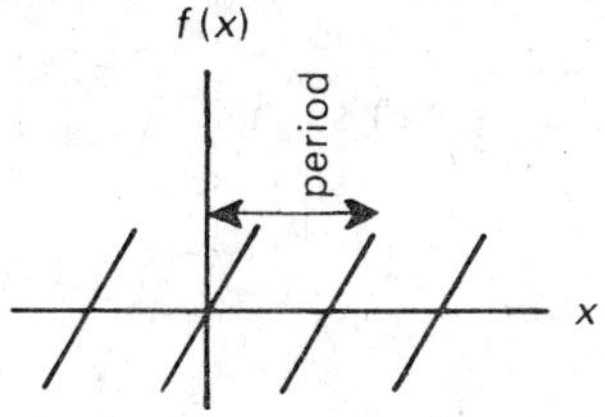

Fig. 15.1

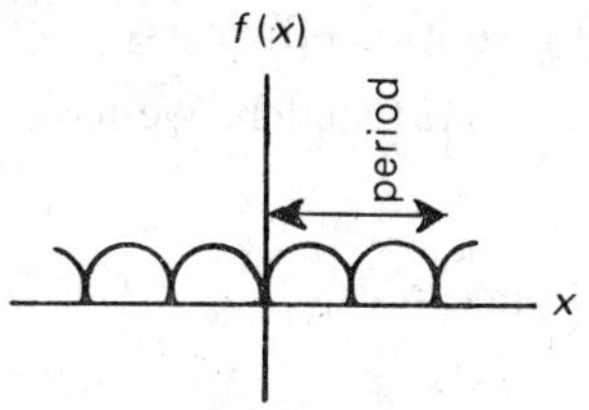

Fig. 15.2

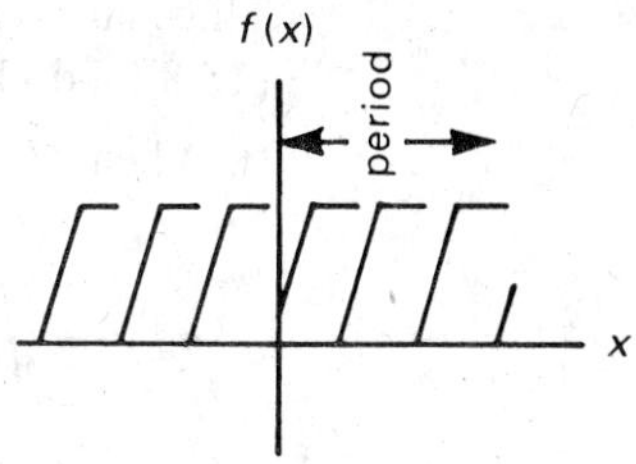

Fig. 15.3

Fig. 15.1 *Odd function*
Fig. 15.2 *Even function*
Fig. 15.3 *Neither odd nor even function.*

$$f(x) = \frac{a_0}{2} + \sum_{n=1}^{\infty} a_n \cos\frac{n\pi x}{p} + \sum_{n=1}^{\infty} b_n \sin\frac{n\pi x}{p}$$

$$\left.\begin{aligned} \text{where } a_0 &= \frac{1}{p}\int_{-p}^{p} f(x)dx \\ a_n &= \frac{1}{p}\int_{-p}^{p} f(x)\cos\frac{n\pi x}{p}dx \\ b_n &= \frac{1}{p}\int_{-p}^{p} f(x)\sin\frac{n\pi x}{p}dx \end{aligned}\right\} \qquad [15.7]$$

1. When $f(x)$ is an *even* function, we have $f(-x) = f(x)$ and therefore

$$a_0 = \frac{1}{p}\int_{-p}^{p} f(x)dx = \frac{2}{p}\int_{0}^{p} f(x)dx. \qquad [15.8]$$

And $f(x)\cos\frac{n\pi x}{p}$ is also an *even* function so that

$$a_n = \frac{1}{p}\int_{-p}^{p} f(x)\cos\frac{n\pi x}{p}dx$$

$$= \frac{2}{p}\int_{0}^{p} f(x)\cos\frac{n\pi x}{p}dx. \qquad [15.9]$$

Again, since $f(x)\sin\frac{n\pi x}{p}$ is an *odd* function we have

$$b_n = \frac{1}{p}\int_{-p}^{p} f(x)\sin\frac{n\pi x}{p}dx = 0.$$

Thus, an *even* periodic function $f(x)$ has only *cosine* terms in its Fourier expansion and [15.8] and [15.9] give its coefficients.

2. When $f(x)$ is an *odd* function, We have $f(-x) = -f(x)$ so that

$$a_0 = \frac{1}{p}\int_{-p}^{p} f(x)dx = 0.$$

Since $\cos\dfrac{n\pi x}{p}$ is an *even* function, we have $f(x)\cos\dfrac{n\pi x}{p}$, which is an odd function.

$$\therefore \quad a_n = \frac{1}{p}\int_{-p}^{p} f(x)\cos\frac{n\pi x}{p}dx = 0.$$

Again, since $\sin\dfrac{n\pi x}{p}$ is an *odd* function, we have $f(x)\sin\dfrac{n\pi x}{p}$, an even function, so that

$$b_n = \frac{1}{p}\int_{-p}^{p} f(x)\sin\frac{n\pi x}{p}dx$$

$$= \frac{2}{p}\int_{0}^{p} f(x)\sin\frac{n\pi x}{p}dx. \qquad [15.10]$$

Thus, an *odd* periodic function $f(x)$ has only *sine* terms in its Fourier expansion and [3] gives its coefficient.

Example 15.3 Expand $f(x) = x^2$ in a Fourier series over the interval $(-p, p)$. Hence show that

$$\frac{1}{1^2} - \frac{1}{2^2} + \frac{1}{3^2}\cdots = \frac{\pi^2}{12} \text{ in } (-\pi, \pi).$$

Solution 15.3 Now $f(x) = x^2$ is an *even* function in $(-p, p)$. Thus, we have

$$f(x) = \frac{a_0}{2} + \sum_{n=1}^{\infty} a_n \cos\frac{n\pi x}{p} \qquad [1]$$

where $a_0 = \dfrac{2}{p}\displaystyle\int_{0}^{p} x^2 dx = \dfrac{2}{p}\dfrac{x^3}{3}\Bigg]_0^p = \dfrac{2p^2}{3}$. [2]

Again $a_n = \dfrac{1}{p}\displaystyle\int_{-p}^{p} f(x)\cos\dfrac{n\pi x}{p}dx = \dfrac{2}{p}\displaystyle\int_{0}^{p} x^2\cos\dfrac{n\pi x}{p}dx$

$$= \frac{2}{p}\left[x^2\frac{\sin(n\pi x/p)}{n\pi/p} - 2x\left(\frac{-\cos(n\pi x/p)}{n^2\pi^2/p^2}\right) + 2\left(\frac{-\sin(n\pi x/p)}{n^3\pi^3/p^3}\right)\right]_0^p$$

or $\quad a_n = \dfrac{4p^2}{n^2\pi^2}(-1)^n$ (since $\cos n\pi = (-1)^n$). Thus, we have

$$a_1 = -\frac{4p^2}{\pi^2}, a_2 = \frac{4p^2}{2^2\pi^2}, a_3 = -\frac{4p^2}{3^2\pi^2}, \dots \qquad [3]$$

Substituting [2] and [3] in [1], we get the required expansion as

$$f(x) = x^2 = \frac{p^2}{3} - \frac{4p^2}{\pi^2}\left[\frac{\cos(\pi x/p)}{1^2} - \frac{\cos(2\pi x/p)}{2^2} + \frac{\cos(3\pi x/p)}{3^2} - \dots\right].$$

Now, taking $p = \pi$ and $x = 0$ in the above expansion, we get

$$f(0) = \frac{\pi^2}{3} - 4\left[\frac{1}{1^2} - \frac{1}{2^2} + \frac{1}{3^2} - \frac{1}{4^2} + \dots\right].$$

But $f(0) = 0$ and hence we have

$$\frac{\pi^2}{3} - 4\left[\frac{1}{1^2} - \frac{1}{2^2} + \frac{1}{3^2} - \frac{1}{4^2} + \dots\right] = 0$$

or $\quad \dfrac{1}{1^2} - \dfrac{1}{2^2} + \dfrac{1}{3^2} - \dfrac{1}{4^2} + \dots = \dfrac{\pi^2}{12}.$

15.5 Half range series

In various physical and engineering problems it is required to obtain a Fourier expansion of a function $f(x)$ for the range $(0, p)$ which is half the period of the Fourier series. Now we can extend the function f from 0 to $-p$ by *reflecting* it in the *vertical axis* so that $f(-x) = f(x)$. Then, the original function together with its extension is *even*. Hence, the Fourier expansion of this extended function will contain only *cosine* terms, including the constant $a_0/2$, whose coefficients are given by

$$a_n = \frac{2}{p}\int_0^p f(x)\cos\frac{n\pi x}{p}dx.$$

On the other hand, if we extend the function f from 0 to $-p$, by *reflecting* it in the *origin*, so that $f(-x) = -f(x)$, the extended function, with 0 or $\pm p$ omitted from its domain according as $f(0) \neq 0$ or $f(p) \neq 0$, is *odd* and hence the Fourier series of this extended function will contain only *sine* terms, and its coefficients are given by

$$b_n = \frac{2}{p}\int_0^p f(x)\sin\frac{n\pi x}{p}\,dx.$$

Thus, by imagining the appropriate extension of a function f defined only from 0 to p, we can obtain expansions containing only cosine terms or only sine terms, as we please. Such series are known as *half range expansions.*

Example 15.4 Expand $f(x) = x$, $0 < x < 2$ in a half range **(a)** sine series and **(b)** cosine series.

Solution 15.4 **(a)** We can extend the definition of the given function to that of the odd function of period 4 shown in Fig. 15.4 below.

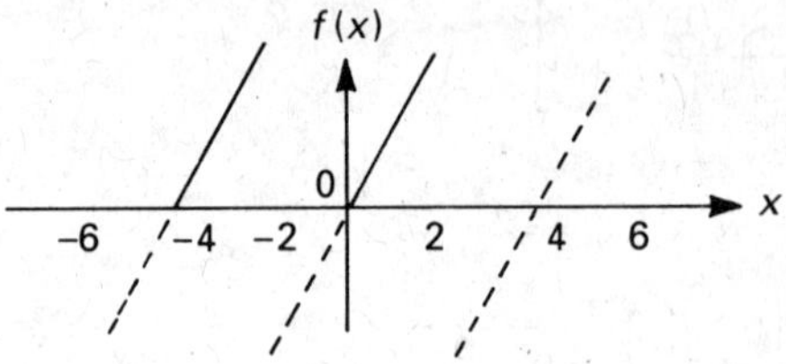

Fig. 15.4

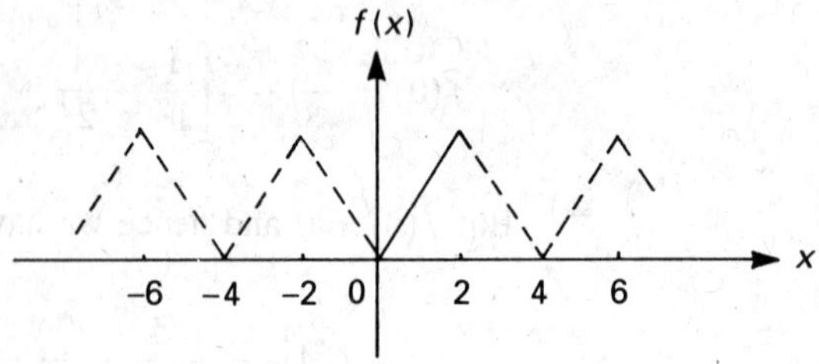

Fig. 15.5

Fig. 15.4 $f(x) = x$ *extended as an odd function.*
Fig. 15.5 $f(x) = x$ *extended as an even function.*

Then $a_n = 0$ and $b_n = \frac{2}{p}\int_0^p f(x)\sin\frac{n\pi x}{p}\,dx = \frac{2}{2}\int_0^2 x\sin\frac{n\pi x}{p}\,dx$

$$= \left[x\left(\frac{-2}{n\pi}\cos\frac{n\pi x}{2}\right) - 1\left(\frac{-4}{n^2\pi^2}\sin\frac{n\pi x}{2}\right)\right]_0^2 = \frac{-4}{n\pi}\cos n\pi$$

so that $f(x) = \sum_{n=1}^{\infty}\frac{-4}{n\pi}\cos n\pi\sin\frac{n\pi x}{2}$

$$= \frac{4}{\pi}\left(\sin\frac{\pi x}{2} - \frac{1}{2}\sin\frac{2\pi x}{2} + \frac{1}{3}\sin\frac{3\pi x}{2}\ldots\right).$$

(b) We can also extend the definition of $f(x)$ to that of the *even* function of period 4 shown in Fig. 15.5. Then $b_n = 0$ and

$$a_n = \frac{2}{p}\int_0^p f(x)\cos\frac{n\pi x}{p}\,dx = \frac{2}{2}\int_0^2 x\cos\frac{n\pi x}{2}\,dx$$

$$= \left[(x)\left(\frac{2}{n\pi} \sin \frac{n\pi x}{2} \right) - 1\left(\frac{-4}{n^2\pi^2} \cos \frac{n\pi x}{2} \right) \right]_0^2 = \frac{4}{n^2\pi^2}(\cos n\pi - 1) \quad \text{if} \quad n \neq 0$$

But if $n = 0$, we have $a_0 = \int_0^2 x dx = 2$; so that

$$f(x) = 1 - \frac{8}{\pi^2}\left(\cos \frac{\pi x}{2} + \frac{1}{3^2} \cos \frac{3\pi x}{2} + \frac{1}{5^2} \cos \frac{5\pi x}{2} + \dots \right).$$

Note: Thus, we can expand a function as a series of sines or cosines, by assuming that it is an odd or even function even though the function may not be odd or even.

Example 15.5 Find the half range expansion of the function

$$f(x) = \begin{cases} 2kx/p & \text{when } 0 < x < p/2 \\ 2k(p - x)/p & \text{when } p/2 < x < p \end{cases}$$

Solution 15.5 This function is the triangular pulse as shown in Fig 15.6 below.

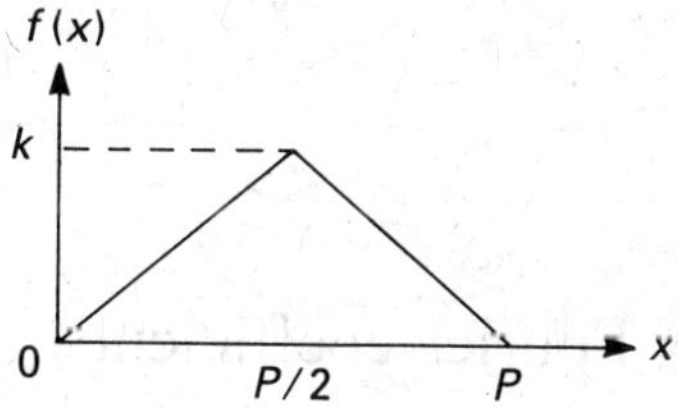

Fig. 15.6 *Triangular pulse.*

We know that

$$a_0 = \frac{1}{p} \int_0^p f(x) dx = \frac{1}{p} \left[\frac{2k}{p} \int_0^{p/2} x dx + \frac{2k}{p} \int_{p/2}^p (p - x) dx \right] \quad \text{or} \quad a_0 = \frac{k}{2}.$$

Also $a_n = \frac{2}{p} \int_0^p f(x) \cos \frac{n\pi x}{p} dx$

$$= \frac{2}{p} \left[\frac{2k}{p} \int_0^{p/2} x \cos \frac{n\pi x}{p} dx + \frac{2k}{p} \int_{p/2}^p (1 - x) \cos \frac{n\pi x}{p} dx. \right. \qquad [1]$$

Integrating by parts, we obtain

$$\int_0^{p/2} x \cos \frac{n\pi x}{p} dx = \frac{px}{n\pi} \sin \frac{n\pi x}{p} \Bigg]_0^{p/2} - \frac{1}{n\pi} \int_0^{p/2} \sin \frac{n\pi x}{p} dx$$

$$= \frac{p^2}{2n\pi}\sin\frac{n\pi}{2} + \frac{p^2}{n^2\pi^2}\left(\cos\frac{n\pi}{2} - 1\right). \qquad [2]$$

Similarly,

$$\int_{p/2}^{p}(1-x)\cos\frac{n\pi x}{p}dx = \frac{-p^2}{2n\pi}\sin\frac{n\pi}{2} - \frac{p^2}{n^2\pi^2}\left(\cos n\pi - \cos\frac{n\pi}{2}\right). \qquad [3]$$

Substituting [2] and [3] in [1], we get $a_n = \frac{4k}{n^2\pi^2}\left(2\cos\frac{n\pi}{2} - \cos n\pi - 1\right).$

We have $a_2 = \frac{-16k}{2^2\pi^2},\ a_6 = \frac{-16k}{6^2\pi^2}$ and $a_{10} = \frac{-16k}{10^2\pi^2}.$

Hence $f(x) = \frac{k}{2} - \frac{16k}{\pi^2}\left(\frac{1}{2^2}\cos\frac{2\pi x}{p} + \frac{1}{6^2}\cos\frac{6\pi x}{p} + \dots\right).$

Similarly we can obtain $b_n = \frac{8k}{n^2\pi^2}\sin\frac{n\pi}{2}$, so that

$$f(x) = \frac{8k}{\pi^2}\left(\frac{1}{p^2}\sin\frac{\pi x}{p} - \frac{1}{3^2}\sin\frac{3\pi x}{p} + \frac{1}{5^2}\sin\frac{5\pi x}{p} - \dots\right).$$

15.6 Complex Fourier coefficients

We know that $e^{i\theta} = \cos\theta + i\sin\theta$, and hence

$$\cos nx = \frac{1}{2}(e^{inx} + e^{-inx}) \quad \text{and} \quad \sin nx = \frac{1}{2i}(e^{inx} - e^{-inx})$$

and the Fourier series $f(x) = a_0 + \sum_{n=1}^{\infty}(a_n\cos nx + b_n\sin nx)$ may be written as

$$f(x) = c_0 + \sum_{n=1}^{\infty}[c_n e^{inx} + c'_n e^{-inx}]$$

where $c_0 = a_0,\ c_n = (a_n - ib_n)/2$ and $c'_n = (a_n + ib_n)/2$ for $n = 1, 2, \dots$

The coefficients c_0, c_n and c'_n can be easily obtained as follows:

$$c_0 = \frac{1}{2}a_0 = \frac{1}{2p}\int_k^{k+2p} f(x)dx$$

$$c_n = \frac{a_n - ib_n}{2} = \frac{1}{2}\left[\frac{1}{p}\int_k^{k+2p} f(x)\cos\frac{n\pi x}{p}dx - i\frac{1}{p}\int_k^{k+2p} f(x)\sin\frac{n\pi x}{p}dx\right]$$

$$= \frac{1}{2p}\int_k^{k+2p} f(x)\left(\cos\frac{n\pi x}{p} - i\sin\frac{n\pi x}{p}\right)dx = \frac{1}{2p}\int_k^{k+2p} f(x)e^{-ni\pi x/p}dx$$

Similarly

$$c_n' = \frac{a_n + ib_n}{2} = \frac{1}{2}\left[\frac{1}{p}\int_k^{k+2p} f(x)\cos\frac{n\pi x}{p}dx + i\frac{1}{p}\int_k^{k+2p} f(x)\sin\frac{n\pi x}{p}dx\right]$$

$$= \frac{1}{2p}\int_k^{k+2p} f(x)\left(\cos\frac{n\pi x}{p} + i\sin\frac{n\pi x}{p}\right)dx = \frac{1}{2p}\int_k^{k+2p} f(x)e^{ni\pi x/p}dx.$$

Writing $c_n' = c_{-n}$, we can see that c_n is given by the single formula

$$c_n = \frac{1}{2p}\int_k^{k+2p} f(x)e^{-ni\pi x/p}dx \; (n = 0, \pm 1, \pm 2, \ldots).$$

The complex form of the Fourier series is then given by

$$f(x) = \sum_{n=-\infty}^{\infty} c_n e^{inx}.$$

Here c_n are called the *complex* Fourier coefficients of $f(x)$.

Example 15.6 Find the complex form of the Fourier series of the function whose definition in one period is $f(x) = e^{-x}, \quad -1 < x < 1.$

Solution 15.6 Since $p = 1$, we have from [15.7], taking $k = -1$:

$$c_n = \frac{1}{2}\int_{-1}^{1} e^{-x}e^{-ni\pi x}dx = \frac{1}{2}\left[\frac{e^{-(1+ni\pi)x}}{-(1+ni\pi)}\right]_{-1}^{1}$$

$$= \frac{e^{-(1+ni\pi)} - e^{(1+ni\pi)}}{-2(1+ni\pi)} = \frac{ee^{ni\pi} - e^{-1}e^{-ni\pi}}{2(1+ni\pi)}.$$

Now $e^{i\pi} = \cos\pi + i\sin\pi = -1$ so that $e^{ni\pi} = e^{-ni\pi} = (-1)^n$. Hence

$$c_n = \frac{(-1)^n(e - e^{-1})}{(1+ni\pi)2} = \frac{(-1)^n(1-ni\pi)\sinh 1}{1+n^2\pi^2}.$$

$$\therefore \; f(x) = \sum_{n=-\infty}^{\infty} (-1)^n \frac{(1-ni\pi)\sinh 1}{1+n^2\pi^2} e^{ni\pi x} \qquad [1]$$

We can easily calculate a_n and b_n using $a_n = c_n + c_{-n}$ and $b_n = i(c_n - c_{-n})$. Thus,

$$a_n = \frac{(-1)^n(1-ni\pi)\sinh 1}{1+n^2\pi^2} + \frac{(-1)^n(1+ni\pi)\sinh 1}{1+n^2\pi^2} = \frac{(-1)^n 2\sinh 1}{1+n^2\pi^2} \qquad [2]$$

and $$b_n = i\left[\frac{(-1)^n(1-ni\pi)\sinh 1}{1+n^2\pi^2} - \frac{(-1)^n(1+ni\pi)\sinh 1}{1+n^2\pi^2}\right]$$

$$= \frac{(-1)^n 2n\pi\sinh 1}{1+n^2\pi^2}. \qquad [3]$$

So, $f(x)$ can also be written as

$$f(x) = \sinh 1 - 2\sinh 1\left(\frac{\cos\pi x}{1+\pi^2} - \frac{\cos 2\pi x}{1+4\pi^2} + \frac{\cos 3\pi x}{1+9\pi^2} - \dots\right)$$

$$-2\pi\sinh 1\left(\frac{\sin\pi x}{1+\pi^2} - \frac{2\sin 2\pi x}{1+4\pi^2} + \frac{3\sin 3\pi x}{1+9\pi^2} - \dots\right).$$

Exercise 15B

15.9 Obtain a Fourier series for the function $f(x)$ given by

$$f(x) = \begin{cases} 1 + (2x/\pi) & \text{for } -\pi \le x \le 0 \\ 1 - (2x/\pi) & \text{for } \ 0 \le x \le \pi \end{cases}$$

and hence deduce that $\dfrac{1}{1^2} + \dfrac{1}{3^2} + \dfrac{1}{5^2} + \dots = \dfrac{\pi^2}{8}$.

15.10 Find the half range sine and cosine expansions of each of the following functions

(i) $f(x) = 1;\quad 0 < x < 2$ (ii) $f(x) = x;\quad 0 \le x < p$

(iii) $f(x) = \sin x;\quad 0 \le x \le 2\pi$ (iv) $f(x) = \sin ax;\quad 0 \le x < \pi,$ a is not an integer

(v) $f(x) = x(\pi - x);\quad 0 \le x \le \pi$ (vi) $f(x) = \begin{cases} 1 & \text{for } \ 0 < x < p/2 \\ 0 & \text{for } \ p/2 \le x \le p \end{cases}$

(vii) $f(x) = \begin{cases} 2kx/p & \text{for } \ 0 \le x \le p/2 \\ 2k(p-x)/p & \text{for } \ p/2 \le x \le p \end{cases}$

(viii) $f(x) = \begin{cases} (1/4) - x & \text{for } 0 < x < 1/2 \\ x - (3/4) & \text{for } 1/2 < x < 1 \end{cases}$

15.11 Graph the following functions and find their corresponding Fourier series

(i) $f(x) = \begin{cases} -x & \text{for } -4 \le x \le 0 \\ x & \text{for } 0 \le x \le 4 \end{cases}$ (ii) $f(x) = \begin{cases} 2x & \text{for } 0 \le x < 3 \\ 0 & \text{for } -3 < x < 0 \end{cases}$

15.12 Find the Fourier series of **(i)** and **(ii)** assuming them to have period 2π.

(i) $f(x) = \begin{cases} 1 & \text{for } -\pi/2 < x < \pi/2 \\ 0 & \text{for } \pi/2 < x < 3\pi/2 \end{cases}$ (ii) $f(x) = x^2/4$ for $-\pi < x < \pi$

15.13 Using problem 15.12, show that

(i) $1 - \frac{1}{3} + \frac{1}{5} - \frac{1}{7} + \ldots = \frac{\pi}{4}$* (ii) $1 + \frac{1}{4} + \frac{1}{9} + \frac{1}{16} + \frac{1}{25} + \ldots = \frac{\pi^2}{6}$

(iii) $1 - \frac{1}{4} + \frac{1}{9} - \frac{1}{16} + \frac{1}{25} - \ldots = \frac{\pi^2}{12}$ (iv) $1 + \frac{1}{3^2} + \frac{1}{5^2} + \frac{1}{7^2} + \ldots = \frac{\pi^2}{8}$

Hint: Use **(ii)** and **(iii)** to solve **(iv)**.

15.14 Find the half range sine expansion of the function $f(x) = x^2$, $0 \le x < 1$.

Use this expansion to show that $1 - \frac{1}{3^3} + \frac{1}{5^3} - \frac{1}{7^3} + \ldots = \frac{\pi^3}{32}$.

By integrating the expansion of $f(x)$, show that $1 + \frac{1}{3^4} + \frac{1}{5^4} + \frac{1}{7^4} + \ldots = \frac{\pi^4}{96}$.

15.15 Show that the complex Fourier coefficients of an odd function are purely imaginary and those of an even function are real.

15.16 Find the complex form of the Fourier series

$$\begin{aligned} f(x) &= e^x \\ f(x + 2\pi) &= f(x) \end{aligned}, \quad -\pi < x < \pi$$

15.17 Prove that for $0 \le x \le \pi$,

(i) $x(\pi - x) = \frac{\pi^2}{6} - \left(\frac{\cos 2x}{1^2} + \frac{\cos 4x}{2^2} + \frac{\cos 6x}{3^2} + \ldots \right)$

*This series is called *Gregory series*.

Solution 15.8 It is possible to find the required deflection curve by solving the simple differential equation

$$EIy^{\text{iv}} = -P_0x/p \tag{1}$$

under the conditions

$$y(0) = y(p) = 0; \quad y''(0) = y''(p) = 0. \tag{2}$$

We can determine the deflection curve by an application of Fourier series for y which satisfies [1] and [2]. So, we can choose a sine series as

$$y = \sum_{n=1}^{\infty} b_n \sin \frac{n\pi x}{p} \tag{3}$$

(where b_n is to be determined) because each term of such a series satisfies the end conditions. Now, we expand the load function $-P_0x/p$ in a half range series in the interval $(0, p)$. So, the required expansion will be

$$-\frac{P_0x}{p} = \frac{2P_0}{\pi}\sum_{n=1}^{\infty}\frac{(-1)^n}{n}\sin\frac{n\pi x}{p}$$

since every $a_n = 0$ and $F_1 = 2P_0$ at $x = -P$. Computing y^{iv} from [3] and substituting the value in [1], we get

$$EI\frac{\pi^4}{p^4}\sum_{n=1}^{\infty} n^4 b_n \sin\frac{n\pi x}{p} = \frac{2P_0}{\pi}\sum_{n=1}^{\infty}\frac{(-1)^n}{n}\sin\frac{n\pi x}{p}.$$

If this equation holds on $0 < x < p$, then the coefficients of like terms on either side must be equal. Thus for $n \geq 1$, we get

$$EI\frac{\pi^4}{p^4}n^4 b_n = \frac{2P_0}{\pi}\frac{(-1)^n}{n} \quad \text{or} \quad b_n = (-1)^n\frac{2P_0p^4}{EI\pi^5 n^5}$$

which is the required solution when we substitute for b_n in [3].

15.8 Finite Fourier transforms

Sometimes we use *finite Fourier transforms* in solving differential equations. The *finite Fourier sine transform* of $F(x)$, $0 < x < p$, is defined as

$$F_S(n) = \int_0^p f(x)\sin\frac{n\pi x}{p}\,dx \tag{15.11}$$

where n is an integer. The function $f(x)$ is then called the *inverse finite Fourier sine transform* of $F_S(n)$ and can be easily shown to be of the form

$$f(x) = \frac{2}{p}\sum_{n=1}^{\infty} F_S(n)\sin\frac{n\pi x}{p}. \qquad [15.12]$$

If $f(x)$ is an odd function in $(-p, p)$, then

$$f(x) = \sum_{n=1}^{\infty} b_n \sin\frac{n\pi x}{p} \quad \left(\text{where } b_n = \frac{2}{p}\int_0^p f(x)\sin\frac{n\pi x}{p}dx\right). \qquad [15.13]$$

If we write $\displaystyle\int_0^p f(x)\sin\frac{n\pi x}{p}dx = F_S(n)$, then $b_n = \dfrac{2}{p}F_S(n)$.

So [15.13] can be written as

$$f(x) = \frac{2}{p}\sum_{n=1}^{\infty} F_S(n)\sin\frac{n\pi x}{p}$$

which is the required result [15.12].

Similarly the finite Fourier *cosine* transform of $f(x), 0 < x < p$ is defined as

$$F_C(n) = \int_0^p f(x)\cos\frac{n\pi x}{p}dx$$

from which we can obtain the *inverse finite Fourier cosine transform* of $F_C(n)$, given by

$$f(x) = \frac{1}{p}F_C(0) + \frac{2}{p}\sum_{n=1}^{\infty} F_C(n)\cos\frac{n\pi x}{p}.$$

Example 15.9 Find: **(a)** the finite Fourier sine transform, and **(b)** the finite Fourier cosine transform of the function

$$F(x) = 2x, \quad 0 < x < 4.$$

Solution 15.9 **(a)** Since $p = 4$, we have

CHAPTER SIXTEEN

Numerical Solution Of Differential Equations

16.1 Introduction

In the preceding chapters we have discussed in detail various methods of solving "*analytically*" different types of differential equations. However, a vast majority of differential equations are such that their closed form solutions cannot be obtained using the available analytical methods.

Further, for even those differential equations that can be solved analytically, the solutions may be very tedious and the expressions obtained may be complicated. In most application-oriented problems, for practical utility, one only needs the numerical values of the unknown function (the dependent variable) corresponding to discrete numerical values of the dependent variables. For obtaining solutions of differential equations, in terms of different discrete numerical values we have numerical methods.

The numerical methods for solving differential equations are essentially *step-by-step* methods. The interval [*a, b*] over which the independent variable x varies, is divided into sub-intervals

$$a = x_0 < x_1 < x_2 < \ldots < x_n = b$$

and the length of each sub-interval is called a *step*. Usually, for convenience, the points of division, $x_1, x_2, \ldots$ are equally spaced so that the length of each sub-interval is the same i.e.,

$$x_{i+1} - x_i = h, i = 0, \ldots, n-1$$

where the fixed length h of each sub-interval is called the *step-length*. In the first step, an approximate value y_1 of the unknown function y of a differential equation

$$\frac{dy}{dx} = f(x, y), \quad y(x_0) = y_0 \qquad [16.1]$$

is evaluated, by any one of the methods to be discussed, at $x = x_1 = x_0 + h$. At the second step, an approximate value y_2 of y of [16.1] is calculated at $x = x_2 = x_0 + 2h$ by the same method. This process is continued till the value y_n is determined at the last point $x_n = x_0 + nh$. Thus, we get a set of approximate values $y_1, y_2, \ldots, y_n$, for the unknown function $y(x)$ of the differential equation [16.1], at the *discrete* values $x_1, x_2, \ldots, x_n$ of the independent variable x. The accuracy of the values of y, thus obtained, depends on the numerical method employed. The amount of error is different in different methods. Depending on the accuracy needed, suitable methods can be employed to solve [16.1].*

Sufficient conditions for the existence of a unique solution function y, satisfying the *initial value* problem [16.1], are given in the following theorem.

Theorem 16.1 Let D be a rectangular domain $a < x < b, A - c < y < A + c$. If

(i) f is *continuous* in D with $|f(x, y)| \leq M$ for all points (x, y) in D and some positive number M (i.e., $f(x, y)$ is *bounded* in D);

(ii) $|f(x, y') - f(x, y'')| \leq L|y' - y''|$ for all points $(x, y'), (x, y'')$ in D and a given positive number L (called *Lipschitz constant);* and

(iii) $d = \min\left\{|b - a|, \dfrac{c}{M}\right\}$, then there exists a unique solution y of the initial problem [16.1] defined for $a \leq x \leq a + d$ (the proof is left out since it is beyond the scope of this book).

Example 16.1 Find suitable values of M and d for the initial value problem

$$\frac{dy}{dx} = 1 + x^2 + y^2, \ y(0) = 0.$$

Solution 16.1 We have $f(x, y) = 1 + x^2 + y^2, a = 0, A = 0$. Assuming $b = 1, c = 1, D$ is the rectangle $0 \leq x < 1, \quad -1 \leq y \leq 1$. For any two points $(x, y'), (x, y'')$ in D, we have

$$|f(x, y') - f(x, y'')| = |(y')^2 - (y'')^2| = |(y' + y'')(y' - y'')| \leq 2|y' - y''|$$

so that $L = 2$. Also for (x, y) in D we get $|f(x, y)| \leq 3$. Therefore $M = 3$ and $d = \min\{1, 1/3\} = 1/3$. Hence the given initial value problem has a unique solution y defined for

$$0 \leq x \leq \frac{1}{3}, \ |y| \leq 1\left(0 \leq x \leq \frac{1}{3}\right).$$

*See *Numerical Methods* (With programs in BASIC, FORTRAN and PASCAL), by S. Balachandra Rao and C.K. Shantha, Universities Press, Hyderabad (1992).

where the initial condition $y(x_0) = y_0$ means $y = y_0$ when $x = x_0$. The known function $f(x, y)$ may be *linear* or *nonlinear*. If $y(x)$ is the exact solution of [16.1], then we can expand $y(x)$ as a Taylor series about the initial point x_0 in the form

$$y(x) = y_0 + (x - x_0)y_0^{(1)} + \frac{(x - x_0)^2}{2!}y_0^{(2)} + \dots \qquad [16.8]$$

where $y_0^{(1)}, y_0^{(2)}, \dots$ are the successive derivatives of y at $x = x_0$. But, these derivatives are not known explicitly. However, if f is differentiable a sufficient number of times, we have, for the first few derivatives, the following expressions in terms of f and its partial derivatives:

$$y^{(1)} = f(x, y); \quad y^{(2)} = f^{(1)}(x, y) = f_x + f_y y^{(1)} = f_x + f_y f;$$

$$y^{(3)} = f_{xx} + f_{xy}f + f_{yx}f + f_{yy}f^2 + f_y f_x + f_y^2 f$$

$$= f_{xx} + 2f_{xy}f + f_{yy}f^2 + f_x f_y + f_y^2 f$$

where the subscripts represent partial derivatives. Continuing this process, a derivative of a higher order can be expressed in terms of f and its partial derivatives. Since the evaluation of high-order total derivatives becomes more and more cumbersome, for practical purposes, we have to restrict the Taylor expansion to a first few convenient number of terms of the series. This *truncation* in the series leads to a restriction on the value of x for which the expansion is a reasonable approximation.

For a suitable small *step-length* h, where $h = x - x_0$, the function $y(x)$ is evaluated at $x = x_0 + h$. The Taylor expansion [16.8] then becomes

$$y_1 = y(x_0 + h) = y_0 + hy_0^{(1)} + \frac{h^2}{2!}y_0^{(2)} + \dots \qquad [16.9]$$

The derivatives $y^{(1)}$, $y^{(2)}$, etc., are then evaluated at $x_1 = x_0 + h$ and substituted in [16.9] to obtain the value of y at $x_0 + 2h$ given by

$$y_2 = y(x_0 + 2h) = y_1 + hy_1^{(1)} + \frac{h^2}{2!}y_1^{(2)} + \dots$$

Continuing in this manner, we obtain a discrete set of values y_n which are approximations to the exact values of y at the points $x_n = x_0 + nh, n = 1, 2, \dots$

16.3.1 Algorithm for Taylor's method

An *algorithm* is a well-defined sequential set of mathematical and logical operations which, when performed, produces the solution of a given mathematical problem.

Suppose we want to find an approximate solution of the initial value problem

$$\frac{dy}{dx} = f(x, y), y(a) = y_0 \text{ over } [a, b].$$

(i) If we need the values of y at N points besides a in $[a, b]$, which are equally spaced, then we choose the *step-length* as $h = (b - a)/N$ so that

$$x_0 = a, x_n = a + nh, y_n = y(a + nh) \quad \text{with} \quad x_N = b, n = 1, 2, 3, \ldots, N - 1.$$

(ii) Approximations y_n to the exact values $y(x_n)$ are evaluated from the *recursion* formula

$$y_{n+1} = y_n + hT_k(x_n, y_n), n = 0, 1, 2, \ldots, N - 1$$

where $T_k(x, y) = f(x, y) + \frac{h}{2!} f^{(1)}(x, y) + \ldots + \frac{h^{k-1}}{k!} f^{(k-1)}(x, y)$ for $k = 1, 2, \ldots$

Note: From the *remainder term,* in the Taylor expansion, the truncation error in the above algorithm for order k is given by

$$E = \frac{h^{k+1}}{(k+1)!} y^{(k+1)}(c), x_n < c < x_n + h.$$

Example 16.2 Solve: $\frac{dy}{dx} = y^2, y(0) = 1$ at $x = 0.01$ using Taylor's series.

Solution 16.2 We have $y^{(1)}(0) = [y(0)]^2 = 1$ and $y^{(2)} = 2yy^{(1)}$, so that

$$y^{(2)}(0) = 2y(0) \cdot y^{(1)}(0) = 2(1)(1) = 2;$$

$$y^{(3)} = 2[y^{(1)}]^2 + 2y \cdot y^{(2)};$$

$$y^{(3)}(0) = 2[1]^2 + 2[1][2] = 6, \text{ etc.}$$

Then, by Taylor's series, we have

$$y(0 + 0.01) = y(0) + (0.01)y^{(1)}(0) + \frac{(0.01)^2}{2!} y^{(2)}(0) + \frac{(0.01)^3}{3!} y^3(0) + \ldots$$

or $\quad y(0.01) \approx 1 + (0.01)(1) + \frac{0.0001}{2}(2) + \frac{0.000001}{6}(6) + \ldots$

or $\quad y(0.01) \approx 1.010101.$

For $n = 2,\ y_2 = y_0 + \int_0^x (1 + xy_1)dx = 1 + \int_0^x \left[1 + \left(x + x^2 + \frac{x^3}{2}\right)\right] dx$

$$= 1 + \left[x + \frac{x^2}{2} + \frac{x^3}{3} + \frac{x^4}{8}\right]_0^x$$

or $y_2 = 1 + x + \frac{x^2}{2} + \frac{x^3}{3} + \frac{x^4}{8}.$

For $n = 3,\ y_3 = y_0 + \int_0^x (1 + xy_2)dx$

$$= 1 + \int_0^x \left[1 + \left(x + x^2 + \frac{x^3}{2} + \frac{x^4}{3} + \frac{x^5}{8}\right)\right] dx$$

or $y_3 = 1 + x + \frac{x^2}{2} + \frac{x^3}{3} + \frac{x^4}{8} + \frac{x^5}{15} + \frac{x^6}{48}.$

Thus, we can go on successively improving the solution.

We shall tabulate (Table 16.1) the values of y for $x = 0(0.1)0.5$, i.e. from $x = 0$ to $x = 0.5$, with step-length 0.1. The values of y are correct to 4 decimal places.

Table 16.1 *Picard's solution.*

x	0	0.1	0.2	0.3	0.4	0.5
y	1.000	1.1053	1.2229	1.3552	1.5053	1.6769

Example 16.4 Solve: $dy/dx = 1 + y^2, y(0) = 0$ by Picard's method.

Solution 16.4 Here, $x_0 = 0, y_0 = 0, f(x, y) = 1 + y^2$. By Picard's iterative formula [16.11], we have

$$y_n = y_0 + \int_{x_0}^x f(x, y_{n-1})dx$$

or $\quad y_n = \int_0^x [1 + y_{n-1}^2]dx \quad$ or $\quad y_n = x + \int_0^x y_{n-1}^2 dx.$

Now, taking $n = 1, 2, 3, \ldots$, we get the following successive approximations to y:

$$y_1 = x + \int_0^x y_0^2 dx = x + \int_0^x (0)dx = x;$$

$$y_2 = x + \int_0^x y_1^2 dx = x + \int_0^x x^2 dx = x + \frac{x^3}{3};$$

$$y_3 = x + \int_0^x y_2^2 dx = x + \int_0^x \left[x + \frac{x^3}{3}\right] dx = x + \int_0^x \left[x^2 + \frac{2}{3}x^4 + \frac{x^6}{9}\right] dx$$

or $y_3 = x + \dfrac{x^3}{3} + \dfrac{2x^5}{15} + \dfrac{x^7}{63}.$

Thus, the solution can be successively improved.

Note: The given differential equation can be solved *analytically* to get the *exact* solution as follows. Separating the variables, the equation can be written as

$$\frac{dy}{1+y^2} = dx.$$

On integration, we get the general solution

$$\tan^{-1} y = x + c \qquad \text{or} \qquad y = \tan(x + c).$$

Initially, when $x = 0$, we have $y = 0$. Substituting these values in the general solution, we get $c = 0$. Therefore, the exact solution of the given initial value problem is $y = \tan x$ which when expanded as a series gives

$$y = x + \frac{x^3}{3} + \frac{2x^5}{15} + \frac{17x^7}{315} + \qquad \text{(converging for } |x| < \pi/2).$$

It may be noted that the first three terms of y_3, obtained by Picard's method, are the same as those of the infinite series expansion for the exact solution.

Example 16.5 Solve: $dy/dx = x + y^2; x = 0, y = 1.$

Solution 16.5 In Picard's iterative formula [16.11]

$$y_n = y_0 + \int_{x_0}^x f(x, y_{n-1}) dx$$

taking $f(x, y) = x + y^2, x_0 = 0, y_0 = 1$ we have

$$y_n = 1 + \int_0^x \left[x + (y_{n-1})\right]^2 dx = 1 + \frac{x^2}{2} + \int_0^x (y_{n-1})^2 dx.$$

Therefore, taking $n = 1, 2, 3, \ldots$, we get

$$y_1 = 1 + \frac{x^2}{2} + \int_0^x (y_0)^2 dx = 1 + \frac{x^2}{2} + \int_0^x (1)^2 dx = 1 + x + \frac{x^2}{2}$$

$$y_2 = 1 + \frac{x^2}{2} + \int_0^x (y_1)^2 dx = 1 + \frac{x^2}{2} + \int_0^x \left[1 + x + \frac{x^2}{2}\right]^2 dx$$

$$= 1 + \frac{x^2}{2} + \int_0^x \left[1 + x^2 + \frac{x^4}{4} + 2x + x^3 + x^2\right] dx$$

$$= 1 + \frac{x^2}{2} + \int_0^x \left[1 + 2x + 2x^2 + x^3 + \frac{x^4}{4}\right] dx$$

$$= 1 + \frac{x^2}{2} + x + x^2 + \frac{2x^3}{3} + \frac{x^4}{4} + \frac{x^5}{20} = 1 + x + \frac{3x^2}{2} + \frac{2x^3}{3} + \frac{x^4}{4} + \frac{x^5}{20}.$$

We see that the evaluation of further successive approximations, $y_3, y_4, \ldots$ becomes more and more complicated.

Example 16.6 Solve $y' = y, y(0) = 1$ by Picard's successive approximation method and compare the solution with the exact solution.

Solution 16.6 We have $f(x, y) = y, x_0 = 0, y_0 = 1.$ Picard's iterative formula takes the form

$$y_n = 1 + \int_0^x y_{n-1}\, dx, n = 1, 2, 3, \ldots$$

Therefore, taking $n = 1, 2, 3, \ldots$, we get

$$y_1 = 1 + \int_0^x y_0\, dx = 1 + \int_0^x (1) dx = 1 + x;$$

$$y_2 = 1 + \int_0^x y_1 dx = 1 + \int_0^x (1 + x) dx$$

$$y_2 = 1 + \left[x + (x^2/2)\right]_0^x = 1 + x + (x^2/2)$$

$$y_3 = 1 + \int_0^x \left[1 + x + (x^2/2)\right] = 1 + x + (x^2/2) + (x^3/3.2);$$

$$y_4 = 1 + \int_0^x \left[1 + x + (x^2/2!) + (x^3/3!)\right] dx$$

or $$y_4 = 1 + x + \frac{x^2}{2!} + \frac{x^3}{3!} + \frac{x^4}{4!}. \qquad [1]$$

Analytical solution: The given differential equation is $dy/dx = y$. By separating the variables, we have $dy/y = dx$. On integrating, we get $\log y = x + c'$ or

$$y = e^{x+c'} = ce^{x} \quad [2]$$

where $c = e^{c'}$ is an arbitrary constant. The *exact* general solution of the given equation is provided by [2]. The initial conditions are $x = 0, y = 1$. Substituting these in [2], we get $c = 1$. Therefore, the *exact* particular solution is $y = e^{x}$ which has an infinite series expansion

$$y = e^{x} = 1 + \frac{x}{1!} + \frac{x^2}{2!} + \frac{x^3}{3!} + \frac{x^4}{4!} + \dots \quad [3]$$

From [2] and [3], we find that the successive approximations by Picard's method give the same terms as the exact solution [3] after 2 terms, 3 terms, etc. In fact, the Picard solution [1] is the same as the first five terms of [3].

16.5 Euler's method

The two methods—Taylor's and Picard's—discussed in sections 16.3 and 16.4 give solutions of a first order differential equation in the form of a power series which can be evaluated at the required values of the independent variable.

Hereafter, we shall consider methods which yield solutions in the form of numerical values of the dependent variable for discrete values of x at specified intervals. In order to solve the initial value problem [16.1]:

$$\frac{dy}{dx} = f(x, y), y(x_0) = y_0.$$

We start from the initial value $y = y_0$ at $x = x_0$ and obtain subsequent values of y in *steps*. In the first step, at $x = x_0 + h$, we compute the value y_1 for the solution of [16.1]. Using the value y_1, the next value y_2 is computed at $x = x_0 + 2h$. In this way, the values of y are computed at the discrete values of x viz., $x_0 + h, x_0 + 2h, \dots, x_0 + nh$. The value of the step-length h is fixed conveniently small, like 0.2 or 0.1 or even smaller if necessary. The computation in *Euler's method* at each step is performed using a formula derived from Taylor's series [16.9] as follows:

We have $y(x + h) = y(x) + hy'(x) + (h^2/2!)y''(x) + \dots$ where the primes denote derivatives with respect to x. But, from [16.1], we have

$$y' \equiv \frac{dy}{dx} = f \text{ and hence } y'' = f', y''' = f''$$

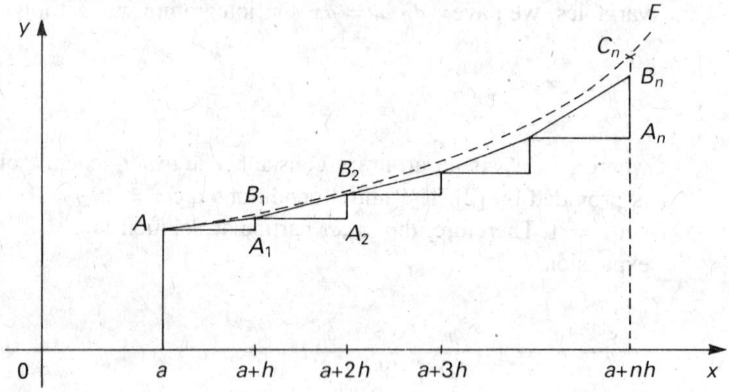

Fig. 16.1 *Euler's method.*

and so on where $f \equiv f(x, y)$ is the known function in the right hand side of [16.1]. We have therefore

$$y(x+h) = y(x) + hf + \frac{h^2}{2!}f' + \frac{h^3}{3!}f'' + \cdots \qquad [16.12]$$

where the derivatives f', f'' etc. are evaluated at the point $[x, y(x)]$.

If the step-length h is taken sufficiently small, then the terms containing h^2, h^3, etc. in [16.12] are still smaller and hence can be neglected. This gives us an approximate solution of [16.1] as

$$y(x+h) = y(x) + hf(x, y).$$

In the first step, using the above formula, we get

$$y_1 = y_0 + hf(x_0, y_0) \quad \text{where} \quad y_1 = y(x_1) = y(x_0 + h).$$

Then, using y_1 instead of y_0 and evaluating f at the next point (x_1, y_1), we now have $y_2 = y_1 + hf(x_1, y_1)$ where $y_2 = y(x_2) = y(x_1 + h) = y(x_0 + 2h)$ and so on. In general, we have the recursive formula:

$$y_{n+1} = y_n + hf(x_n, y_n) \text{ (for } n = 0, 1, 2, \ldots). \qquad [16.13]$$

16.5.1 Algorithm for Euler's method

Suppose we want an approximate solution of the differential equation [16.1]:

$$dy/dx = f(x, y) \text{ with } y = y_0 \text{ when } x = x_0.$$

(i) If, besides at x_0, the values of y are required at N points which are equally spaced with step-length h, then let

$$x_1 = x_0 + h, x_2 = x_0 + 2h, \ldots, x_N = x_0 + Nh.$$

(ii) Approximating y_n to the exact values, $y(x_n)$ are evaluated using the recursive formula

$$y_{n+1} = y_n + hf(x_n, y_n) \text{ for } n = 0, 1, 2, \ldots, N-1.$$

Example 16.7 Obtain the solution of $y' = x - y;\ y(0) = 1$ at $x = 0(0.2)0.6$.

Solution 16.7 Here, we have $f(x, y) = x - y; x_0 = 0, y_0 = 1$ and $h = 0.2$. We have to find the values of y at $x_1 = 0.2, x_2 = 0.4$ and $x_3 = 0.6$. Now,

$$f(x_0, y_0) = f(0, 1) = 0 - 1 = -1.$$

By Euler's iterative formula [16.13], $y_{n+1} = y_n + hf(x_n, y_n)$. For $n = 0$,

$$y_1 = y_0 + hf(x_0, y_0) = 1 + (0.2)(-1) = 0.8.$$

For the next step, consider $f(x_1, y_1) = x_1 - y_1 = 0.2 - 0.8 = -0.6$. Therefore,

$$y_2 = y_1 + hf(x_1, y_1) = 0.8 + (0.2)(-0.6) \text{ or } y_2 = 0.8 - 0.12 = 0.68.$$

For the third step, consider $f(x_2, y_2) = x_2 - y_2 = 0.4 - 0.68 = -0.28$. Therefore,

$$y_3 = y_2 + hf(x_2, y_2) = 0.68 + (0.2)(-0.28) = 0.68 - 0.056 = 0.624.$$

The solution is tabulated below:

Table 16.2 *Euler's solution.*

x	0	0.2	0.4	0.6
y	1.000	0.800	0.680	0.624

Let us compare the solution obtained with the exact solution which we get analytically.

Analytical solution: The given equation is

$$\frac{dy}{dx} = x - y. \qquad [1]$$

We shall reduce [1] to the variables separable form by the transformation $z = x - y$ so that $dz/dx = 1 - (dy/dx)$ and hence [1] takes the form

$$1 - \frac{dz}{dx} = z \quad \text{or} \quad \frac{dz}{dx} = 1 - z \quad \text{or} \quad \frac{dz}{1-z} = dx.$$

On integrating, we get $-\log(1-z) = x + c'$ or $\log(1-z) = -x - c'$. So, $1 - z = e^{-x-c'} = ce^{-x}$ where $c = e^{-c'}$, an arbitrary constant. Therefore, the general solution of [1] is $1 - (x - y) = ce^{-x}$ or

$$y = x - 1 + ce^{-x}. \qquad [2]$$

Initially, for $x = 0, y = 1$. Substituting these values in [2], we get $c = 2$ so that [2] becomes

$$y = x - 1 + 2e^{-x}. \qquad [3]$$

The values of y obtained numerically by Euler's method are compared with the corresponding values from the exact solution [3] in the following table:

Table 16.3 *Solutions by Euler's and analytical methods.*

x	0	0.2	0.4	0.6
y_{Euler}	1.000	0.800	0.680	0.624
y_{Exact}	1.00000	0.83746	0.74064	0.69762

The values in the last row are correct to five decimal places. We observe that as the value of x is farther from the initial value, $x = 0$, the corresponding value of y obtained numerically by Euler's method deviates more from the exact value.

Example 16.8 The motion of a simple pendulum is governed by the differential equation

$$\frac{d^2\theta}{dt^2} + \frac{g}{l}\sin\theta = 0 \qquad [1]$$

where θ is the angle made by the string with the vertical at an instant of time t, l is the constant length of the thin, light and inextensible string and g is the constant acceleration due to gravity.

Obtain a numerical solution of [1] by Euler's method taking $g = 980\text{cm/sec}^2$, $l = 100$cm and the maximum value of θ as 4°.

Solution 16.8 For the derivation of [1] governing the motion of the simple pendulum, see Example 5.4 in Chapter 5. It is important to recognise that [1] is of the second order and *nonlinear* because of $\sin\theta$. In Example 5.4, the equation was solved by rendering it linear with the approximation $\sin\theta \approx \theta$ when θ is small. Now, without restricting θ to be small, we shall obtain a numerical solution of the nonlinear equation [1]. Equation [1] can be integrated by multiplying it by $2(d\theta/dt)$ which gives

$$2\frac{d\theta}{dt}\frac{d^2\theta}{dt^2} + 2\frac{g}{l}\sin\theta\frac{d\theta}{dt} = 0.$$

On integrating with respect to t, we get

$$\left[\frac{d\theta}{dt}\right]^2 - 2\frac{g}{l}\cos\theta = c \qquad [2]$$

where c is a constant of integration. Taking the initial conditions as $d\theta/dt = 0$ when $\theta = \alpha$, where $d\theta/dt$ is the angular velocity and α is the maximum value of the angular displacement θ, we get $c = (-2g/l)\cos\alpha$. Substituting this value of c in [2], the differential equation becomes

$$\left[\frac{d\theta}{dt}\right]^2 = 2\frac{g}{l}\left[\cos\theta - \cos\alpha\right]$$

or $$\frac{d\theta}{dt} = \sqrt{2\frac{g}{l}\left[\cos\theta - \cos\alpha\right]}. \qquad [3]$$

Differential equation [3] is of the *first order* but *nonlinear* and hence cannot be solved analytically. It is given that $g = 980\text{cm/sec}^2$, $l = 100$ cm and $\alpha = 4°$. On substituting these values of the constants in [3], the differential equation becomes

$$\frac{d\theta}{dt} = \sqrt{19.6\left[\cos\theta - \cos 4°\right]}. \qquad [4]$$

We shall find the solution of [4], by Euler's method, with the initial conditions $\theta = 0$ when $t = 0$. By Euler's algorithm we have

$$\theta_{n+1} = \theta_n + hf(t_n, \theta_n), n = 0, 1, 2, \ldots \left(\text{where } f(t,\theta) = \sqrt{19.6\left[\cos\theta - \cos 4°\right]}\right).$$

Taking the initial values as $\theta_0 = 0, t_0 = 0$ and $h = 0.25$ we get

$$\theta_1 = \theta_0 + (0.25)\sqrt{19.6\left[\cos\theta_0 - \cos 4°\right]}$$

$$= 0 + 0.25\sqrt{19.6\left[1 - \cos 4°\right]} = 0°.0546263.$$

Continuously applying Euler's algorithm, we get the solution of the simple pendulum problem as shown in Table 16.4. The values of θ are correct to 6 decimal places.

Table 16.4 *Simple pendulum solution.*

n	t (sec)	θ (deg.)
0	0.00	0.000000
1	0.25	0.054626
2	0.50	0.1092147
3	0.75	0.163853
4	1.00	0.218434
5	1.25	0.272979
6	1.50	0.327478
7	1.75	0.381922
8	2.00	0.436298
9	2.25	0.490599
10	2.50	0.544813

16.5.2 Program for Euler's method

The algorithm for Euler's method is set to a computer program. The program is given in BASIC, FORTRAN and Pascal languages. The print-outs of the source as well as output are given below.

Program 16.1 (PASCAL)
Euler's method

```
program euler;
uses crt;
var
      x0, y0, h, y1: real;
      i, n: integer;
function fnf (p, q: real):real;
begin
      fnf:=-p*p*q*q;
end;
procedure start;
begin
      clrscr;
      writeln ('************************************************************* ');
      writeln ('*                                                            *');
      writeln ('*   EULER'S METHOD FOR SOLVING FIRST ORDER O.D.E.            *');
      writeln ('*                          y' = f(x, y)                      *');
      writeln ('*                                                            *');
```

```
        writeln ('*************************************************************');
        gotoxy (26, 8); write ('Initial Value of x: x0 = '); readln (x0);
        gotoxy (26, 9); write ('Initial Value of y: y0 = '); readln (y0);
        gotoxy (26, 10); write ('Number of points : n = '); readln (n);
        gotoxy (26, 11); write ('Step-length : h = '); readln (h);
        writeln; writeln ('NUMBER' :12, 'x' : 20, 'y' : 30);
        writeln;
end;

procedure workout;
begin
        for i: =1 to n do
        begin
            y1: =y0+h*fnf (x0, y0);
            y0: = y1;
            writeln (i:10, x0:25:6, y0:32:11);
            x0:=x0+h;
        end;
end;

procedure choice;
begin
        writeln; writeln;
        write ('Press <Esc> to exit or any other key to try again' :64);
end;

begin
        repeat
            start;
            workout;
            choice;
        until readkey =#27;
end.
```

Program 16.1 (FORTRAN)
Euler's method

```
C    *******************************
            REAL FUNCTION F(P, Q)
            REAL P, Q
            F=- P*P*Q*Q
            RETURN
            END
C    *******************************
            EXTERNAL F
            REAL X0, Y0, H, Y1
```

```
      INTEGER I, N
      PRINT *, ' ****************************************'
      PRINT *, '*                                        *'
      PRINT *, '*    EULER'S METHOD FOR SOLVING          *'
      PRINT *, '*       FIRST ORDER O.D.E. y'=f(x, y)    *'
      PRINT *, '*                                        *'
      PRINT *, ' ****************************************'
      PRINT *, 'INITIAL VALUES:'
      PRINT *, 'X0='
      READ (*, *) X0
      PRINT *, 'Y0='
      READ (*,*) Y0
      PRINT *, 'NUMBER OF POINTS:'
      READ (*,*) N
      PRINT *, 'STEP LENGTH:'
      READ (*,*) H
      WRITE (*, 70)
70    FORMAT ('NUMBER', 17X, 'x', 22X, 'Y')
      DO 280 I=1, N
      Y1=Y0+H*F (X0, Y0)
      Y0=Y1
      WRITE (*, 80) I, X0, Y0
      X0=X0+H
280   CONTINUE
80    FORMAT (2X, I2, 15X, F10.6, 14X, F10.6)
      STOP
      END
```

```
INITIAL VALUES:
x0= 0
y0 =1
NUMBER OF POINTS= 10
STEP LENGTH: .1
```

NUMBER	x	y
1	0	1
2	.1	.9989999
3	.2	.9950079
4	.3	.9860976
5	.4	.9705393
6	.5	.9469907
7	.6	.9147062
8	.7000001	.8737085
9	.8000001	.824853
10	.900001	.769742

16.6 Improved Euler's method

The errors introduced by the use of Euler's method and the build-up of these errors by the repeated application of the method are reduced in the *improved* Euler's method for solving [16.1]. This method starts off exactly in the same way as Euler's method (Section 16.5) i.e., using the initial point (x_0, y_0), y_1 is calculated from the formula $y_1 = y_0 + hf(x_0, y_0)$ and then $f(x_1, y_1)$ is evaluated. In Fig. 16.2, the initial condition, $y = y_0$ when $x = x_0$ is given. This means (x_0, y_0) is a known point on the integral curve $y = F(x)$. Let the step-length be

$$C_0C_1 - x_1 - x_0 = h.$$

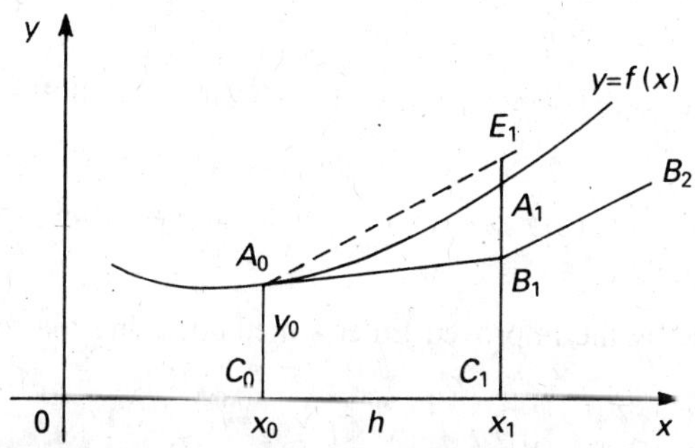

Fig. 16.2 *Improved Euler's method.*

The true value of y at $x = x_1$ is C_1A_1 and the estimated value, by Euler's method, is represented by C_1B_1. Now, A_0B_1 is the tangent to the curve $y = F(x)$ at $A_0(x_0, y_0)$. A line is drawn through A_0 and parallel to B_1B_2 which has slope equal to $f(x_1, y_1)$. Let this line meet C_1A_1 at E_1. The height C_1E_1 is given by

$$C_1E_1 = y_0 + hf(x_1, y_1)$$

while $C_1B_1 = y_0 + hf(x_0, y_0)$.

Now, C_1B_1 given by Euler's method is an *under-estimate* of the true value C_1A_1 while C_1E_1 is an *over-estimate*. It is therefore reasonable to take the *mean* of these two estimates as being a better approximation to the true value C_1A_1 than either of them separately. The mean of C_1B_1 and C_1E_1 is given by

$$y_{est} = y_0 + \frac{h}{2}\left[f(x_0, y_0) + f(x_1, y_1)\right] \qquad [16.14]$$

where y_{est} is an improved estimate for y at $x = x_1$.

Example 16.9 Solve: $dy/dx = x - y, y(0) = 1$ at $x = 0.2$.

Solution 16.9 In Example 16.7, the given initial value problem was solved using Euler's method and the values of y thus obtained for $x = 0(0.2)0.6$ were compared with the exact values of y. Now, we apply the improved Euler's method to the same problem and evaluate y at $x = 0.2$. For the first interval, with $h = 0.2$, we have the relationship

$$f(x_0, y_0) = x_0 - y_0 = 0 - 1 = -1$$

so that, by Euler's method, we get

$$y_1 = y_0 + hf(x_0, y_0) = 1 + (0.2)(-1) = 0.8$$

and $f(x_1, y_1) = x_1 - y_1 = 0.2 - 0.8 = -0.6.$

Now, by the improved Euler's method, using the formula [16.14], we get

$$y_{est} = 1 + \frac{0.2}{2}[-1 - 0.6] = 0.84.$$

This value is much closer to the exact value $y = 0.83746$ given in Example 16.7. Similarly, improved estimates of y can be obtained at $x = 0.4$ and $x = 0.6$ also.

Note: **(i)** Euler's method is also referred to as the *constant slope* method and the improved Euler's method as the *average slope* method.

(ii) The improved Euler's method itself is referred to as the *modified* Euler's method by some authors. However, we consider a slightly different method as the *modified Euler's method* and discuss it in Section 16.8.

(iii) In Euler's method, the *error* is given by the remainder term in Taylor's expansion:

$$E = \frac{h^2}{2!}y''(c), x_0 < c < x_0 + h.$$

Since the error E is proportional to h^2, i.e. of order h^2 [written as $O(h^2)$], the error may considerably reduce the step-length h. It can be shown that the *cumulative error*, in evaluating y through n steps, is of order h.

(iv) In the case of the *improved* Euler's method, the cumulative error is of order h^2.

16.7 Euler's predictor – corrector method

Using the improved Euler's method (Section 16.6), to solve the initial value problem Example 16.1 starting from (x_0, y_0), it is required to find an initial estimate of y_1 by using the formula

$$y_1 = y_0 + hf(x_0, y_0).$$

But this formula is relatively less accurate than the estimate [16.14] of the *improved* Euler's method and hence used as a "*predictor*" formula for an approximate value of y_1. The value of y_1 thus obtained from the *predictor* formula is now used for a more accurate estimation of y_1 using [16.14] of the improved Euler's method. This estimate of y_1 is a corrected value and hence [16.14] is considered a *corrector* formula.

To distinguish between the two different values of y_1 obtained from different formuale, these are denoted by y_1^p and y_1^c representing respectively the *predicted* and the *corrected* values. Thus, we have the *predictor* and *corrector* formulae respectively as

$$y_1^p = y_0 + hf(x_0, y_0) \quad [16.15]$$

$$y_1^c = y_0 + \frac{h}{2}\left[f(x_0, y_0) + f(x_1, y_1^p)\right]. \quad [16.16]$$

After finding the corrected estimate of y_1, we can proceed to evaluate similarly the corrected values of y_2, y_3, etc. We have the *generalized* predictor and corrector formulae as follows:

$$y_{n+1}^p = y_n^c + hf(x_n, y_n^c) \quad [16.17]$$

$$y_{n+1}^c = y_n^c + \frac{h}{2}\left[f(x_n, y_n^c) + f(x_{n+1}, y_{n+1}^p)\right]. \quad [16.18]$$

16.8 Modified Euler's method

The solution obtained from the *improved* Euler's method, for the initial value problem Example 16.1, can be made more accurate by an *iterative process*.

After calculating y_1^c using [16.16], this value can now be used instead of y_1^p in [16.16]. This gives a second value of y_1^c. This second value is now used in place of the first value of y_1^c in [16.16] to obtain a third value of y_1^c. Thus, each time we obtain a better corrected value of y_1^c. This iterative process is repeated until the absolute difference between two successive values of y_1^c is within the prescribed limit of accuracy.

The modification of the improved Euler's method is embodied in the formula

$$\left[y_1^c\right]_{i+1} = y_0 + \frac{h}{2}\left[f(x_0, y_0) + f(x_1, (y_1^c)_i)\right] \quad [16.19]$$

where i denotes the order of the iterate (i.e. first, second, etc.). The final value of y_1^c obtained from [16.19], is used as the best approximation of y_1 in evaluating y_2, the value of y at $x = x_2$. The same iterative process can be applied to get the best values of y_2, y_3, etc. Generalising [16.19], for getting the best values of y_1^c, y_2^c, y_3^c, etc., we have the *modified Euler's method* given as

$$\left[y_{n+1}^c\right]_{i+1} = y_n^c + \frac{h}{2}\left[f(x_n, y_n^c) + f(x_{n+1}, (y_{n+1}^c)_i)\right] \quad [16.20]$$

where the best value of y_n^c and the earlier iterate $(y_{n+1}^c)_i$ are used to obtain $(y_{n+1}^c)_{i+1}$.

Example 16.10 Solve: $dy/dx = x - y, y(0) = 1$ at $x = 0.2, 0.4$.

Solution 16.10 In Examples 16.7 and 16.9, we obtained the solution of the equation by using Euler's and *improved* Euler's methods. Now, we shall solve the initial value problem by the *modified* Euler's method. In Example 16.9, we obtained at $x = 0.2$, $y_1^p = 0.8$ and $y_1^c = 0.84$. We now take this value of y_1^c as the *first* corrected value and go on improving the same iteratively. We have

$$F(x_1, y_1^c) = x_1 - y_1^c = 0.2 - 0.84 = -0.64.$$

We have the iterative formula [16.19]:

$$[y_1^c]_{i+1} = y_0 + \frac{h}{2}\left[f(x_0, y_0) + f(x_1, (y_1^c)_i)\right].$$

With $i = 1$, we have the *second* corrected value

$$[y_1^c]_2 = y_0 + \frac{h}{2}\left[f(x_0, y_0) + f(x_1, (y_1^c)_1)\right] \quad \text{(where } (y_1^c)_1 = 0.84\text{)}$$

so that $[y_1^c]_2 = 1 + \frac{0.2}{2}[-1 - 0.64]$ with $h = 0.2$ and $f(x_0, y_0) = -1$. Therefore

$$[y_1^c]_2 = 0.836.$$

Now, we have $f(0.2, 0.836) = 0.2 - 0.836 = -0.636$ and hence the *third* corrected value of y_1^c from [16.20] with $i = 2$ is given by

$$[y_1^c]_3 = y_0 + \frac{h}{2}\left[f(x_0, y_0) + f(x_1, (y_1^c)_2)\right] = 1 + \frac{0.2}{2}[-1 - 0.636] = 0.8364.$$

The iterative process can be continued till there is no difference between two successive iterates upto the required number of decimal places. In the current example, y_1^c settles down at $y_1^c = 0.8364$ correct to 4 decimal places after 3 iterations.

Now, taking this best value of y_1 viz., $y_1^c = 0.8364$ as y_1, we proceed to evaluate y_2. We have $y_2^p = y_1 + hf(x_1, y_1)$ where $h = 0.2$ and $x_1 = 0.2$. Therefore

$$y_2^p = 0.8364 + 0.2(0.2 - 0.8364) = 0.70912.$$

Then, $f(x_2, y_2^p) = x_2 - y_2^p = 0.4 - 0.70912 = -0.30912$ so that

$$y_2^c = y_1^c + (h/2)\left[f(x_1, y_1^c) + f(x_2, y_2^p)\right]$$

$$= 0.8364 + 0.1\left[-0.6364 - 0.30912\right] = 0.741848$$

which is the *first* corrected value of y_2^c. The further corrected (iteratively) values of y_1^c are obtained as follows:

$$f(0.4, 0.741848) = -0.341848$$

$$(y_2^c)_2 = 0.8364 + 0.1[-0.6364 - 0.341848] \quad \text{or} \quad (y_2^c)_2 = 0.7385752.$$

Now $f(0.4, 0.7385752) = -0.3385752$

$$(y_2^c)_3 = 0.8364 + 0.1[-0.6364 - 0.3385752] \quad \text{or} \quad (y_2^c)_3 = 0.7389024.$$

Similarly, we have $(y_2^c)_4 = 0.7388697, (y_2^c)_5 = 0.738873, (y_2^c)_6 = 0.7388727$ etc. We notice that y_2^c settles down, correct to four decimal places, to $y_2^c = 0.7389$.

16.9 Runge–Kutta methods

The Runge–Kutta methods score over the earlier methods in securing greater accuracy for the solution and at the same time avoiding the need for higher-order derivatives. These methods are named after two German mathematicians, Carl Runge (1856–1927) and Wilhelm Kutta (1867–1944). The methods were devised by Runge in 1894 and later extended by Kutta in 1901. In the simplest of these methods, a formula of the following form is sought:

$$y_{n+1} = y_n + ak_1 + bk_2 \qquad [16.21]$$

where $k_1 = hf(x_n, y_n)$, $k_2 = hf(x_n + \alpha h, y_n + \beta k_1)$ and a, b, α and β are constants to be determined so that [16.21] will coincide with the Taylor algorithm (Section 16.3) for as high an order as possible.

On expanding $y(x_{n+1})$ in a Taylor series upto the term containing h^3, we have

$$y(x_{n+1}) = y(x_n) + hy'(x_n) + \frac{h^2}{2}y''(x_n) + \frac{h^3}{3!}y'''(x_n) + \dots$$

$$= y(x_n) + hf(x_n, y_n) + \frac{h^2}{2!}(f_x + ff_y)_n$$

$$+ \frac{h^3}{3!}(f_{xx} + 2ff_{xy} + f_{yy}f^2 + f_xf_y + f_y^2 f)_n + O(h^4) \qquad [16.22]$$

where we have used the expansions for y'' and y''' in terms of partial derivatives given that $y' = f(x, y)$ from [16.1]. The use of the subscript n in [16.22] means that all functions involved are evaluated at (x_n, y_n).

From Taylor's expansion for a function of two variables we have

$$\frac{k_2}{h} = f(x_n + \alpha h, y_n + \beta k_1) = f(x_n, y_n) + \alpha h f_x$$

$$+ \beta k_1 f_y + \frac{\alpha^2 h^2}{2} f_{xx} + \alpha h \beta k_1 f_{xy} + \frac{\beta^2 k_1^2}{2} f_{yy} + O(h^3)$$

where the subscripts stand for partial derivatives and all the derivatives are evaluated at (x_n, y_n). Now, on substituting for k_2 in [16.21] and noting that $k_1 = hf(x_n, y_n)$ on rearrangement of terms of powers of h, we get

$$y_{n+1} = y_n + (a + b)hf + bh^2(\alpha f_x + \beta ff_y)$$

$$+bh^3\left[\frac{\alpha^2}{2}f_{xx} + \alpha\beta ff_{xy} + \frac{\beta^2}{2}f^2 f_{yy}\right] + O(h^4). \qquad [16.23]$$

On comparing [16.23] with [16.22] for the coefficients of h and h^2, we get

$$a + b = 1, \; b\alpha = b\beta = \frac{1}{2}. \qquad [16.24]$$

In [16.24] we have three equations in four unknowns: a, b, α and β. Since there is still one degree of freedom, there are many solutions for [16.24]. We choose the simplest solution

$$a = b = \frac{1}{2}, \alpha = \beta = 1$$

so that [16.21] becomes

$$y_{n+1} = y_n + \frac{1}{2}(k_1 + k_2) \qquad [16.25]$$

where $k_1 = hf(x_n, y_n), k_2 = hf(x_n + h, y_n + k_1)$. The method of solution of the initial value problem [16.1] provided by the formula [16.25] is called the Runge–Kutta method of *second order* (or RK method of second order).

16.9.1 Algorithm for RK second order method

Suppose the given initial value problem [16.1] is in the form

$$\frac{dy}{dx} = f(x, y), y(x_0) = y_0.$$

For the approximations y_{n+1} to the exact solution $y[x_0 + (n+1)h]$, for a fixed step-length h, we have the recursion formula [16.25] viz.,

$$y_{n+1} = y_n + \frac{1}{2}(k_1 + k_2) \text{ (with } k_1 = hf(x_n, y_n), k_2 = hf(x_n + h, y_n + k_1))$$

where $n = 0, 1, 2, 3, \ldots$

16.9.2 Error estimation in RK second order method

The error in the approximate solution y_{n+1}, from the algorithm [16.25] is given by

$$y(x_{n+1}) - y_{n+1} = \frac{h^3}{12}\left[f_{xx} + 2ff_{xy} + f_{yy}f^2 - 2f_xf_y - 2f_y^2 f\right] + O(h^4).$$

The actual estimation of error is quite complicated because of the complexity of the coefficient of h^3 in the error term. However, the error is of the order h^3 while that of Euler's method is lower. Therefore, the RK method of second order is expected to give a more accurate solution, even with a slightly bigger step-length h, than Euler's method.

Example 16.11 Solve: $y' = x - y, y(0) = 1, x = 0.2, 0.4.$

Solution 16.11 For the convenience of comparing the accuracy of the different methods of solution, we have taken the same example as discussed in Examples 16.7, 16.9 and 16.10. We shall take the step-length, $h = 0.2$ although a smaller step-length like 0.1 would give a more accurate solution. We have the algorithm of the RK second order method given by [16.25]:

$$y_{n+1} = y_n + \frac{1}{2}(k_1 + k_2) \text{ with } k_1 = hf(x_n, y_n), k_2 = hf(x_n + h, y_n + k_1).$$

Substituting $n = 0$, we have

$$y_1 = y_0 + \frac{1}{2}(k_1 + k_2) \qquad [1]$$

where $k_1 = hf(x_0, y_0), k_2 = hf(x_0 + h, y_0 + k_1)$. Here, $x_0 = 0, y_0 = 1, f(x, y) = x - y, h = 0.2$. Therefore, at $x_1 = 0.2$, we have

$$k_1 = 0.2(0 - 1) = -0.2; k_2 = 0.2f(0.2, 1 - 0.2) = 0.2(0.2 - 0.8) = -0.12,$$

so that substituting for k_1 and k_2 in [1], we have

$$y_1 = 1 + \frac{1}{2}(-0.2 - 0.12) = 0.84.$$

Next, taking $n = 1$ in the recursion formula [16.25]

$$y_2 = y_1 + \frac{1}{2}(k_1 + k_2) \qquad [2]$$

where $k_1 = hf(x_1, y_1)$ and $k_2 = hf(x_1 + h, y_1 + k_2)$. Here, at $x_2 = 0.4$, we have

$$k_1 = hf(x_1, y_1) = 0.2f(0.2, 0.84) = 0.2(0.2 - 0.84) = -0.128$$

$$k_2 = hf(x_1 + h, y_1 + k_1) = 0.2[0.4 - (0.84 - 0.128)]$$

$$= 0.2[-0.312] = -0.0624;$$

$$\therefore \quad y_2 = 0.84 + \frac{1}{2}[-0.128 - 0.0624] = 0.84 + \frac{1}{2}[-0.1904] = 0.84 - 0.0952 = 0.7448.$$

In Table 16.5, we compare the solutions of the given problem obtained from Euler's method, the RK second order method and the analytical method.

Table 16.5 *Comparison of Euler's and RK second order solutions*

x	0	0.2	0.4
y_{Euler}	1.0000	0.800	0.680
$y_{\text{RK-II}}$	1.0000	0.8400	0.7448
y_{Exact}	1.0000	0.83746	0.74064

We notice that the RK second order method gives a far more accurate result than Euler's.

16.9.3 Program for the RK second order method

Program 16.2 (PASCAL)
Runge–Kutta second order method

```
program runge2;
uses crt;
var
      x0, x1, y0, h, k1, k2: real;
      function fnf (x, y:real): real;
begin
      fnf:=-x*x*y*y;
end;

procedure start;
begin
      clrscr;
      writeln ('*****************************************************************');
      writeln ('*                                                               *');
      writeln ('*          RUNGE-KUTTA METHOD (SECOND ORDER)                    *');
      writeln ('*                    TO SOLVE AN O.D.E                          *');
      writeln ('*****************************************************************');
      gotoxy (27, 10); write ('Initial Value of x: x0 = '); read (x0);
      gotoxy (27, 11); write ('Final Value of x: x1 = '); read (x1);
      gotoxy (27, 12); write ('Initial Value of y: y0= '); read (y0);
      gotoxy (27, 13); write ('Step Length : h = '); read (h);
      writeln ('_______________________________________________________________');
      writeln ('X': 34, 'Y' : 20);
      writeln ('_______________________________________________________________');
end;

procedure workout;
begin
      while (x0<=x1-1E-11) do
      begin
         writeln (x0:39:11, y0:20:11);
         k1:=h× fnf (x0, y0);
```

```
        k2:=h× fnf (x0+h, y0+k1);
        x0:=x0+h;
        y0:=y0+(k1+k2)/2;
      end;
  end;

  procedure choice;
  begin
      writeln;
      gotoxy (16,23); write ('Press <Esc> to exit or any other key to try again');
  end;

  begin
      repeat
        start;
        workout;
        choice;
      until readkey =#27;
  end.
```

Program 16.2 (FORTRAN)
Runge–Kutta second order method

```
C     *******************************
      REAL FUNCTION F(X, Y)
      REAL X, Y
      F=- X*X*Y*Y
      RETURN
      END
C     *******************************
      EXTERNAL F
      REAL X0, X1, Y0, K1, K2, H
      INTEGER I, N
      PRINT *, '*************************************************'
      PRINT *, '*                                               *'
      PRINT *, '* RUNGE-KUTTA METHOD (SECOND ORDER)             *'
      PRINT *, '* TO SOLVE FIRST ORDER O.D.E.                   *'
      PRINT *, '*                                               *'
      PRINT *, '*************************************************'
      PRINT *, 'INITIAL VALUE OF X:'
      READ (*, *) X0
      PRINT *, 'FINAL VALUE OF X:'
      READ (*, *) X1
      PRINT *, 'INITIAL VALUE OF Y (AT X0):'
      READ (*,*) Y0
```

```
      PRINT *, 'STEP LENGTH, H:'
      READ (*,*) H
      WRITE (*, 50)
50    FORMAT (18X, 'X', 24X, 'Y')
70    WRITE (*, 60) X0, Y0
      IF (X0 .GT. (X1-(1E-7))) GOTO 100
      K1=H*F (X0, Y0)
      K2=H*F (X0+h, Y0+K1)
      X0=X0+H
      Y0=Y0+(K1+K2)/2
      GOTO 70
60    FORMAT (10X, F12.8, 15X, F12.8)
      STOP
      END
```

```
INITIAL VALUE OF X: X0=0
FINAL VALUE OF X: X1=1
INITIAL VALUE OF Y: Y0= 1
STEP LENGTH:       H= .1
```

X	Y
0	1
.1	.9985
.2	.9970065
.3	.9905809
.4	.9784546
.5	.9592002
.6	.931923
.7000001	.8964163
.8000001	.8532244
.9000001	.8035771
.1	.7492042

16.10 Runge–Kutta fourth order method

We shall seek a more accurate solution of the initial value problem [16.1] than provided by the RK second order method discussed in Section 16.9. Let us consider additional terms in the solution [16.21] in the form

$$y_{n+1} = y_n + ak_1 + bk_2 + ck_3 + dk_4 \qquad [16.26]$$

where $k_1 = hf(x_n, y_n); k_2 = hf(x_n + \alpha_2 h, y_n + \beta_2 k_1);$

$$k_3 = hf(x_n+\alpha_3 h, y_n+\beta_3 k_1+\gamma_3 k_2); k_4 = hf(x_n+\alpha_4 h, y_n+\beta_4 k_1+\gamma_4 k_2+\delta_4 k_3)$$

where the constants have to be determined by expanding [16.26] by Taylor's series and then equating the coefficients of powers of h, upto h^4, to the corresponding coefficients in Taylor's algorithm (Section 16.3). Here also, because of the degree of freedom several solutions are possible to obtain the constants. Of these, the popularly used Runge–Kutta fourth order method is obtained with the following choice of coefficients:

$$\alpha_2 = \alpha_3 = \beta_2 = \gamma_3 = \frac{1}{2}; \beta_3 = \beta_4 = \gamma_4 = 0; \alpha_4 = \delta_4 = 1; a = d = \frac{1}{6}, b = c = \frac{1}{3}.$$

16.10.1 Algorithm for the RK fourth order method

With the choice of coefficients shown above, [16.26] becomes

$$y_{n+1} = y_n + \frac{1}{6}[k_1 + 2k_2 + 2k_3 + k_4] \qquad [16.27]$$

where $k_1 = hf(x_n, y_n); k_2 = hf\left(x_n + \frac{h}{2}, y_n + \frac{k_1}{2}\right);$

$$k_3 = hf\left(x_n + \frac{h}{2}, y_n + \frac{k_2}{2}\right); k_4 = hf(x_n + h, y_n + k_3)$$

for $n = 0, 1, 2, 3, \ldots$ given that $y = y_0$ when $x = x_0$ and the step-length is h.

Note: **1.** If $f(x, y)$ is a function of x only, then the fourth order Runge–Kutta method reduces to *Simpson's one-third rule* for integration.

2. The advantages of the RK fourth order method are the following:

(i) No special procedure is needed for starting.

(ii) It is easy to change the step-length h for greater accuracy.

(iii) More computer memory is available for progressive computations since earlier values need not be stored.

Example 16.12 For the initial value problem $dy/dx = x - y, y(0) = 1$, find the value of y when $x = 0.2$ and $x = 0.4$, taking $h = 0.2$.

Solution 16.12 We have $f(x, y) = x - y, x_0 = 0, y_0 = 1$ and $h = 0.2$.

(i) At $x = 0.2$ the RK fourth order formula [16.27] becomes, with $n = 0$:

$$y_1 = y_0 + \frac{1}{6}[k_1 + 2k_2 + 2k_3 + k_4] \qquad [1]$$

where $k_1 = hf(x_0, y_0), k_2 = hf\left(x_0 + \frac{h}{2}, y_0 + \frac{k_1}{2}\right);$

$$k_3 = hf\left(x_0 + \frac{h}{2}, y_0 + \frac{k_2}{2}\right), k_4 = hf(x_0 + h, y_0 + k_3).$$

Here we get $k_1 = 0.2f(0, 1) = 0.2(0 - 1) = -0.2;$

$$k_2 = 0.2f(0.1, 1 - 0.1) = 0.2(0.1 - 0.9) = -0.16;$$

$$k_3 = 0.2f(0.1, 1 - 0.08) = 0.2(0.1, 0.92) = -0.164; \quad \text{and}$$

$$k_4 = 0.2f(0.2, 1 - 0.164) = 0.2(0.2 - 0.836) = -0.1272$$

so that we get, on substituting these in [1]:

$$y_1 = 1 + \frac{1}{6}[-0.2 - 0.32 - 0.328 - 0.1272] = 1 - \frac{1}{6}[0.9752] = 1 - 0.16253 = 0.83747.$$

(ii) At $x = 0.4$, we shall evaluate y_2, the value of y at $x_2 = 0.4$. Taking $n = 1$ in [16.27], we have

$$y_2 = y_1 + \frac{1}{6}[k_1 + 2k_2 + 2k_3 + k_4] \qquad [2]$$

where $k_1 = hf(x_1, y_1), k_2 = hf\left(x_1 + (h/2), y_1 + (k_1/2)\right);$

$$k_3 = hf\left(x_1 + (h/2), y_1 + (k_2/2)\right), k_4 = hf(x_1 + h, y_1 + k_3) \quad \text{and} \quad x_1 = 0.2.$$

Here, $k_1 = 0.2f(0.2, 0.83747) = 0.2(-0.63747) = -0.127494;$

$$k_2 = 0.2f(0.2 + 0.1, 0.83747 - 0.063747) = 0.2(0.3 - 0.773723) = -0.0947446;$$

$$k_3 = 0.2f(0.3, 0.83747 - 0.0473723) = 0.2(0.3 - 0.7900977) = -0.0980195; \quad \text{and}$$

$$k_4 = 0.2f(0.4, 0.83747 - 0.0980195) = 0.2(0.4 - 0.7394505) = -0.0678901.$$

Therefore, on substituting these values in [2], we get

$$y_2 = 0.83747 + \frac{1}{6}[-0.127494 - 2(0.0947446)$$

$$-2(0.0980195) - 0.0678901] = 0.83747 - \frac{1}{6}[0.5809123] = 0.7406512.$$

For a comparison, we tabulate (Table 16.6) the values of y at $x = 0.2$ and $x = 0.4$ as obtained from Euler's, RK second order, RK fourth order and the analytical methods. We notice that the RK fourth order method is accurate to the fourth decimal place.

Table 16.6 *Comparison of Euler's, RK second and RK fourth order solutions.*

x	**0**	**0.2**	**0.4**
y_{Euler}	1.000	0.800	0.680
$y_{\text{RK–II}}$	1.0000	0.8400	0.7448
$y_{\text{RK–IV}}$	1.00000	0.83747	0.74065
y_{Exact}	1.00000	0.83746	0.74064

Example 16.13 Infusion of glucose into the blood stream is an important medical technique. The amount of glucose in the blood stream of a patient at time t is denoted by $G(t)$. Suppose glucose is infused at a constant rate of k millimoles/litre in a minute. At the same time, the glucose is converted and removed from the blood stream at a rate proportional to the amount of glucose present. The function $G(t)$ satisfies the equation

$$\frac{dG}{dt} = k - \alpha G$$

where α is a positive constant. Given that initially, at $t = 0$, the amount of glucose present is 3.5 millimoles/litre, solve the equation using the Runge–Kutta fourth order method. Compare the solution thus obtained with the exact solution.

Solution 16.13 We shall first obtain the numerical solution using the RK fourth order method. The given equation is

$$\frac{dG}{dt} = 0.5 - (0.11)G \qquad [1]$$

with $G = 3.5$ when $t = 0$. In [1], we have chosen the constant of proportionality α as 0.11 on empirical basis and the same value will be used later to obtain the exact solution by the analytical method. Here, $t_0 = 0; G_0 = 3.5$; and let $h = 1$ minute. Let the right hand side of [1] be denoted by $f(t,G)$. Then, we have

$$f(t,G) = 0.5 - (0.11)G.$$

The RK fourth order algorithm [16.27], with t and G in place of x and y, becomes

$$G_{n+1} = G_n + \frac{1}{6}[k_1 + 2k_2 + 2k_3 + k_4] \qquad [2]$$

where, with $n = 0$, we have

$$k_1 = hf(t_0, G_0) = (1)[0.5 - (0.11)3.5] = 0.115;$$

$$k_2 = hf\left(t_0 + \frac{h}{2}, G_0 + \frac{k_1}{2}\right) = (1)[0.5 - (0.11)(3.5575)] = 0.108675;$$

$$k_3 = hf\left(t_0 + \frac{h}{2}, G_0 + \frac{k_2}{2}\right) = (1)[0.5 - (0.11)(3.5543375)] = 0.1090228;$$

$$k_4 = hf(t_0 + h, G_0 + k_3)$$

$$= (1)[0.5 - 0.11(3.6090228)]$$

$$= 0.1030074.$$

Substituting these values in [2], with $n = 0$, we get $G_1 = G_0 + (1/6)[k_1 + 2k_2 + 2k_3 + k_4]$

or $\quad G_1 = 3.5 + \dfrac{1}{6}[0.115 + 2(0.108675) + 2(0.1090228) + (0.1030074)]$

$$= 3.5 + \frac{1}{6}[0.653403] = 3.6089005.$$

Similarly, we can evaluate G for further values of t. The values of G_n are tabulated at the end of this example.

Analytical solution: The given differential equation [1] is a linear first order equation and can be written in the form

$$\frac{dG}{dt} + (0.11)G = 0.5.$$

The integrating factor is $e^{\int (0.11)dt} = \frac{e^{(0.11)t}}{0.11}$ so that the general solution of [1] is

$$G\left[\frac{e^{(0.11)t}}{0.11}\right] = \int (0.5)\left[\frac{e^{(0.11)t}}{0.11}\right]dt$$

or $\quad Ge^{(0.11)t} = [((0.5)/(0.11))\,e^{(0.11)t}] + c$

or $\quad G = 4.54 + ce^{-(0.11)t}$ [3]

Initially, at $t = 0$, $G = 3.5$ so that, from [3] we get $3.5 = 4.\dot{5}\dot{4} + c$ or $c = -1.0\dot{4}\dot{5}$. Substituting the value of c in [3], the exact solution of the initial value problem is

$$G(t) = 4.\dot{5}\dot{4} - (1.0\dot{4}\dot{5})e^{-(0.11)t}. \qquad [4]$$

As $t \to \infty$, the glucose content approaches a constant 4.54. We compare the numerical solution (from the RK fourth order method) with the exact solution [4] in Table 16.7. The values of G are given correct to six decimal places.

Table 16.7 *Glucose content by RK fourth order and exact solution.*

t	G RK fourth order method	G Exact solution
0	3.5	3.5
1	3.608901	3.608901
2	3.706458	3.706458
3	3.793852	3.793852
4	3.872143	3.872144
⋮	⋮	⋮
10	4.197453	4.197453
⋮	⋮	⋮
20	4.429615	4.429615
⋮	⋮	⋮
30	4.506895	4.506895

We observe in the above table that the solution given by the RK fourth order method coincides with the exact solution even upto the sixth decimal place in the case of the given problem. The variation of the glucose content in blood stream is shown in Fig. 16.3.

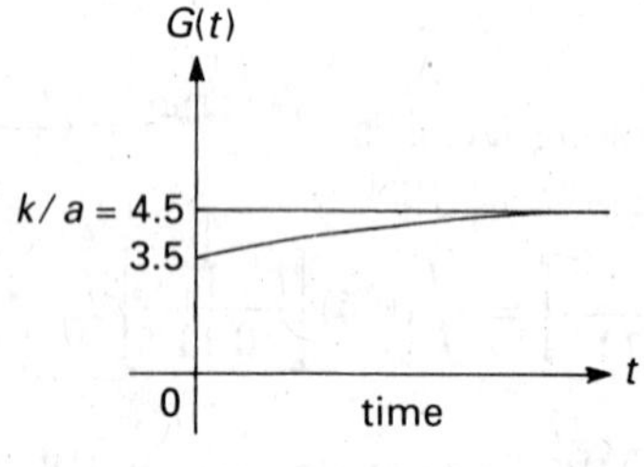

Fig. 16.3 *Variation of glucose in blood stream.*

Example 16.14 Suppose a source of light is placed at the origin O (see Fig. 2.6) and that a beam of light from the source O, after reflection at P on a mirror of a certain shape, travels parallel to the x-axis. Find the shape of the mirror.

Solution 16.14 For the formulation of the differential equation for this problem and its *analytical* solution, see Example 2.20 in Section 2.5.

The differential equation governing this problem of reflection in a curved reflector is shown to be

$$\frac{dy}{dx} = \frac{-x \pm \sqrt{x^2 + y^2}}{y} \quad \text{(see [1] in Example 2.20).} \qquad [1]$$

We shall apply the Runge–Kutta fourth order method with the initial condition $y = 4$ when $x = -3$ and step-length $h = 0.5$ (see Fig. 16.4). The solution is given after Program 16.3. The numerical solution obtained by the RK fourth order method is tabulated, for comparison, along with those obtained analytically as well as from the extensions of the Runge–Kutta methods at the end of Section 16.11. With the given initial condition, the equation to the cross section of the curved reflector is $y^2 = 16(x+4)$ which is a parabola symmetrical about the x-axis and passing through the point $(-3,4)$. The vertex of the parabola is $(-4,0)$ lying 4 units to the left of the origin on the x-axis.

16.10.2 Error estimation in the RK fourth order method

The truncation error in the Runge–Kutta fourth order method is of the order of h^5. Although it is possible to obtain a precise expression for the error it is highly complicated involving high order partial derivatives. We have a more practical method, albeit less precise, of estimating the error in this case.

In solving the initial value problem [16.1] viz.

$$\frac{dy}{dx} = f(x,y), y(x_0) = y_0,$$

suppose we proceed to (x_1,y_1) and then to (x_2,y_2) using the same step-length h. In that case, if E is the truncation error, we obtain the value $(y_1 - E)$ instead of the expected correct value y_1. In the second step, we start with the value $(y_1 - E)$ instead of y_1. If we assume that the truncation error in the second step is still approximately E, then we obtain the next value of y as $(y_2 - 2E)$ instead of the correct value y_2.

On the other hand, suppose we go directly from (x_0,y_0) to (x_2,y_2) using the step-length $2h$. Since the truncation error is of the order h^5, its value is now approximately $32E$. We thus obtain the value of y approximately as $(y_2 - 32E)$ instead of y_2. The value of E is roughly one-thirtieth of the difference between the two computed values of y_2. In other words, the error in the first value of y_2 is roughly one-fifteenth of this difference.

Consider, for example, the equation

$$\frac{dy}{dx} = -x^2y^2 \text{ with } y(0) = 1.$$

With $h = 0.2$, we obtain, by the RK fourth order method

$$y(0.2) = 0.997339983; \quad y(0.4) = 0.979111274.$$

On the other hand, with $h = 0.4$, we get $y(0.4) = 0.979084908$. The error in the first value of $y(0.4)$ is approximately

$$\frac{1}{15}[0.979111274 - 0.979084908] = 0.000001758 \quad \text{(which is of the order } 10^{-6}\text{)}.$$

In fact, the correct value of $y(0.4)$ obtained analytically is 0.979112271 so that the actual error in the first value of $y(0.4)$ is 0.000000997, also of order 10^{-6}.

Even though this method does not provide an accurate error estimate, it does indicate as to how many decimal places are correct.

16.10.3 Comparison between the RK second and fourth order methods

We have already observed, in Sections 16.9.2 and 16.10.2, that the truncation error in the second order method is of the order h^3 while that in the fourth order method is of the order h^5. Consider for example

$$\frac{dy}{dx} = y, y(0) = 1.$$

The exact solution, from the separation of variables is $y = e^x$. Now, if the values of y, obtained from the RK second and fourth order methods with step-length $h = 0.1$, are compared, we find that the error occurs in the *third* decimal place in the second order method while the error is only in the *seventh* decimal place in the fourth order method for x upto the value of 0.6. The relative errors for $x = 1$ in the RK second and fourth order solutions are respectively of orders 10^{-3} and 10^{-7}.

16.10.4 Program for RK fourth order method

Program 16.3 (PASCAL)
Runge–Kutta fourth order method

```
program runge4;
uses crt;
var
        x0, x1, y0, h, k1, k2, k3, k4: real;
        function fnf (x, y:real): real;
begin
        fnf:=-x*x*y*y;
end;

procedure start;
begin
        clrscr;
        writeln ('*************************************************************');
        writeln ('*                                                           *');
        writeln ('*          RUNGE-KUTTA METHOD (FOURTH ORDER)                *');
```

```
    writeln ('*                    TO SOLVE AN O.D.E                         *');
    writeln ('****************************************************************');
    gotoxy (27, 10); write ('Initial Value of x: x0 = '); read (x0);
    gotoxy (27, 11); write ('Final Value of x: x1 = '); read (x1);
    gotoxy (27, 12); write ('Initial Value of y: y0= '); read (y0);
    gotoxy (27, 13); write ('Step Length : h = '); read (h);
    writeln ('——————————————————————————————————————————————————');
    writeln ('X': 34, 'Y' : 20);
    writeln ('——————————————————————————————————————————————————');
end;

procedure workout;
begin
    while (x0<=x1-1E-11) do
    begin
        writeln (x0:39:11, y0:20:11);
        k1:=h*fnf (x0, y0);
        k2:=h*fnf (x0+h/2, y0+k1/2);
        k3:=h*fnf (x0+h/2, y0+k2/2);
        k4:=h*fnf (x0+h, y0+k3);
        x0:=x0+h;
        y0:=y0+(k1+2× (k2+k3)+k4)/6;
    end;
end;

procedure choice;
begin
    writeln;
    gotoxy (16,23); write ('Press <Esc> to exit or any other key to try again');
end;

begin
    repeat
        start;
        workout;
        writeln;
        choice;
    until readkey =#27;
end.
```

Program 16.3 (FORTRAN)
Runge–Kutta fourth order method

```
C    ********************************
           REAL FUNCTION F(X, Y)
           REAL X, Y
           F=-X*X*Y*Y
```

```
      RETURN
      END
C     ******************************
      EXTERNAL F
      REAL X0, X1, Y0, K1, K2, K3, K4, H
      PRINT *, '*************************************************'
      PRINT *, '*                                               *'
      PRINT *, '* RUNGE-KUTTA METHOD (FOURTH ORDER)  *'
      PRINT *, '* TO SOLVE FIRST ORDER O.D.E.                   *'
      PRINT *, '*                                               *'
      PRINT *, '*************************************************'
      PRINT *, 'INITIAL VALUE OF X:'
      READ (*, *) X0
      PRINT *, 'FINAL VALUE OF X:'
      READ (*, *) X1
      PRINT *, 'INITIAL VALUE OF Y (at X0):'
      READ (*,*) Y0
      PRINT *, 'STEP LENGTH, H:'
      READ (*,*) H
      WRITE (*, 50)
50    FORMAT (18X, 'X', 24X, 'Y')
70    WRITE (*, 60) X0, Y0
      IF (X0 .GT. (X1-0.0000001)) GOTO 100
      K1=H*F (X0, Y0)
      K2=H*F (X0+H/2, Y0+K1/2)
      K3=H*F (X0+H/2, Y0+K2/2)
      K4=H*F (X0+H, Y0+K3)
      X0=X0+H
      Y0=Y0+(K1+2*(K2+K3)+K4)/6
      GOTO 70
60    FORMAT (10X, F12.8, 15X, F12.8)
100   STOP
      END
```

INITIAL VALUE OF X: X0=0

FINAL VALUE OF X: X1=1

INITIAL VALUE OF Y: Y0= 1

STEP LENGTH: H= .1

X	Y
0	1
.1	.9996668
.2	.9973404
.3	.9910802
.4	.9791122
.5	.9599999
.6	.9328357
.7000001	.8973973
.8000001	.8542138
.9000001	.8045049
.1	.7499996

16.11 Extensions of Runge–Kutta methods

The Runge–Kutta methods can be extended to higher orders. The fourth order method discussed in Section 16.10 can be modified to provide still greater accuracy in the solution. Some of these extensions and modifications are discussed briefly in this section.

16.11.1 Runge–Kutta–Gill method

In the solution [16.26], discussed in Section 16.10, Gill introduced a modification in 1951 and hence the modified method is called the Runge–Kutta–Gill (RKG) method.

In the solution [16.26], let us set

$$\alpha_2 = \alpha_3 = \beta_2 = \frac{1}{2}; \quad \alpha_4 = 1; \quad \beta_3 = \frac{\sqrt{2}-1}{2}; \quad \beta_4 = 0;$$

$$\gamma_3 = 1 - \frac{1}{\sqrt{2}}; \quad \gamma_4 = -\frac{1}{\sqrt{2}}; \quad \delta_4 = 1 + \frac{1}{\sqrt{2}};$$

$$a = d = \frac{1}{6}; \quad b = \frac{1}{3}\left(1 - \frac{1}{\sqrt{2}}\right); \quad c = \frac{1}{3}\left(1 + \frac{1}{\sqrt{2}}\right).$$

With this modification, the differential equation

$$\frac{dy}{dx} = f(x,y), y(x_n) = y_n$$

has the next iterate of its solution given by

$$y_{n+1} = y_n + \frac{1}{6}[k_1 + (2-\sqrt{2})k_2 + (2+\sqrt{2})k_3 + k_4] \quad [16.28]$$

where $k_1 = hf(x_n, y_n)$; $k_2 = hf\left(x_n + \frac{h}{2}, y_n + \frac{k_1}{2}\right)$;

$$k_3 = hf\left[x_n + \frac{h}{2}, y_n + \frac{(\sqrt{2}-1)k_1}{2} + \frac{(2-\sqrt{2})k_2}{2}\right];$$

$$k_4 = hf\left[x_n + h, y_n - \frac{\sqrt{2}k_2}{2} + \left(1+\frac{\sqrt{2}}{2}\right)k_3\right].$$

The Runge–Kutta–Gill method is applied to the problem given in Example 16.14 and the values of y_n are tabulated at the end of this section (Table 16.8).

16.11.2 Program for Runge–Kutta–Gill method

Program 16.4 (PASCAL)
Runge–Kutta–Gill method

```
program gill;
uses crt;
var
      x0, x1, y0, h, k1, k2, k3, k4: real;
function fnf (x, y:real): real;
begin
      fnf:=-x*x*y*y;
end;
procedure start;
begin
      clrscr;
      writeln ('*************************************************************');
      writeln ('*                                                           *');
      writeln ('*      RUNGE-KUTTA-GILL METHOD TO SOLVE AN O.D.E            *');
      writeln ('*                                                           *');
      writeln ('*************************************************************');
      gotoxy (27, 10); write ('Initial Value of x: x0 = '); read (x0);
      gotoxy (27, 11); write ('Final Value of x: x1 = '); read (x1);
      gotoxy (27, 12); write ('Initial Value of y: y0= '); read (y0);
      gotoxy (27, 13); write ('Step Length : h = '); read (h);
      writeln ('————————————————————————————————————————————————————————————');
      writeln ('X': 34, 'Y' : 20);
      writeln ('————————————————————————————————————————————————————————————');
end;

procedure workout;
begin
```

```
     while (x0<=x1-1E-11) do
     begin
        writeln (x0:39:11, y0:20:11);
        k1:=h*fnf (x0, y0);
        k2:=h*fnf (x0+h/2, y0+k1/2);
        k3:=h*fnf (x0+h/2, y0+(sqrt(2)-1)*k1/2 +(2-sqrt(2))*k2/2);
        k4:=h*fnf (x0+h, y0-sqrt(2)*k2/2+(1+sqrt(2)/2*k3);
        x0:=x0+h;
        y0:=y0+(k1+(2-sqrt(2)) * k2+(2+sqrt(2)) * k3+k4)/6;
     end;
end;

procedure choice;
begin
     writeln;
     gotoxy (16,23); write ('Press <Esc> to exit or any other key to try again');
end;

begin
     repeat
        start;
        workout;
        writeln;
        choice;
     until readkey =#27;
end.
```

Program 16.4 (FORTRAN)
Runge–Kutta–Gill method

```
C    ********************************
           REAL FUNCTION F(X, Y)
           REAL X, Y
           F=-X*X*Y*Y
           RETURN
           END
C    ********************************
           EXTERNAL F
           REAL X0, X1, Y0, K1, K2, K3, K4, H
           PRINT *, ' **************************************************'
           PRINT *, '*                                                  *'
           PRINT *, '*          RUNGE-KUTTA-GILL METHOD                 *'
           PRINT *, '*         TO SOLVE A FIRST ORDER O.D.E.            *'
           PRINT *, '*                                                  *'
           PRINT *, ' **************************************************'
           PRINT *, 'INITIAL VALUES OF X:'
```

```
      READ (*, *) X0
      PRINT *, 'FINAL VALUE OF X:'
      READ (*, *) X1
      PRINT *, 'INITIAL VALUE OF Y (at X0):'
      READ (*,*) Y0
      PRINT *, 'STEP LENGTH, H:'
      READ (*,*) H
      WRITE (*, 50)
50    FORMAT (18X, 'X', 24X, 'Y')
70    WRITE (*, 60) X0, Y0
      IF (X0 .GT. (X1-0.0000001)) GOTO 100
      K1=H*F (X0, Y0)
      K2=H*F (X0+H/2, Y0+K1/2)
      K3=H*F (X0+H/2, Y0+(SQRT(2.0)-1)*K1/2+(2-SQRT(2.0))*K2/2)
      K4=H*F (X0+H, Y0-SQRT(2.0)*K2/2+(1+SQRT(2.0)/2)*K3)
      X0=X0+H
      Y0=Y0+(K1+(2-SQRT(2.0))*K2+(2+SQRT (2.0))*K3+K4)/6
      GOTO 70
60    FORMAT (10X, F12.3, 15X, F12.8)
100   STOP
      END
```

```
****************************************
*                                      *
*      RUNGE-KUTTA-GILL METHOD         *
*   TO SOLVE A FIRST ORDER O.D.E.      *
*                                      *
****************************************

INITIAL VALUES OF X: 0
FINAL VALUE OF X: 1
INITIAL VALUE OF Y (at X0): 1
STEP LENGTH, H: 0.1
```

X	Y
0.000	1.00000000
0.100	0.99966680
0.200	0.99734050
0.300	0.99108030
0.400	0.97911230
0.500	0.96000000
0.600	0.93283580
0.700	0.89739750
0.800	0.85421400
0.900	0.80450510
1.000	0.74999980

16.11.3 The Runge–Kutta–Merson method

This is also a fourth order method which involves an additional derivative calculation and also provides an error estimate. This error estimate will be exact if the derivative function is linear in x and y. In fact, the method also provides a technique for automatic adjustment of the step-length h to ensure good convergence of the solution. The algorithm of the Runge–Kutta–Merson (RKM) method is as follows:

$$y_{n+1} = y_n + \frac{1}{6}(k_1 + 4k_4 + k_5) \tag{16.29}$$

where $k_1 = hf(x_n, y_n)$; $k_2 = hf\left[x_n + \frac{h}{3}, y_n + \frac{k_1}{3}\right]$;

$$k_3 = hf\left[x_n + \frac{h}{3}, y_n + \frac{k_1}{6} + \frac{k_2}{6}\right];$$

$$k_4 = hf\left[x_n + \frac{h}{2}, y_n + \frac{k_1}{8} + \frac{3k_3}{8}\right];$$

$$k_5 = hf\left[x_n + h, y_n + \frac{k_1}{2} - \frac{3k_3}{2} + 2k_4\right].$$

The RKM method is applied for the problem given in Example 16.14 and the values of y_n are given in Table 16.8 and also after Program 16.5.

16.11.4 Program for Runge–Kutta–Merson method

Program 16.5 (PASCAL)

Runge–Kutta–Merson method

```
program merson;
uses crt;
var
   x0, x1, y0, h, k1, k2, k3, k4, k5: real;
function fnf (x, y:real): real;
begin
   fnf:=-x*x*y*y;
end;
procedure start;
begin
   clrscr;
   writeln ('*************************************************************');
   writeln ('*                                                           *');
   writeln ('*   RUNGE-KUTTA-MERSON METHOD TO SOLVE AN O.D.E             *');
   writeln ('*                                                           *');
   writeln ('*************************************************************');
   gotoxy (27, 10); write ('Initial Value of x: x0 = '); read (x0);
   gotoxy (27, 11); write ('Final Value of x: x1 = '); read (x1);
   gotoxy (27, 12); write ('Initial Value of y: y0= '); read (y0);
   gotoxy (27, 13); write ('Step Length : h = '); read (h);
   writeln ('___________________________________________________________ ');
   writeln ('X': 34, 'Y' : 20);
   writeln ('___________________________________________________________ ');
end;
procedure workout;
begin
   while (x0<=x1-1E-11) do
   begin
      writeln (x0:39:11, y0:20:11);
      k1: = h*fnf (x0, y0);
      k2: = h*fnf (x0+h/3, y0+k1/3);
      k3: = h*fnf (x0+h/3, y0+(k1+k2)/6);
      k4: = h*fnf (x0+h/2, y0+(k1+3*k3)/8);
      k5 = h*fnf (x0+h, y0+(k1-3k3)/2 + 2*k4);
      x0: = x0+h;
      y0: = y0+(k1+4×k4+k5)/6;
   end;
end;
procedure choice;
begin
   writeln;
   gotoxy (16,23); write ('Press <Esc> to exit or any other key to try again');
end;
begin
```

```
  repeat
    start;
    workout;
    writeln;
    choice;
  until readkey =#27;
end.
```

Program 16.5 (FORTRAN)
Runge–Kutta–Merson method

```
C     ******************************
      REAL FUNCTION F(X, Y)
      REAL X, Y
      F=-X*X*Y*Y
      RETURN
      END
C     ******************************
      EXTERNAL F
      REAL X0, X1, Y0, K1, K2, K3, K4, K5, H
      PRINT *, '**********************************************'
      PRINT *, '*                                            *'
      PRINT *, '*      RUNGE-KUTTA-MERSON METHOD             *'
      PRINT *, '*      TO SOLVE A FIRST ORDER O.D.E.         *'
      PRINT *, '*                                            *'
      PRINT *, '**********************************************'
      PRINT *, 'INITIAL VALUES OF X:'
      READ (*, *) X0
      PRINT *, 'FINAL VALUE OF X:'
      READ (*, *) X1
      PRINT *, 'INITIAL VALUE OF Y (at X0):'
      READ (*,*) Y0
      PRINT *, 'STEP LENGTH, H:'
      READ (*,*) H
      WRITE (*, 50)
50    FORMAT (18X, 'X', 24X, 'Y')
70    WRITE (*, 60) X0, Y0
      IF (X0 .GT. (X1-0.0000001)) GOTO 100
      K1=H*F (X0, Y0)
      K2=H*F (X0+H/3, Y0+K1/3)
      K3=H*F (X0+H/3, Y0+(K1+K2)/6)
      K4=H*F (X0+H/2, Y0+(K1+3× K3)/8)
      K5=H* F (X0+H/2, Y0+(K1-3× K3)/2+2×K4)
```

```
        X0=X0+H
        Y0=Y0+(K1+4*K4+K5)/6
        GOTO 70
60      FORMAT (10X, F12.3, 15X, F12.8)
100     STOP
        END
```

```
****************************************
*                                      *
*  RUNGE-KUTTA-MERSON METHOD           *
*   TO SOLVE A FIRST ORDER O.D.E.      *
*                                      *
****************************************
INITIAL VALUES OF X: 0
FINAL VALUE OF X: 1
INITIAL VALUE OF Y (at X0): 1
STEP LENGTH, H: 0.1
```

X	Y
0.000	1.00000000
0.100	0.99966680
0.200	0.99734050
0.300	0.99108030
0.400	0.97911230
0.500	0.96000040
0.600	0.93283640
0.700	0.89739820
0.800	0.85421480
0.900	0.80450580
1.000	0.75000050

16.11.5 Runge–Kutta–Butcher method

J.C. Butcher (1965) enhanced the order of the solution to five so that the error term is now of the order h^6. This method requires six functional values. Butcher showed that this method involves minimum computing time and ensures greater accuracy.

In the Runge–Kutta–Butcher (RKB) method, we have

$$y_{n+1} = y_n + \frac{1}{90}[7k_1 + 32k_3 + 12k_4 + 32k_5 + 7k_6] \tag{16.30}$$

where $k_1 = hf(x_n, y_n); k_2 = hf\left[x_n + \frac{h}{4}, y_n + \frac{k_1}{4}\right];$

$$k_3 = hf\left[x_n + \frac{h}{4}, y_n + \frac{(k_1+k_2)}{8}\right];$$

$$k_4 = hf\left[x_n + \frac{h}{2}, y_n - \frac{k_2}{2} + k_3\right]; \quad k_5 = hf\left[x_n + \frac{3h}{4}, y_n + \frac{(3k_1+9k_4}{16}\right];$$

$$k_6 = hf\left[x_n + h, y_n - \frac{1}{7}(3k_1 - 2k_2 - 12k_3 + 12k_4 - 8k_5)\right].$$

The solution obtained for the problem in Example 16.14, from the RKB method is tabulated at the end of this section (Table 16.8) and also after Prog. 16.6.

16.11.6 Program for Runge–Kutta–Butcher method

Program 16.6 (PASCAL)
Runge–Kutta–Butcher method

```
program butcher;
uses crt;
var
     x0, x1, y0, h, k1, k2, k3, k4, k5, k6: real;
function fnf (x, y:real): real;
begin
     fnf:=-x*x*y*y;
end;
procedure start;
begin
     clrscr;
     writeln ('**************************************************************');
     writeln ('*                                                            *');
     writeln ('* RUNGE-KUTTA-BUTCHER METHOD TO SOLVE AN O.D.E              *');
     writeln ('*                                                            *');
     writeln ('**************************************************************');
     gotoxy (27, 10); write ('Initial Value of x: x0 = '); read (x0);
     gotoxy (27, 11); write ('Final Value of x: x1 = '); read (x1);
     gotoxy (27, 12); write ('Initial Value of y: y0= '); read (y0);
     gotoxy (27, 13); write ('Step Length: h = '); read (h);
     writeln ('-------------------------------------------------------------');
     writeln ('X': 34, 'Y' : 20);
     writeln ('-------------------------------------------------------------');
end;
procedure workout;
begin
```

```
        while (x0<=x1-1E-11) do
        begin
           writeln (x0:39:11, y0:20:11);
           k1:=h*fnf (x0, y0);
           k2:=h*fnf (x0+h/4, y0+k1/4);
           k3:=h*fnf (x0+h/4, y0+(k1+k2)/8);
           k4:=h*fnf (x0+h/2, y0-k2/2+k3);
           k5:=h*fnf (x0+3*h/4, y0+(3*k1+9*k4)/16);
           k6:=h*fnf (x0+h, y0-(3*k1-2*k2-12*k3+12*k4-8*k5)/7);
           x0:=x0+h;
           y0:=y0+(7*k1+32*k3+12*k4+32*k5+7*k6)/90;
        end;
end;
procedure choice;
begin
        writeln;
        gotoxy (16,23); write ('Press <Esc> to exit or any other key to try again');
end;
begin
        repeat
           start;
           workout;
           writeln;
           choice;
        until readkey =#27;
end.
```

Program 16.6 (FORTRAN)
Runge–Kutta–Butcher method

```
C     ******************************
            REAL FUNCTION F(X, Y)
            REAL X, Y
            F=-X*X*Y*Y
            RETURN
            END
C     ******************************
            EXTERNAL F
            REAL X0, X1, Y0, K1, K2, K3, K4, K5, K6, H
            PRINT *, ' ********************************************'
            PRINT *, '*                                            *'
            PRINT *, '*      RUNGE-KUTTA-BUTCHER METHOD            *'
            PRINT *, '*       TO SOLVE A FIRST ORDER O.D.E.        *'
            PRINT *, '*                                            *'
```

```
      PRINT *, ' ********************************************'
      PRINT *, 'INITIAL VALUE OF X:'
      READ (*, *) X0
      PRINT *, 'FINAL VALUE OF X:'
      READ (*, *) X1
      PRINT *, 'INITIAL VALUE OF Y (at X0):'
      READ (*,*) Y0
      PRINT *, 'STEP LENGTH, H:'
      READ (*,*) H
      WRITE (*, 50)
50    FORMAT (18X, 'X', 24X, 'Y')
70    WRITE (*, 60) X0, Y0
      IF (X0 .GT. (X1-0.0000001)) GOTO 100
      K1=H*F (X0, Y0)
      K2=H*F (X0+H/4, Y0+K1/4)
      K3=H*F (X0+H/4, Y0+(K1+K2)/8)
      K4=H*F (X0+H/2, Y0-K2/2+K3)
      K5=H*F (X0+3*H/4, Y0+(3*K1+9*K4)/16)
      K6=H*F (X0+H, Y0-(3*K1-2*K2-12* K3+12*K4-8*K5)/7)
      X0=X0+H
      Y0=Y0+(7*K1+32*K3+12*K4+32*K5+7*K6)/90
      GOTO 70
60    FORMAT (10X, F12.8, 15X, F12.8)
100   STOP
      END
```

```
*****************************************
*                                       *
*  RUNGE-KUTTA-BUTCHER METHOD           *
*   TO SOLVE A FIRST ORDER O.D.E.       *
*                                       *
*****************************************

INITIAL VALUE OF X: 0
FINAL VALUE OF X: 1
INITIAL VALUE OF Y (at X0): 1
STEP LENGTH, H: 0.1
```

X	Y
0.00000000	1.00000000
0.10000000	0.99966680
0.20000000	0.99734050
0.30000000	0.99108040
0.40000000	0.97911240
0.50000000	0.96000020
0.60000000	0.93283610
0.70000000	0.89739780
0.80000010	0.85421430
0.90000010	0.80450540
1.00000000	0.75000020

16.11.7 Runge–Kutta–Nyström method

E.J. Nyström, a Finnish mathematician introduced an extension in the RK method, in his 1925 paper, by raising the order to 5, the error term being of order h^6. The algorithm involves six functional values. The Kutta–Nyström or Runge–Kutta–Nyström (RKN) method provides the following algorithm:

$$y_{n+1} = y_n + \frac{1}{192}[23k_1 + 125k_2 - 81k_5 + 125k_6] \qquad [16.31]$$

where $k_1 = hf(x_n, y_n)$; $k_2 = hf\left[x_n + \frac{h}{2}, y_n + \frac{k_1}{3}\right]$;

$$k_3 = hf\left[x_n + \frac{2h}{5}, y_n + \frac{(6k_2 + 4k_1)}{25}\right];$$

$$k_4 = hf\left[x_n + h, y_n + \frac{15k_3 - 12k_2 + k_1}{4}\right];$$

$$k_5 = hf\left[x_n + \frac{2h}{3}, y_n + \frac{8k_4 - 50k_3 + 90k_2 + 6k_1}{81}\right];$$

$$k_6 = hf\left[x_n + \frac{4h}{5}, y_n + \frac{8k_4 + 10k_3 + 36k_2 + 6k_1}{75}\right].$$

The RKN method is applied to the initial value problem given in Example 16.14 and the values of y are tabulated at the end of this section (Table 16.8) and also after Prog. 16.7.

16.11.8 Program for Runge–Kutta–Nyström method

Program 16.7 (PASCAL)
Runge–Kutta–Nyström method

```
program nystrom;
uses crt;
var
      x0, x1, y0, h, k1, k2, k3, k4, k5, k6: real;
function fnf (x, y:real): real;
begin
      fnf:=-x*x*y*y;
end;
procedure start;
begin
      clrscr;
      writeln ('******************************************************************');
      writeln ('*                                                                *');
      writeln ('* RUNGE-KUTTA-NYSTROM METHOD TO SOLVE AN O.D.E                   *');
      writeln ('*                                                                *');
      writeln ('******************************************************************');
      gotoxy (27, 10); write ('Initial Value of x: x0 = '); read (x0);
      gotoxy (27, 11); write ('Final Value of x: x1 = '); read (x1);
      gotoxy (27, 12); write ('Initial Value of y: y0= '); read (y0);
      gotoxy (27, 13); write ('Step Length: h = '); read (h);
      writeln ('————————————————————————————————————————————————————————————');
      writeln ('X': 34, 'Y' : 20);
      writeln ('————————————————————————————————————————————————————————————');
end;
procedure workout;
begin
      while (x0<=x1-1E-11) do
      begin
         writeln (x0:39:11, y0:20:11);
         k1:=h*fnf (x0, y0);
         k2:=h*fnf (x0+h/2, y0+k1/3);
         k3:=h*fnf (x0+2*h/5, y0+(4*k1+6*k2)/25);
         k4:=h*fnf (x0+h, y0+(k1-12*k2+15*k3)/4);
         k5:=h*fnf (x0+2*h/3, y0+(6*k1+90*k2-50*k3+8*k4)/81);
         k6:=h*fnf (x0+4*h/5, y0+(6*k1+36*k2+10*k3+8*k4)/75);
         x0:=x0+h;
         y0:=y0+(23*k1+125*k3-81*k5+125*k6)/192;
      end;
end;
procedure choice;
```

```
begin
    writeln;
    gotoxy (16,23); write ('Press <Esc> to exit or any other key to try again');
end;

begin
    repeat
        start;
        workout;
        writeln;
        choice;
    until readkey =#27;
end.
```

Program 16.7 (FORTRAN)
Runge–Kutta–Nyström method

```
C   ********************************
         REAL FUNCTION F(X, Y)
         REAL X, Y
         F=-X*X*Y*Y
         RETURN
         END
C   ********************************
         EXTERNAL F
         REAL X0, X1, Y0, K1, K2, K3, K4, K5, K6, H
         PRINT *, '**********************************************'
         PRINT *, '*                                            *'
         PRINT *, '*        RUNGE-KUTTA-NYSTROM METHOD          *'
         PRINT *, '*        TO SOLVE A FIRST-ORDER O.D.E.       *'
         PRINT *, '*                                            *'
         PRINT *, '**********************************************'
         PRINT *, 'INITIAL VALUE OF X:'
         READ (*, *) X0
         PRINT *, 'FINAL VALUE OF X:'
         READ (*, *) X1
         PRINT *, 'INITIAL VALUE OF Y (at X0):'
         READ (*,*) Y0
         PRINT *, 'STEP LENGTH, H:'
         READ (*,*) H
         WRITE (*, 50)
  50     FORMAT (18X, 'X', 24X, 'Y')
  70     WRITE (*, 60) X0, Y0
         IF (X0 .GT. (X1-0.0000001)) GOTO 100
```

```
      K1=H*F (X0, Y0)
      K2=H*F (X0+H/2, Y0+K1/3)
      K3=H*F (X0+2*H/5, Y0+(4*K1+6*K2)/25)
      K4=H*F (X0+H, Y0+(K1-12*K2+15*K3)/4)
      K5=H*F (X0+2*H/3, Y0+(6*K1+90*K2-50*K3+8*K4)/81)
      K6=H*F (X0+4*H/5, Y0+(6*K1+36*K2+10*K3+8*K4)/75)
      X0=X0+H
      Y0=Y0+(23*K1+125*K3-81*K5+125*K6)/192
      GOTO 70
60    FORMAT (10X, F12.8, 15X, F12.8)
100   STOP
      END
```

```
****************************************
*                                      *
*  RUNGE-KUTTA-NYSTROM METHOD          *
*   TO SOLVE A FIRST ORDER O.D.E.      *
*                                      *
****************************************

INITIAL VALUE OF X: 0
FINAL VALUE OF X: 1
INITIAL VALUE OF Y (at X0): 1
STEP LENGTH, H: 0.1
```

X	Y
0.00000000	1.00000000
0.10000000	0.99966680
0.20000000	0.99734050
0.30000000	0.99108040
0.40000000	0.97911250
0.50000000	0.96000040
0.60000000	0.93283640
0.70000000	0.89739850
0.80000010	0.85421540
0.90000010	0.80450700
1.00000000	0.75000230

Example 16.15 Solve the equation

$$\frac{dy}{dx} = \frac{-x + \sqrt{x^2 + y^2}}{y}, y(-3) = 4$$

for $-3 \le x \le 7$ taking the step-length $h = 0.5$ using the Runge–Kutta fourth order, RK–Gill, RK–Merson, RK–Butcher and RK–Nyström methods and compare the solution with the exact solution.

Solution 16.15 The solution of the given initial-value problem using the numerical methods given in Sections 16.10, 16.11.1, 16.11.3, 16.11.5 and 16.11.7 is tabulated below (Table 16.8) and also compared with the exact solution (see Example 2.20 in Section 2.5).

Table 16.8 *RK fourth order, RKG, RKM, RKB, RKN and Exact solutions compared.*

*F(X, Y) = (−X +SQR (X*X + Y* Y)) /Y*

Initial Value of X: − 3 Final Value of X: 7

Initial Value of Y: 4 Step Length : 0.5

	RKM	RK	RKG	RKN	RKB	Exact
X	*Y*	*Y*	*Y*	*Y*	*Y*	*Y*
−3	4	4	4	4	4	4
−2.5	4.899017	4.899018	4.899007	4.898979	4.898979	4.89898
−2	5.656899	5.656898	5.656886	5.656855	5.656853	5.656854
−1.5	6.324601	6.3246	6.324589	6.324556	6.324555	6.324556
−1	6.92825	6.928248	6.928237	6.928204	6.928202	6.928203
−.5	7.483362	7.48336	7.483349	7.483316	7.483314	7.483315
0	8.000048	8.000045	8.000033	8.000001	7.999999	8
.5	8.485329	8.485327	8.485316	8.485283	8.48528	8.485281
1	8.944321	8.944318	8.944306	8.944273	8.94427	8. 944272
1.5	9.380881	9.380878	9.380866	9.380832	9.38083	9.380832
2	9.798008	9.798005	9.797994	9.79796	9.797958	9.797959
2.5	10.19809	10.19809	10.19807	10.19804	10.19804	10.19804
3	10.58306	10.58305	10.58304	10.58301	10.583	10.58301
3.5	10.9545	10.9545	10.95449	10.95445	10.95445	10.95445
4	11.31376	11.31376	11.31375	11.31371	11.31371	11.31371
4.5	11.66196	11.66195	11.66194	11.6619	11.6619	11.6619
5	12.00005	12.00005	12.00004	12	12	12
5.5	12.32888	12.32888	12.32887	12.32883	12.32883	12.32883
6	12.64916	12.64916	12.64915	12.64911	12.64911	12.64911
6.5	12.96154	12.96153	12.96152	12.96148	12.96148	12.96148
7	13.26655	13.26655	13.26654	13.2665	13.2665	13.2665

From Table 16.8, we see that the RK–Butcher and the RK–Nyström methods give solutions close to that of the analytical method correct to 6 decimal places. The RK–Gill solution is slightly better than the RK fourth order. The RK–Merson values are almost the same as the RK fourth order values. The solution y is plotted against x in Fig. 16.4.

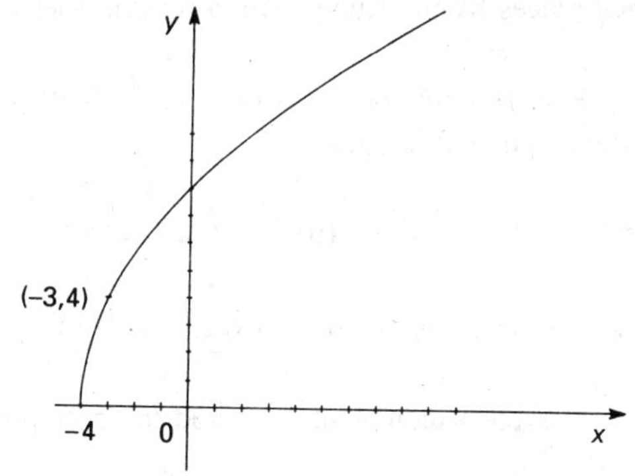

Fig. 16.4 *Parabolic shape of the mirror.*

Exercise 16

16.1 Determine the value of y at the given point for each of the initial value problems by taking the indicated number of sub-division n using Euler's method and improved Euler's method:

(i) $y' = 2x + y; y(0) = 0$. Find $y(0.5); n = 5$.
(ii) $y' = (1 + y)/x; y(2) = 3$. Find $y(2.8); n = 4$.
(iii) $y' = x^2 - y; y(1) = 0$. Find $y(1.6); n = 6$.
(iv) $y' = x^2 + y^2; y(1) = 2$. Find $y(0.5); n = 5$.
(v) $y' = (x - y)/(x + y); y(3) = 2$. Find $y(1); n = 5$.
(vi) $y' = \sin y; y(0) = 1$. Find $y(1); n = 10$.
(vii) $y' = \sqrt{x + y}; y(5) = 4$. Find $y(4); n = 5$.

16.2 Use the modified Euler's method to find y at $x = 0.1$ and $x = 0.2$ for the equation

$$\frac{dy}{dx} = \sqrt{x + y}, y(0) = 0.36.$$

16.3 Solve the equation $dy/dx = x - y^2, y(0) = 1$ for $x = 0.2$ and $x = 0.4$ correct to three decimal places.

16.4 Solve the equation $dy/dx = 2x + y, y(0) = 2$ at $x = 0.2$ taking **(i)** single step of length 0.2, and **(ii)** two steps of length 0.1; using Euler's method, the improved Euler's method and the modified Euler's method (correct to 4 decimal places).

16.5 Solve the equation $dy/dx = x - y^2, y(0) = 1$ for $x = 0.2$ and $x = 0.4$ to 3 decimal places by the Runge–Kutta fourth order method.

16.6 For the equation $dy/dx = 3x + (y/2)y(0) = 1$ find y at the following points with the given step-lengths:

(i) $x = 0.1, h = 0.1$ **(ii)** $x = 0.2, h = 0.2$ **(iii)** $x = 0.4, h = 0.4$.

Compare these values with the exact values of y.

16.7 Use the Runge–Kutta method to approximate y at $x = 0.1$ and $x = 0.2$ for the equation

$$\frac{dy}{dx} = x + y, y(0) = 1.$$

16.8 Solve by the RK fourth order method

$$\frac{dy}{dx} = \frac{1}{x+y}, y(0) = 1 \text{ for } x = 0.5(0.5)2.$$

16.9 Find y at $x = 0.2(0.2)0.6$ for the equation

$$\frac{dy}{dx} = 1 + y^2, y(0) = 0.$$

16.10 Find the value of y at $x = 1$ for the initial value problem

$$\frac{dy}{dx} = \frac{y-x}{y+x}.$$

16.11 The response x of a given hydraulic wave subject to sinusoidal input variation is given by

$$\frac{dx}{dt} = \sqrt{2\left[1 - \frac{x^2}{\sin^2 t}\right]}$$

with $x = 0$ at $t = 0$. Show that $dx/dt = \sqrt{2}/3$ at $x = 0$ and hence use the Runge–Kutta fourth order method to obtain a solution at $t = 0.2$. [**Hint:** For getting dx/dt at $x = 0$, rearrange the given expression without surds or fractions and differentiate twice with respect to t.]

16.12 The rate of emission of a radio active substance is proportional to the amount N of the substance remaining at any instant t, governed by the equation

$$\frac{dN}{dt} = -kN$$

(where the negative sign indicates that the radioactivity decreases with time). Taking $k = 0.01$ and at $t = 0, N = 100$ gm, find how much of the material remains when $t = 100$ sec. using the Runge–Kutta fourth order method with **(i)** $h = 100$ **(ii)** 50. Compare the results with the exact values.

Answers To Exercises

Chapter One

1.1 (i) $\frac{dy}{dx} = -5$; one, one (ii) $\frac{dy}{dx} = \frac{y}{x+y}$; one, one (iii) $r\frac{d\theta}{dr} = \tan\ \theta$; one, one

1.2 (i) $\frac{dx}{dt} = \frac{k}{t}$ (ii) $L\frac{dI}{dt} = E$ (iii) $F = k\left(\frac{dx}{dt}\right)^2$

1.3 (i) $y = x^3 - x^2 + c$ (ii) $s = \sin\ t + \cos\ t + c$ (iii) $y = \cos\ x + Ax + B$

(iv) $y = \frac{x^6}{120} + \frac{x^5}{60} + \frac{x^3}{6} + \frac{Ax^2}{2} + Bx + C$

1.4 (i) $y = \sin\ x + 1$ (ii) $y = xe^x - e^x + 3$ (iii) $y = -\sin\ x + 2$

Chapter Two

2.1 $\frac{1+x^2}{1+y^2} = c$ **2.2** $y^2 = -\frac{2}{3}x^3 + c$ **2.3** $(1+y^3)^{1/3} = \frac{cx}{\sqrt{1+x^2}}$

2.4 $\cos\ y = ce^{-x\tan\ x}$ **2.5** $y = ce^{(\log\ x)^2/2}$ **2.6** $\sin\ y = \cos\ x + c$

2.7 $y - 2 = ce^{-\tan\ x}$ **2.8** $x\sqrt{1-y^2} + y\sqrt{1-x^2} = c$

2.9 $\sqrt{1+x^2} + \sqrt{1+y^2} = \log\ [(1+\sqrt{1+x^2})(1+\sqrt{1+y^2})/xy] + c$

2.10 $1 - e^{-3x} = ce^{x^2}$ **2.11** $\frac{a}{2}\log\left[\frac{x-y-a}{x-y+a}\right] = y + c$

2.12 $(x+y) = \tan\ (x+c)$ **2.13** $\tan\ \frac{x+y}{2} = x + c$

2.14 $3(y-x) + 2\ \log\ (3x+6y-1) = c$ **2.15** $2(x-y) = \sin\ 2(x+y) + c$

2.16 $4\ \log\ (1+x-2y) = (2y-2x-c)$

2.17 $2x + 8y = \tan\left(\frac{\pi}{4} + 2x\right)$ **2.18** $2x - y + 1 = \log\ (x - y + 2)$

2.19 $\frac{1}{\sqrt{6}} \tan^{-1}\left[\sqrt{\frac{2}{3}}(3x + 2y + 4)\right] = x + c$ **2.20** $2x - y - 15\log\ (3x - y + 19) = c$

2.21 $x + e^{-(x+y)} = c$ **2.22** $x^4 - 4\cos\ xy = c$

2.23 $y = \frac{x}{2 + x}$ **2.24** $y = \log\ \sec\ x$ **2.25** $y + \log\left[\frac{9}{4} - \sqrt{1 + x^2}\right] = 0$

2.26 $e^x = e^y(x + 1)$ **2.27** $y = (1 + x^2)^2$ **2.29** 32/3 **2.31** 22.78 minutes

2.32 5.7 months

2.33 Approximately 2.4 billion in 1960 and 2.9 billion in 2000.

2.34 (i) $\frac{dp}{dt} = 0.003p - 0.001p^2 - 0.002$

(ii) $p(t) = \frac{1,999,998 - 999,998e^{-0.001t}}{999,999 - 999,998e^{-0.001t}}$. As $t \to \infty, p(t) \to 2$.

2.35 $y^2 = x^2 \log\ (cx^2)$ **2.36** $y = c\ \exp\ (x^2/2y^2)$ **2.37** $y^2 = x^2 + cx^4$

2.38 $y = cx^2(x + y)$ **2.39** $y = x\tan\ (cx^3)$

2.40 $\cos\left(\frac{y}{x}\right) + \log\ (cx) = 0$ **2.41** $x + y = 0$ **2.42** $x + y = 0$

2.43 $x^2 = y^2$ **2.44** $4(y - x)^3 = x^2y^2$ **2.45** $x^2 + y^2 = 1$

2.46 $\tan^{-1}\frac{y + 5}{x - 1} = \log\ \sqrt{(x - 1)^2 + (y + 5)^2} + c$ **2.47** $y - x = 5\ \log\ (x + y - 1) + c$

2.48 $2\ \tan^{-1}\left(\frac{y}{x + 1}\right) = \log\ (x^2 + y^2 + 2x + 1) + c$ **2.49** $2x + y - 3 = c(x + y - 1)^4$

2.50 $(8x^2 - 2tx - 14x + t^2 + 7) = c$ **2.51** $(2x + y - 2)^2 = (2x - 2y + 1)^5$

2.52 $(x + 2t - 3)^2(t - 4x + 3) = 112$ **2.53** $(x + 2y)(x - y)^2 = c$ **2.54** $x^8y^4(x^2 + y^2) = c$

2.55 $x^2 - 2xy = c$ **2.56** $x^5 - 5x^3y = c$ **2.57** $xy = y^2 + \log\ y + c$

2.58 $x^2\cos\ 3x + 3y = cx^2$ **2.59** $2y\sin\ x - \sin^2 x = c$

2.60 $I = e^{-2t} + 4e^{-3t}$ **2.61** $y = \frac{x}{4} + \frac{9}{7} - \frac{15}{28x^7}$

2.62 $y = x^2 \sin x + x \cos x + x$ **2.63** $y = \frac{\sin x - \cos x}{1+x} + \frac{3e^{-x}}{1+x}$

2.64 $y = x \sin x - x$

2.65 $y(x+4)^2(x-1) = \frac{x^3}{3} + \frac{3x^2}{2} - 4x - 4$ **2.68** $y = \frac{3}{4}[2x - 1 + e^{-2x}]$

2.67 $v = v_1(1 - e^{-kt})$ if the body falls from rest initially

2.69 $y = \sqrt{cx^2+1}$ **2.70** $y^3 = c \exp(-x^2/2) + 1$ **2.71** $\frac{1}{y^2} = -x^4 + cx^2$

2.72 $y^3 = 3 \sin x + \frac{9 \cos x}{x} - \frac{18 \sin x}{x^2} - \frac{18 \cos x}{x^3} + \frac{c}{x^3}$

2.73 $y = \frac{1}{cx - (1 + \log x)}$

2.74 $y = \frac{1}{\sec x(\tan x + c)}$ **2.75** $y^3 = cx^3 + \sin x + \frac{3 \cos x}{x} - \frac{6 \sin x}{x^2} - \frac{6 \cos x}{x^3}$

2.76 $\frac{x}{\sin y} + 1 = c\sqrt{x}$ **2.77** $y = 2x^3 - \sqrt{yx^3}$ $\left[\textbf{Hint}: \text{Consider} \frac{dx}{dy} = \frac{x(2x^3+y)}{6y^2}\right]$

2.78 $\frac{1}{y^3} = \frac{1}{7}\left[3x^4 - \frac{10}{x^3}\right]$ **2.79** $y^2 = x(x-1)$ **2.81** Not exact

2.82 Not exact **2.83** $(x^3 + \log x)y = c$ **2.84** $e^{xy} + y^2 = c$ **2.85** $xy + \log y^2 = c$

2.86 Not exact **2.87** $4xy - x^4 + y^4 = c$ **2.88** Not exact **2.89** $xy + \sin(xy) = c$

2.90 $(x-3)^2 + 9(y-1)^2 = 9$ **2.91** $y = \tan x$ **2.92** $x^2 - xy + y^2 = 3$

2.93 $x^2 y + y + 6 = 0$

2.94 $x^2 - x \sin y = 4$ **2.95** $2x \tan y + \cos 2x = 2\pi + 1$ **2.96** $x^2 y + y^2 e^{2x} = 1$

2.97 $x^3 + x^2 y^2 = c$ **2.98** $y = 2x - x^3$ **2.99** $y^2 - x = y(c - \sin x)$

2.100 $x^2(c + \cos 2y) = 2y$ **2.101** $x = \left(\frac{\pi}{2} - y\right) \sin y$ **2.102** $y = (x+c) \cos^2 x$

2.103 $6x^4y - x^6 = c$ **2.104** $6xy^3 - y^6 = c$ **2.105** $x^2y^3 = c$ **2.106** $y = \dfrac{c}{\log x}$

2.107 $y = c\sqrt{x}$ **2.108** $\sin y = cx$ **2.109** $x^2y + 4 = 0$ **2.110** $e^x \cos y + 1 = 0$

2.111 $\dfrac{x^2}{y^3} - \dfrac{1}{y} = c$ **2.112** $xy - \log x = \dfrac{1}{2}y^2 + c$ **2.113** $\dfrac{3}{2}y^2 + \dfrac{1}{2x^2y^2} = c$

2.114 $\dfrac{x}{y} - \dfrac{1}{y} + y + c = 0$ **2.115** $\log \dfrac{x}{y} = \dfrac{1}{3}y^3 + c$ **2.116** $\tan^{-1}\dfrac{x}{y} + \dfrac{x^4}{4} = c$

Chapter Three

3.1 $(y-x-c)(xy-c) = 0$ **3.2** $(y-cx)(x^2-y^2-c) = 0$ **3.3** $(y-cx^2)(x^3y-c) = 0$

3.4 $(x^2+y^2-c)(y-x-c) = 0$ **3.5** $(xy-c)(x^2y-c) = 0$

3.6 $(2y-x^2-c)(y^2-2x-c) = 0$ **3.7** $(y-e^x-c)(y+e^{-x}-c) = 0$

3.8 $\sin^{-1}\left(\dfrac{y}{x}\right) = \pm \log cx$ **3.9** $(y-cx^2)(y^2+3x^2-c) = 0$

3.10 $(y^2-x^2-c)(y^2-x^2-cx^4) = 0$ **3.12** $x^2+y^2 = cx$

3.11 $\left[y - \dfrac{x^2}{2} - c\right][y - e^{c+x}][y + x - 1 - ce^{-x}] = 0$

3.13 $\left[y + \dfrac{1}{2}x^2 - c\right][x + \log y - c] = 0$

3.14 $(y-c)(x+y-c)(x^2+xy+y^2-c) = 0$ **3.15** $y^2 = 2cx - c^2$

3.16 $x = \left[\dfrac{c - a\cosh^{-1}p}{\sqrt{p^2-1}}\right]p,\ y = \dfrac{c - a\cosh^{-1}p}{\sqrt{p^2-1}} - ap$ **3.17** $y^2 = 2cy + c^2$

3.18 $x = -\dfrac{12}{7}p^2 + \dfrac{c}{3}p^{-3/2},\ y = -\dfrac{8}{7}p^3 + cp^{-1/2}$ **3.19** $\log y = cx + c^2$

3.20 $2cx = c^3 + y^2$ **3.21** $c(c-4x)^2 = 64y$

3.22 $(x-a)^2 + (y-c)^2 = 1$ **3.23** $x = \dfrac{p}{1+p^2} + \tan^{-1}p,\ y = -\dfrac{1}{1+p^2} + c$

3.24 $x = c - 2[p + \log(p-1)],\ \ y = c - 2\left[\dfrac{1}{2}p^2 + p + \log(p-1)\right]$

3.25 $y^2 - a = (x+c)^2$ **3.26** $y = -\frac{8}{7}p^3 + \frac{c}{\sqrt{p}}, x = -\frac{12}{7}p^2 + \frac{c}{3p^{3/2}}$

3.27 $y = p \sin p + \cos p, x = \sin p + c$

3.28 $x = \log[(1-p)/\sqrt{1+p^2}] - \tan^{-1} p + c, \quad y = \log[(1-p)/\sqrt{1+p^2}] + \tan^{-1} p + c$

3.29 $2y = cx^2 + \frac{a}{c}$ **3.30** $2cy + 16 = cx^2$ **3.31** $y^2 = 2cx + c^2$

3.32 $x = (\log p - p + c)(p-1)^2, y = xp^2 + p$

3.33 $x = \frac{a}{p^3} + \frac{b}{p^2}, y = \frac{3a}{2p^2} + \frac{2b}{p} + c$ **3.34** $y = 3x + \log\left[\frac{3}{1 - ce^{3x}}\right]$

3.35 $y = cx + \log c;\ xe^{y+1} + 1 = 0 (x < 0)$

3.36 $(y - cx)(c - 1) = c; (x - y)^2 + 1 = 2(x + y)$

3.37 $y = cx + c^2; x^2 + 4y = 0$ **3.38** $y = cx + e^c; e^{(x+y)/x} + x = 0 (x < 0)$

3.39 $y^2 = cx^2 - \frac{a^2 c}{1+c}; a^4 - (x^2+y^2)^2 = 2a^2(y^2 - x^2 + a^2)$

3.40 $y = cx + (1 - c^2); x^2 = 4(y - 1)$

3.41 $y^2 = cx + (c^3/8)$; no singular solution

3.42 $\sin y = c \sin x + c^2; 4 \sin y - \sin^2 x = 0$

Chapter Four

4.1 $x^2 - y^2 = c$ **4.2** $x^2 + 2y^2 = c^2$ **4.3** $y^2 = -2x + c$ **4.4** $4x + y = c$

4.5 $(c - 2x)y = 1$ **4.6** $y = \exp\left(\frac{2}{3}cx - A\right)$ **4.7** $(x^2+y^2)^2 = c(2x^2 + y^2)$

4.8 $y + 1 = x + ce^{-x}$ **4.9** $x^2 + ny^2 = c^2$ **4.10** $y = c - \frac{x}{2} - \frac{\sin 2x}{4}$

4.11 $y^2 = 2(c - \sin x)$ **4.12** $(x+1)^2 + y^2 = a^2$

4.13 $(\log r)^2 + \theta^2 = c$ **4.14** $r^n = c^n \sin n\theta$

4.15 $r(1 - \cos\theta) = 2c$ **4.16** $r + \theta = c + \frac{1}{r}$ **4.17** $r = ce^{-\theta^2/2}$ **4.18** $r^n \cos n\theta = c$

Chapter Five

5.1 $y = c_1e^{3x} + c_2e^{-3x}$ **5.2** $y = c_1 + c_2e^{x}$ **5.3** $y = A\cos x + B\sin x$

5.4 $y = e^{-2x}[A\cos x + B\sin x]$ **5.5** $y = c_1e^{6x} + c_2e^{3x}$

5.6 $y = c_1e^{(\sqrt{7}-1)x} + c_2e^{-(\sqrt{7}+1)x}$

5.7 $y = c_1e^{2x/3} + c_2e^{-7x}$ **5.8** $y = (Ax + B)e^{3x}$ **5.9** $y = (c_1 + c_2x)e^{(3/2)x}$

5.10 $y = e^{4x}(A\cos 3x + B\sin 3x)$ **5.11** $y = \dfrac{1}{2}e^{x} + \dfrac{5}{2}e^{-x}$ **5.12** $y = e^{2x} - 2e^{x}$

5.13 $y = \dfrac{1}{2}(5e^{2x} + 3e^{-2x})$ **5.14** $y = 3e^{x/2} - 2e^{(3/4)x}$ **5.15** $y = 5\cos 6x + 7\sin 6x$

5.16 $q = \dfrac{Q}{2}\left[2\cos\left(\dfrac{R}{L}t\right) + \sin\left(\dfrac{R}{L}t\right)\right]\exp\left(-\dfrac{Rt}{2L}\right)$

5.17 $y = c_1e^{-2x} + c_2e^{-x} + \dfrac{1}{15}e^{-4x}$

5.18 $y = e^{-2x}(\cos x + \sin x) + 3e^{-2x}$ **5.19** $x = (At + B)e^{t} + \dfrac{2}{9}e^{-2t}$

5.20 $y = c_1e^{3x} + c_2e^{-2x} - 4xe^{-2x}$ **5.21** $y = c_1e^{-5x} + c_2xe^{-5x} + 7x^2e^{-5x}$

5.22 $y = c_1\sin 2x + c_2\cos 2x + \sin x$ **5.23** $y = c_1e^{x} + c_2e^{2x} + 2\sin 2x + 3\cos 2x$

5.24 $y = c_1\sin x + c_2\cos x + x\sin x$

5.25 $y = e^{x}(c_1\cos 2x + c_2\sin 2x) + (5x^2 + 4x + 2)$

5.26 $y = c_1 + c_2e^{2x} + 2x - 3x^2$ **5.27** $N = \dfrac{n}{m}\left[t - \dfrac{1}{m}(1 - e^{-mt})\right]$

5.29 (i)$y = e^{-t}\cos t$ (ii)$y = e^{-t}(\cos t + \sin t) = \sqrt{2}e^{-t}\cos\left(t - \dfrac{\pi}{4}\right)$

(iii) $y - e^{-t}(\cos t + 2\sin t) = \sqrt{5}e^{-t}\cos(t - 1.11)$

5.30 Minimum at $t = \frac{3\pi}{4}$; maximum at $t = \frac{7\pi}{4}$ etc. being the positive solutions of $\tan t = -1$.

5.31 $y = c_1 e^{-2x} + c_2 e^{-x} + \frac{1}{170}[11 \sin\ x - 7\cos\ x]e^{2x}$

5.32 $y = e^{x/2}\left[A\cos\ \frac{\sqrt{3}}{2}x + B\sin\ \frac{\sqrt{3}}{2}x\right] + \frac{e^{2x}}{3} + e^{x}$

5.33 $y = (Ax + B)e^{x} + \frac{e^{-x}}{2} + \frac{e^{-2x}}{9} + 1$

5.34 $y = e^{-x/2}\left[A\cos\ \frac{\sqrt{3}}{2}x + B\sin\ \frac{\sqrt{3}}{2}x\right] - 2\cos\ x + \frac{3}{26}\cos\ 2x - \frac{1}{13}\sin\ 2x + \frac{3}{2}$

5.35 $y = e^{x}[A\cos\ 2x + B\sin\ 2x] + \frac{e^{-x}}{20}[\sin\ 2x + 2\cos\ 2x] - \frac{e^{x}}{8}\sin\ 2x$

5.36 $y = c_1 e^{5x/2} + c_2 e^{-x} - \frac{31}{205}\sin\ 2x - \frac{33}{205}\cos\ 2x$

5.37 $y = c_1 e^{2x} + c_2 e^{-2x} - \frac{1}{4}x^2 - \frac{1}{8}$

5.38 $y = c_1 e^{2x} + c_2 e^{3x} + \frac{1}{6}x + \frac{5}{36}$ **5.39** $y = c_1\cos\ 2x + c_2\sin\ 2x + \frac{1}{4}\left(x^2 + \frac{5}{2}\right)$

5.40 $y = c_1 e^{-2x} + e^{x}[c_2\cos\ \sqrt{3x} + c_3\sin\ \sqrt{3x}] + \frac{1}{8}(x^4 - x + 1)$

5.41 $y = (c_1 + c_2 x)e^{x} + x + 1$ **5.42** $y = c_1 + (c_2 + c_3)xe^{-x} + \frac{x}{6}(2x^2 - 9x + 24) + \frac{1}{19}e^{2x}$

5.43 $y = e^{x}\left[c_1 + c_2 x + c_3 x^2 + \frac{1}{6}x^3 + \frac{1}{24}x^4\right]$

5.44 $y = c_1 e^{x} + c_2 e^{-x} + \frac{1}{3}xe^{x}(x^2 - 3x + 9)$

5.45 $y = c_1 + c_2 x + c_3 e^{-2x} + \frac{1}{48}[10x^4 - 20x^3 - 54x^2 + 54x - 27]$

5.46 $y = (c_1 + c_2 x)e^{2x} + c_3 e^{-4x} + \frac{1}{4}$

5.47 $y = c_1 e^{-x} + c_2\cos\ 2x + c_3\sin\ 2x - \frac{1}{16}[\cos\ 2x - \sin\ 2x]xe^{x}$

5.48 $y = c_1 e^{-x} + c_2 e^{3x} + c_3 e^{-2x} - \frac{1}{12}\left(x + \frac{17}{12}\right)e^{2x}$

5.49 $y = c_1 + (c_2 + c_3x + c_4x^2)e^{-x} + \frac{20}{3}x^3 - 59x^2 + 227x + 367$

5.50 $y = c_1e^x + c_2e^{2x} + c_3e^{-2x} - \frac{1}{3}xe^x$

Chapter Six

6.1 $y = [c_1 \cos(\log x^3) + c_2 \sin(\log x^3)]/x$ **6.2** $y = [c_1 + c_2 \log x]x^2$

6.3 $y = c_1x^2 + \frac{c_2}{x^2} + \frac{1}{5}x^3$

6.4 $y = \frac{1}{x}[c_1 \cos \log x + c_2 \sin \log x] + \frac{x^2}{10} + \frac{3x}{5}$

6.5 $y = c_1x^{8/3} + \frac{c_2}{x^2} + \frac{x^3}{5}(7\log x + 15)$

6.6 $y = x(x^2 - 1)$ **6.7** $y = \frac{c_1}{x^2} + c_2x^3 + c_3x - \frac{x\log x}{6} + \frac{\log x}{5x^2}$

6.8 $y = -4x(\log x)^2 + \left(8\log 2 - \frac{8}{3}\right)x\log x + \left[\frac{8}{3}\log 2 - 4(\log 2)^2 - \frac{8}{9}\right]x + \frac{x^4}{9}$

6.9 $y = c_1(x+2)^{-3+\sqrt{21}} + c_2(x+2)^{-3-\sqrt{21}} + \frac{3}{4}x^2 + 5x + \frac{23}{4}$

6.10 $y = \frac{c_1}{2x+1} + c_2(2x+1)^3 - \frac{3}{8}x + \frac{1}{16}$

6.11 $y = c_1 \cos[\log(1+x)] + [c_2 + 2\log(1+x)][\sin \log(1+x)]$

6.12 $y = (2x-1)[c_1 + c_2(2x-1)^{\sqrt{3/2}} + c_3(2x-1)^{\sqrt{3/2}}]$

6.13 $y = x^2 + \sqrt{x}\left[c_1 \cos\left(\frac{\sqrt{3}}{2}\log x\right) + c_2 \sin\left(\frac{\sqrt{3}}{2}\log x\right)\right]$

6.14 $y = x^2[c_1 \cos(\log x) + c_2 \sin(\log x)] - \frac{x^2}{2}\log x \cos(\log x)$

6.15 $y = c_1(3x+2)^2 + c_2(3x+2)^{-2} + \frac{1}{100}[(3x+2)^2 \log(3x+2) + 1]$

Chapter Seven

7.1 $x = \cos t, y = \sin t$ **7.2** $x = 2c_1e^{-t} + c_2e^t, y = c_1e^{-t} + c_2e^t$

7.3 $x = e^{3t}(2c_1 \cos 3t + 2c_2 \sin 3t)$, $y = e^{3t}[c_1(\cos 3t + 3\sin 3t) + c_2(\sin 3t - 3\cos 3t)]$

7.4 $x = -2c_1e^{3t} + c_2(1+2t)e^{3t}$, $y = c_1e^{3t} - c_2te^{3t}$

7.5 $x = 3c_1 + c_2e^{-2t}, y = 4c_1 + 2c_2e^{-2t}$

7.6 $x = \left(c_1 - \frac{c_2}{2} + c_2t\right)e^{-3t} + \frac{t}{9} + \frac{1}{27}$, $y = (c_1 + c_2t)e^{-3t} - \frac{2}{9}t + \frac{4}{27}$

7.7 $x = c_1e^t + c_2e^{-5t} + \frac{6}{7}e^{2t}$, $y = c_2e^{-5t} + c_1e^t + \frac{8}{7}e^{2t}$

7.8 $x = c_1e^t + c_2e^{-2t} + 2e^{-t}$, $y = 3c_1e^t + 2c_2e^{-2t} + 3e^{-t}$

7.9 $x = (c_1 + c_2t)e^{-t} + c_3e^{2t} - \frac{1}{4}e^t$, $y = -2(c_1 + c_2 + c_2t)e^{-t} + \frac{c_3}{2}e^{2t} - \frac{1}{4}e^t$

7.10 $x = (c_1 + c_2t)e^t + (c_3 + c_4t)e^{-t}$, $y = \frac{1}{2}(c_2 - c_1 - c_2t)e^t - \frac{1}{2}(c_4 + c_3 + c_4t)e^{-t}$

7.11 $x = c_1e^{\sqrt{2}t} + c_2e^{-\sqrt{2}t} + 3\cos t$, $y = (\sqrt{2}+1)c_1e^{\sqrt{2}t} + (1-\sqrt{2})c_1e^{-\sqrt{2}t} + 2\sin t$

7.12 $x = c_1 \sin t - c_2 \cos t + 2$, $y = c_1 \cos t + c_2 \sin t - t$

7.13 $x = c_1e^{(3+\sqrt{10})t} + c_2e^{(3-\sqrt{10})t}$, $y = \frac{1}{2}\sqrt{10}\,[c_1e^{(3+\sqrt{10})t} - c_2e^{(3-\sqrt{10})t}]$

7.14 $x = c_1e^{-5t} + c_2e^{-t/3} + \dfrac{8\sin t + \cos t}{65}$,

$$y = -\frac{4}{3}c_1e^{-5t} + c_2e^{-t/3} + \frac{61\sin t - 33\cos t}{130}$$

7.15 $x = (c_1 - c_2)\cos t + (c_1 + c_2)\sin t + (3e^{2t}/5)$,

$$y = c_1 \cos t + c_2 \sin t + (2e^{2t}/5) + (e^t/2)$$

7.16 $x = e^t + e^{-t}, y = e^{-t} - e^t + \sin t$

7.17 $y = c_1e^x + c_2e^{-2x} + 2e^{-x}$, $z = 3c_1e^x + 2c_2e^{-2x} + 3e^{-x}$

7.18 $x = (c_1 + c_2t)e^{-4t} + \frac{4}{25}e^t - \frac{1}{36}e^{2t}$, $y = -(c_1 + c_2 + c_2t)e^{-4t} + \frac{7}{36}e^{2t} + \frac{1}{25}e^t$

7.19 $y = (c_1 + c_2x)e^x - c_3e^{-3x/2} - \frac{x}{2}$, $z = 2(3c_2 - c_1 - c_2x)e^x - c_3e^{-3x/2} - \frac{1}{3}$

7.20 $x = \frac{8}{9}\left(1 - \cos\frac{3}{2t}\right)^{-1}$, $y = \frac{4}{3t} - \frac{8}{9}\sin\frac{3}{2t}$

Chapter Eight

8.1 **(i)** $y = c_1 \cos x + c_2 \sin x + x \sin x + \cos x \log \cos x$

(ii) $y = c_1e^x + c_2e^{3x}(4x^3 - 42x^2 + 150x - 183)$ **(iii)** $y = e^x[c_1 + (c_2x^2/2) + xe^x]$

(iv) $y = c_1xe^{2x} + c_2x - x^2/2$ **(v)** $y = c_2e^{2x} - \frac{1}{4}c_1(2x+5) - e^x$

8.2 $y = [x^2 + (c_1 - 1)x]e^x + c_2x$ **8.3** $y = c_1e^x - x^2 + c_2x - 1$

8.4 $y = c_1 \operatorname{cosec} x + c_2 \cot x$

8.5 $y = c_1 \sin x - c_2(x \sin x + \cos x)$ **8.6** $y = e^x[c_1 \log x + c_2]$

8.7 $y = c_2\left[x + \frac{1}{x}\right] - c_1\frac{1}{2x}$

8.8 $y = 1 - x^2$ **8.9** $y = \frac{2}{3}x^2 - x + \frac{1}{3x^2}$

8.10 General solution : $y = -\frac{1}{2}e^x \cos x + \frac{c_1}{\sqrt{5}}e^{-x}\sin\left(x + \tan^{-1}\frac{1}{2}\right) + c_2e^x$.

Particular solution of the initial-value problem: $y = -\frac{1}{2}e^x \cos x$ since $c_1 = 0$ and $c_2 = 0$.

8.11 $y = c_1e^{2x} + c_2(2x+5) - e^x$ **8.12** $k = -\frac{1}{2}; y = x^{-(1/2)}[2c_1 \sin^{-1}\sqrt{x} + c_2]$

8.13 $y = c_1 \sin x + c_2 \cos x + \sin x \log(\operatorname{cosec} x - \cot x)$

8.14 $y = c_1 \sin ax + c_2 \cos ax + \frac{x}{a}\sin ax + \frac{1}{a^2}\cos ax \log \cos ax$

8.15 $y = c_1 \sin 2x + c_2 \cos 2x + \frac{1}{4}\sin 2x \log \sin 2x - \frac{x}{2}\cos 2x$

8.16 $y = c_1e^{-x} + c_2e^{-2x} + \frac{x}{2} - \frac{3}{4} - 3xe^{-2x}$ **8.17** $y = (c_1 + c_2x - \log x)e^{3x}$

8.18 $y = c_1 x + c_2 x^2 + \dfrac{1}{12x^2}$ **8.19** $y = \dfrac{1}{1-x^2}[c_1 \sin\ x + c_2 \cos\ x + x]$

8.20 $y = c_1 x + c_2 e^x + x^2 + x + 1$ **8.21** $y = \dfrac{3}{2}x - 9(x^2 - 1) + x^4 + 3x^2$

8.22 $y = -\dfrac{2}{\pi}x^3 + 2x - x\cos\ x$ **8.23** $y = c_1 x + \dfrac{c_2}{x} + \left(1 - \dfrac{1}{x}\right)e^x$

8.24 $y = c_1 \sin\left(2\log\ \tan\ \dfrac{x}{2} + c_2\right)$

8.25 $y = c_1 \sin\ (x^2 + c^2) + \dfrac{x^2}{4}$ **8.26** $y = c_1 \cos\left(\dfrac{a}{2x^2}\right) + c_2 \sin\left(\dfrac{a}{2x^2}\right) + \dfrac{1}{a^2x^2}$

8.27 $y = c_1 \cos\left(2\ \log\ \tan\dfrac{x}{2}\right) + c_2 \sin\left(2\log\ \tan\ \dfrac{x}{2}\right)$

8.28 $y = (c_1 + c_2 \sin\ x)e^{-\sin\ x} + (4\sin\ x - \sin^2 x - 5)$ **8.29** $y = \sin^2 x$

8.30 $y = c_1\ \exp\left(\dfrac{x^3}{\sqrt{3}} - 2\right) + c_2\ \exp\left(-\dfrac{x^3}{\sqrt{3}} - 2\right)$

8.31 $y = c_1 \cos\ (\tan^{-1} x) + c_2 \sin\ (\tan^{-1} x)$

8.32 $y = c_1 \cos\ (1/x) - c_2 \sin\ (1/x) + 1/x^2$

8.33 $y = \left[\dfrac{e}{e-1}\right]e^{-\cos\ x} - \left[\dfrac{1}{e-1}\right]e^{\cos\ x} - \cos\ x$

8.34 $y = \left[c_1 - \dfrac{1}{4}\sin^{-1} x\right](1 - 2x^2) + 2c_2 x\sqrt{1-x^2}$

8.35 $y = c_1 e^z + c_2 e^z - \sin\ z$ where $z = x^2$

8.36 $y = [c_1 \cos\ \sqrt{3}x + c_2 \sin\ \sqrt{3}x + (e^x/4)]\ \exp\ (x^2/2)$

8.37 $y = [c_1 \sin\ \sqrt{6}x + c_2 \cos\ \sqrt{6}x] \sec x$

8.38 $y = (c_1 x^2 + c_2 x)e^x$ **8.39** $y = (\sec x)[c_1 \cos\ \sqrt{6}x + c_2 \sin\ \sqrt{6}x + e^x/7]$

8.40 $y = [c_1 \cos\ 2x + c_2 \sin\ 2x + (x/4)]\ \exp\ (-x^2/2)$

8.41 $y = \dfrac{1}{x^2+1}[c_1 \cos\ 3x + c_2 \sin\ 3x]$

8.42 $y = \sqrt{x}e^{x^2}\left[c_1 \cos\left(\frac{\sqrt{19}}{2}\log x\right) + c_2 \sin\left(\frac{\sqrt{19}}{2}\log x\right)\right]$

8.43 $2y = xe^{x^2}[c_1 \cos(\sqrt{2}x) + c_2 \sin(\sqrt{2}x)]$ **8.44** $2y = x[e^{2x-2} - x + 1]$

8.45 $y = x\left[c_1 \cos x + c_2 \sin x + \frac{1}{2}e^x\right]$

8.46 $xy = \log\frac{x}{1-x} + c_1 \log x + c_2$ [see Example 8.19]

8.47 $y = e^x \log x + c_2 e^x [1 + \int x^{-1}e^{-x}dx]$

8.48 $y\text{cosec}^2 x = c_1\left[\log\tan(x/2) - \text{cosec}x \cot x\right] + c_2$

8.49 $(x+1)x^2y = c_1 + c_2\left[\frac{x^4}{4} + \frac{x^3}{3}\right]$ **8.50** $(1+x^2)y = c_1 + \log\sec x + c_2 x$

8.51 $(1+x+x^2)y = \frac{kx^2}{2} + c_1 x + c_2$ **8.52** $x^2 y = \sin x - x\cos x + \frac{c_1 x^2}{2} + c_2$

8.53 $y = x^2 + c_1(1+2x+x^2) + c_2 x^2 e^{(1/x)}$ where $c_1 = \frac{e^2-1}{e^2+5}$ and $c_2 = \frac{6e}{e^2+5}$.

8.54 $y\sqrt{1+x^2} = \frac{1}{3}(1+x^2)^{3/2} + c_1 \sinh^{-1}x + c_2$

8.55 $xy = c_1 \tan^{-1} x + c_2$

Chapter Nine

9.1 **(i)** $y = c_1\left(1 + \frac{3}{5}x + \frac{3.5}{5.7}x^2 + \frac{3.5.7}{5.7.9}x^4 + \ldots\right) + c_2 x^{-3/2}$

(ii) $y = c(x+1)\left\{1 + (x+1) + \frac{(x+1)^2}{2!} + \frac{(x+1)^3}{3!} + \ldots\right\}$

9.2 $y_1(t) = 1 + 2t + t^2 + \frac{4t^3}{15} + \frac{t^4}{14} + \ldots$

$y_2(t) = t^{1/2}\left[1 + \frac{5t}{6} + \frac{17t^2}{60} + \frac{89t^3}{60(21)} + \frac{941t^4}{36(21)(60)} + \ldots\right]$

9.3 $y(t) = 1 - \frac{\lambda t}{(1!)^2} - \frac{\lambda(1-\lambda)t^2}{(1!)^2(2!)^2} + \ldots + \frac{(-\lambda)(1-\lambda)\ldots(n-1-\lambda)t^n}{(n!)^2}$

9.4 $y_1(x) = 1 - \frac{p.p}{2!}x^2 + \frac{p(p-2)p(p+2)}{4!}x^4 - \ldots$

$$y_2(x) = x - \frac{(p-1)(p+1)}{3!}x^3 + \frac{(p-1)(p-3)(p+1)(p+3)}{5!}x^5 - \ldots$$

9.6 $y_1 = x + \frac{x^4}{2.3} + \frac{x^7}{2.3.5.6} + \frac{x^{10}}{2.3.5.6.8.9} + \ldots$

$$y_2 = x^2 + \frac{x^5}{3.4} + \frac{x^8}{3.4.6.7} + \frac{x^{11}}{3.4.6.7.9.10} + \ldots$$

Chapter Twelve

12.1 $2xz = y^2(z^2 + c)$ **12.2** $(y+z)e^x = c$ **12.3** $x - cy - y - y \log\ z = 0$

12.4 $(x + y + z^2)\ \exp\ (x^2) = c$ **12.5** $y(x+z) = c(y+z)$ **12.6** $2y - z = ce^{-x/2}$

12.7 $(x+y)z^2 + \sin\ z - z\cos\ z = 0$ **12.8** $xy = c(z-a)$ **12.9** $1 + xy = c(z-x)$

12.10 $xy + yz + zx = c(x+y+z)$ **12.11** $x^2 + y^2 + z^2 = cx$

12.12 $(x-a)^2 + (y-c)^2 + z^2 = b^2$ **12.13** $(zx + zy - y^2) = cz^2$

12.14 $(zx + yz + x^2 + y^2) = cxy$ **12.15** $\log\ xyz + (a-1)\log\ (x+y+z) = c$

Chapter Thirteen

13.1 $x^4 - y^4 = c_1,\ x^2 - z^2 = c_2$ **13.2** $x + y = c_1, z^2 + (x+y)^2 = c_2 e^{-2y}$

13.3 $x^{-1} - y^{-1} = c_1, \frac{x-y}{z} = c_2$ **13.4** $\frac{1}{x} = \frac{1}{y} + c_1, z = c_2 - \frac{nxy}{y-x}\log\left(x/y\right)$

13.5 $(x+y)(x+y+z) = c_1, xy = c_2$ **13.6** $x^2 - y^2 - z^2 = c_1, 2xy - z^2 = c_2$

13.7 $\frac{xy}{z^2} = c_1, \frac{1}{(x+y)^2} - \frac{1}{(x-y)^2} = c_2$ **13.8** $(x-y) = c_1, z = xy + x^2y^2 + x^3y^3 + c_2$

13.9 $xyz^2 = c_1, x^4 + y^4 + z^4 = c_2$

13.10 $px^2 + qy^2 + rz^2 = c_1,\ p^2x^2 + q^2y^2 + r^2z^2 = c_2$

13.11 $(y-x) = c_1(z-x), (y-x) = c_2(z-y)$

13.12 $y = c_1 x, x^{1-a} = c_2[z + \sqrt{x^2+y^2+z^2}]$

Chapter Fourteen

14.1 (i) $\dfrac{3}{s+2}; s > -2$ (ii) $\dfrac{3s}{s^2+25}; s > 0$ (iii) $\dfrac{1+e^{-\pi s}}{s^2+1}; s > 0$

(iv) $\dfrac{4e^{-2s}}{s}; s > 0$

14.2 (i) $\dfrac{2}{(s+4)^3}$ (ii) $\dfrac{s^2-1}{(s^2+1)^2}$ (iii) $\dfrac{1-5s}{s^2+2s-3}$ (iv) $\dfrac{e^b}{s-a}$

(v) $\dfrac{1}{s} + \dfrac{3}{(s+1)^2} + \dfrac{6}{(s+2)^3} + \dfrac{6}{(s+3)^4}$

14.3 (i) Yes (ii) No (iii) Yes (iv) No (v) Yes **14.5** $\dfrac{2s-1+\tan^{-1}(1/s)}{s^2}$

14.6 (i) $\dfrac{1}{4}\log\left\{\dfrac{s^2+4}{s^2}\right\}$ (ii) $\cos t^{-1}s - \dfrac{1}{2}s\log\left(1+\dfrac{1}{s^2}\right)$

(iii) $\dfrac{\sqrt{\pi}}{2s^{3/2}}$ (iv) $\dfrac{s}{(s^2+a^2)^{3/2}}$

(v) $\dfrac{24s(s^2-1)}{(s^2+1)^4}$ (vi) $\log\left(1+\dfrac{1}{s}\right)$ (vii) $\dfrac{\pi}{2s} - \dfrac{1}{s}\tan^{-1}\left(\dfrac{s+4}{3}\right)$

(viii) $\dfrac{2}{(s^2-1)^2}$ (ix) $\dfrac{3s+2}{2s^2(s^2+1)^{3/2}}$ (x) $\dfrac{1-e^{-2s}}{s^2}$

14.8 $\dfrac{2-2e^{-2s}-4se^{-2s}-4s^2e^{-2s}}{s^3(1-e^{-2s})}$ **14.9** $\dfrac{1}{s^2+s+1}$

14.10 $\dfrac{1}{(s-1)(s+1)(s+2)}$

14.13 $\dfrac{1}{s} + \dfrac{e^{-s}}{s} + \dfrac{2e^{-3s}}{s} - \dfrac{6e^{-4s}}{s}$ **14.14** $\dfrac{1}{s} + e^{-\frac{\pi s}{6}}\left(\dfrac{\sqrt{3}}{2}\dfrac{1}{s^2+1} + \dfrac{1}{2}\dfrac{s}{s^2+1} - \dfrac{1}{s}\right)$

14.15 (i) $\dfrac{6-2e^{-\pi s}-4e^{-2\pi s}}{s}$ (ii) $\dfrac{5^{2-s}}{s-2}$ (iii) $\dfrac{e^{-as}}{s^2}$ (iv) $\dfrac{e^{-2s}(2+4s+4s^2)}{s^3}$

(v) $\dfrac{1}{s}$

14.16 $4e^{-8}$ **14.17** $\dfrac{e^{-s}(s^2+s+1)}{s^2}$

14.18 **(i)** $3e^{-4t}$ **(ii)** $e^{-t}\cos\sqrt{5}t - \dfrac{1}{\sqrt{5}}e^{-t}\sin\sqrt{5}t$ **(iii)** $1+e^{-t}$

(iv) $\dfrac{1}{t}(e^{-2t}-e^{-3t})$ **(v)** $e^{-3t}\cos 2t - 4e^{-3t}\sin 2t$ **(vi)** $4(1-e^{2-t})(t-2)$

(vii) $\dfrac{1}{7}(4\sin 4t - 3\sin 3t)$ **(viii)** $e^{4t}+e^{-t}(\cos 2t + 2\sin t)$ **(ix)** $\dfrac{2}{t}(1-\cos t)$

(x) $\dfrac{1}{8}\sin 2t\sinh 2t$

14.20 **(i)** $\dfrac{e^{-t}-e^{-2t}}{t}$ **(ii)** $\displaystyle\int_0^t \frac{e^{-u}-e^{-2u}}{u}du$ **14.26** $y=\dfrac{1}{2}(t^2-3t+2)e^t$

14.27 $y=\left(\dfrac{1}{2}k\right)\sin kt$ **14.28** $y=\sinh t - 1$ **14.29** $y=e^{-t}\sin 2t + e^{-t}\sin t$

14.30 $y=\dfrac{1-e^{3t}\cos t + 3e^{3t}\sin t}{10}$ **14.31** $y=e^{-t}(\sin 2t+\cos 2t)+2\sin t$

14.32 $y=4(t-2)\left[1-\dfrac{e^{t-2}}{2}-\dfrac{e^{-t+2}}{2}\right]$ **14.33** $y=\dfrac{1}{2\sqrt{2}}e^{-t}\sin\dfrac{2}{\sqrt{2}t}$

14.34 $y=\dfrac{3}{2}e^t-\dfrac{1}{2}e^{3t}+\dfrac{1}{2}\displaystyle\int_0^t(e^{3u}-e^u)f(t-u)du$

14.35 $y=\begin{cases}(1/2)\sin 2t \text{ if } t<2\\ (1/2)\{\sin 2t+\sin(2t-4)\}^2 \text{ if } t>2\end{cases}$

14.36 $y(t)=1+2t$ **14.37** $y^{(t)}=2t+\pi\sin 3t$

Chapter Fifteen

15.1 If $k=0$, the fundamental period is 1.

15.2 $e^{-x}=\dfrac{1-e^{-2\pi}}{\pi}\left\{\dfrac{1}{2}+\left(\dfrac{1}{2}\cos x+\dfrac{1}{5}\cos 2x+\dfrac{1}{10}\cos 3x+\dots\right)\right.$

$\left.+\left(\dfrac{1}{2}\sin x+\dfrac{2}{5}\sin 2x+\dfrac{3}{10}\sin 3x+\dots\right)\right\}$

15.3 **(i)** $\dfrac{4}{\pi}\left[\sin\pi x+\dfrac{\sin 3\pi x}{3}+\dfrac{\sin 5\pi x}{5}+\dots\right]$

(ii) $\frac{2}{\pi}\left[\sin\ \pi x - \frac{\sin\ 2\pi x}{2} + \frac{\sin\ 3\pi x}{3} + \ldots\right]$

(iii) $-1 + \pi \sin\ x - \frac{1}{2}\cos\ x + 2\sum_{n=2}^{\infty}\frac{\cos\ nx}{n^2 - 1}$

(iv) $\frac{1}{2}\sin\ \frac{\pi x}{2} + \frac{4}{\pi}\sum_{n=1}^{\infty}(-1)^{n+1}\frac{n\ \sin\ n\pi t}{4n^2 - 1}$.

(v) $a_0 = \frac{1}{2}, a_n = \begin{cases} 4/n^2\pi^2 & n = 1, 3, 5, \ldots \\ 8/n^2\pi^2 & n = 2, 6, 10, \ldots; \\ 0 & n = 4, 8, 12, \ldots \end{cases} \quad b_n = 0$

15.5 $f(x) = \frac{\pi^2}{12} + \sum_{n=1}^{\infty}\frac{\cos\ nx}{n^2}$

15.6 $\frac{4}{\pi}\left(\sin\ x - \frac{1}{9}\sin\ 3x + \frac{1}{25}\sin\ 5x \ldots\right)$ **15.7** $\pi/4$

15.8 $f(x) = \frac{-8}{\pi^2}\sum_{n=1}^{\infty}\frac{1}{n^2}\cos\ \frac{2n\pi x}{3} + \frac{3}{\pi}\sum_{n=1}^{\infty}+\frac{1}{n}\sin\ \frac{2n\pi x}{3}$

15.10 (ii) $\frac{2\pi}{p}\sum_{n=1}^{\infty}\frac{1}{n}(-1)^{n+1}\sin\ \frac{n\pi x}{p};\ \frac{p}{2} - \frac{4p}{\pi^2}\sum_{n=0}^{\infty}\frac{1}{(2n+1)^2}\cos\ \frac{(2n+1)\pi x}{p}$

(iii) $\sin\ x;\ \frac{8}{\pi}\sum_{n=0}^{\infty}\frac{\cos\ (2n+1)\pi x}{3 - 4n(n+1)}$ (v) $\frac{8}{\pi}\sum_{n=0}^{\infty}\frac{\sin\ (2n+1)x}{(2n+1)^3};\ \frac{\pi^2}{6} - \sum_{n-1}^{\infty}\frac{\cos\ 2nx}{n^2}$

(iv) $a_n = 0; b_n = \frac{(-1)^{n+1}2n\sin\ a\pi}{\pi(n^2 - a^2)};\quad a_n = \frac{2a[1 - (-1)^n\cos\ a\pi]}{\pi(a^2 - n^2)}; b_n = 0$

(vii) $\frac{8k}{\pi^2}\sum_{n=0}^{\infty}\frac{(-1)^n}{(2n+1)^2}\frac{\sin\ (2n+1)\pi x}{p};$

$\frac{k}{2} - \frac{4k}{\pi^2}\sum_{n=0}^{\infty}\frac{1}{(2n+1)^2}\cos\ \frac{2(2n+1)\pi x}{p}$

(viii) $\left(\frac{1}{\pi} - \frac{4}{\pi^2}\right)\sin\ \pi x + \left(\frac{1}{3\pi} + \frac{4}{3^2\pi^2}\right)\sin\ 3\pi x + \left(\frac{1}{5\pi} - \frac{1}{5^2\pi^2}\right)\sin\ 5\pi x \ldots$

15.11 (i) $2 - \dfrac{8}{\pi^2}\sum_{n=1}^{\infty}\dfrac{(1-\cos n\pi)}{n^2}\cos\dfrac{n\pi x}{4}$

(ii) $\dfrac{3}{2} + \sum_{n=1}^{\infty}\left\{\dfrac{6(\cos n\pi - 1)}{n^2\pi^2}\cos\dfrac{n\pi x}{3} - \dfrac{6\cos n\pi}{n\pi}\sin\dfrac{n\pi x}{3}\right\}$

15.12 (i) $\dfrac{1}{2} + \dfrac{2}{\pi}\left(\cos x - \dfrac{1}{3}\cos 3x + \dfrac{1}{5}\cos 5x - \ldots\right)$

(ii) $\dfrac{\pi^2}{12} - \cos x + \dfrac{1}{4}\cos 2x - \dfrac{1}{9}\cos 3x + \dfrac{1}{16}\cos 4x - \ldots$

15.14 $b_n = \begin{cases}(2/n\pi) - (8/n^3\pi^3) & n \text{ odd}\\ -2/n\pi & n \text{ even}\end{cases}$

15.16 $e^x = \dfrac{\sinh \pi}{\pi}\sum_{n=-\infty}^{\infty}(-1)^n\dfrac{1+in}{1+n^2}e^{inx}$

Chapter Sixteen

16.1 (a) 1.95 (b) 4.60 (c) 0.81 (d) 0.80 (e) 2.00 (f) 1.96 (g) 1.33

16.2 0.4263, 0.5045 **16.3** 0.851, 0.780

16.4

Method	One step y	Two steps y
Euler	2.4	2.44
Improved Euler	2.48	2.4841
Modified Euler	2.4889	2.4864

[**Note**: By the analytical method, $y = 2.4856$ correct to 4 decimal places]

16.5 0.851, 0.780

16.6

x	h	y_{RK}	y (exact)	Error
0.1	0.1	1.06652422	1.06652425	3×10^{-8}
0.2	0.2	1.16722083	1.16722194	1.11×10^{-6}
0.4	0.4	1.4782	1.478232	3.2×10^{-5}

Exact solution : $y = 13e^{x/2} - 6x - 12$

16.7 1.1103, 1.2428

16.7

x	0.5	1.0	1.5	2.0
y	1.3571	1.5837	1.7565	1.8950

16.9 0.2027, 0.4228, 0.6841 **16.10** 1.4983

16.11 $\left(\dfrac{dx}{dt}\right)^2 \sin^2 t = 2\sin^2 t - 2x^2.$

For one step from $t = 0$ to 0.2, $k_1 = 0.16330, k_2 = 0.16274, k_3 = 0.16386, k_4 = 0.15993, x = 0.1627.$
16.12 (i) 37.50 gm (ii) 36.82 gm
Exact solution: $N = 100e^{-(0.01)t}$ which gives $N = 36.788$ gm at $t = 100$ sec.

Bibliography

Balachandra Rao, S. and C.K. Shantha, *Differential Calculus*, Wiley Eastern Ltd., 1992.

Balachandra Rao, S. and C.K. Shantha, *Numerical Methods*, Universities Press, 1992.

Batchelet, E., *Introduction to Mathematics for Life Scientists*, Springer-Verlag, 1973.

Betz et. al., *Differential Equations with Applications*, Harper International Edn., 1964.

Biracre, F.F.P., *Applied Calculus*, Blackie & Son Ltd., 1947.

Birkhoff, G. and G. Rota, *Ordinary Differential Equations*, Ginn, 1962.

Bittinger, M.L., *Calculus: A Modelling Approach*, Addison-Wesley, 1976.

Braun, Martin, *Differential Equations and Their Applications* (Short Version), Springer-Verlag, 1978.

Coddington, E.A., *An Introduction to Ordinary Differential Equations*, Prentice-Hall, 1961.

Coddington, E.A. and Levinson, N., *Theory of Ordinary Differential Equations*, TMH Edn., 1977.

Collatz, L., *Numerical Treatment of Differential Equations*, Springer-Verlag, 1966.

Conte, S.D. and Carle de Boor, *Elementary Numerical Analysis*, McGraw-Hill, 1981.

Eastham, *Theory of Ordinary Differential Equations*, Van Nostrand Reinhold Co., 1970.

Ferziger, J.H., *Numerical Methods for Engineering Applications*, John Wiley & Sons, 1981.

Granville, W.A., *Elements of Differential and Integral Calculus*, Ginn & Co., 1941.

Ince, E.L., *Ordinary Differential Equations*, Longmans, 1929.

Krasnov, M.L., *Ordinary Differential Equations*, Mir Publishers, 1983.

Kreyszig, E., *Advanced Engineering Mathematics*, Fifth Edition, Wiley Eastern Ltd., 1990.

Martin Jr., Rabert H., *Ordinary Differential Equations*, McGraw-Hill International Student Edition, 1983.

Mizrahi, A. and M. Sullivan, *Calculus with Applications for Business and Life Sciences*, John Wiley & Sons, 1976.

Murray, Daniel, *Introductory Course in Differential Equations*, Orient Longman, 1991.

Petrovski, I.G., *Ordinary Differential Equations*, Prentice-Hall, 1966.

Piaggio, H.T.H., *Differential Equations (and their Applications)*, G. Bell and Sons Ltd., 1956.

Pontryagin, L.S., *Ordinary Differential Equations*, Addison-Wesley, 1962.

Rabenstein, Albert L., *Introduction to Ordinary Differential Equations*, Academic Press, 1966.

Simmons, G.F., *Differential Equations (with Applications and Historical Notes)*, Tata McGraw-Hill, 1972.

Smith, C., *Biomathematics*, Charles Griffin & Co. Ltd., 1966.

Stibitz, G., *Mathematics in Medicine and the Life Sciences*, Year Book Medical Publishers, 1976.

Index